81

Fortschritte der Chemie organischer Naturstoffe

Progress in the Chemistry of Organic Natural Products

Founded by
L. Zechmeister

Edited by
W. Herz, H. Falk,
G. W. Kirby, and R. E. Moore

Author:
S. Huneck

Springer-Verlag Wien GmbH

Prof. W. Herz, Department of Chemistry,
The Florida State University, Tallahassee, Florida, U. S. A.

Prof. H. Falk, Institut für Chemie,
Johannes-Kepler-Universität, Linz, Austria

Prof. G. W. Kirby, Chemistry Department,
The University of Glasgow, Glasgow, Scotland

Prof. R. E. Moore, Department of Chemistry,
University of Hawaii at Manoa, Honolulu, Hawaii, U. S. A.

Originally published by Springer-Verlag Wien New York 2001
Softcover reprint of the hardcover 1st edition 2001
Library of Congress Catalog Card Number AC 39-1015

Typesetting: Scientific Publishing Services (P) Ltd., Madras

Printed on acid-free and chlorine-free bleached paper

SPIN: 10774546

With 11 Figures

ISSN 0071-7886

ISBN 978-3-7091-7249-0 ISBN 978-3-7091-6196-8 (eBook)
DOI 10.1007/978-3-7091-6196-8

Contents

List of Contributors

S. Huneck, Fliederweg 34a, D-06179 Langenbogen/Saalkreis, Germany

New Results on the Chemistry of Lichen Substances

S. Huneck
Fliederweg 34a, Langenbogen/Saalkreis, Germany

Contents

Abbreviations

The following abbreviations are used in this article: A = "Alcohol" part of a depside, ABIBN=AIBN=2,2'-Azabisisobutyronitrile, Ac=Acetyl, BPO=Benzoylperoxide, Bu=Butyl, Bn=Benzyl, Bz=Benzoyl, CDI=1,1'-Carbonyldiimidazole, CPBA=Chloroperoxybenzoic acid, DBU=1,5-Diazabicyclo [5.4.0] undec-5-ene, DCBCl=2,5-Dichlorobenzoylchloride, DCCD=Dicyclohexylcarbodiimide, DCDME=Dichlorodimethyl ether, DHP=3,4-Dihydro-2-H-pyran, DIBAH=DIBAL=Diisobutylaluminiumhydride, DIPA=Diisopropylamide, DiPy=Dipyridyl, DMPA=Dimethylaminopyridine, DMF=Dimethylformamide, DMS=Dimethylsulphate, DMSO=Dimethylsulfoxide, EE=Ethoxyethyl, Et=Ethyl, HMA=HMTA=Hexamethylenetetramine, HMDS=Hexamethyldisilazide, HMPA=Hexamethylphosphoric acid triamide, LAH=Lithiumaluminiumhydride, LDA=Lithiumdiisopropylamine, MCPBA=*m*-Chloroperbenzoic acid, Me=Methyl, MEM=Methoxy-ethoxy-ethyl, MOM=Methoxymethyl, Mp=Melting point, MTPA=Methyl-α-trifluoromethoxy-phenylacetyl, NBS=N-Bromosuccinimide, NCS=NCSI=N-Chlorosuccinimide, PCC=Pyridinium chlorochromate, PDC=Pyridinium dichromate, PDM=Phenyldiazomethane, Ph=Phenyl, Piv=Pivaloyl, PPTS=Pyridinium-p-toluenesulphonate, Py=Pyridine, RAMP=(*R*)-1-Amino-2-(methoxymethyl)pyrrolizidine, RT=Room temperature, S="Säure" part of a depside, SA=Sulfamic acid, SAMP=(*S*)-1-mino-2-(methoxymethyl)pyrrolizidine, TBA=tert.-Butanol, TBDM=TBS=tert.-Butyldimethylsilyl, TBOK=K-tert.-butylate, 2,4,6,-TCB=2,4,6-Trichlorobenzoyl, TFA=Trifluoroacetic acid, TFAA=Trifluoroacetic anhydride, TFE=Trifluoroethanol, THF=Tetrahydrofuran, THP=Tetrahydropyranyl, TMDA=Tetramethylenediamine, TMS=Trimethylsilyl, TPM=Triphenylmethyl, Ts=Tosyl.

1. Introduction

Lichen chemistry continues to be a flourishing branch of natural product chemistry. Since the last major review by Elix *et al.* in volume 35 of this series (*274*) many new and interesting lichen substances have been isolated and synthesized. This suggests that an update is desirable. As in the previous articles new methods for structure elucidation and synthesis, biosynthesis, chemotaxonomy and biological activities of lichen substances will be discussed. Some other relatively recent treatments of the subject need to be mentioned: A short chapter by

Culberson and Elix (*100*) in Harborne's Methods in Plant Biochemistry treated lichen constituents. Several minor reviews were published in 1984 (*382*), 1991 (*385*) and 1994 (*386*). Volume III of Galun's Handbook of Lichenology (*311*) contains a chapter on chemical constituents of lichens including secondary metabolic products, storage products of lichens, carotenoids and phycobiliproteins.

Schöller's booklet on "Flechten" (*640*) deals with the history, biology, systematics, ecology, protection and cultural significance of lichens and contains numerous colourful photos. Two other popular articles on lichens have been published by Sharnoff and Sharnoff (*645*) and by Ahmadjian (*5*) who also discussed the problems of lichen symbiosis (*6*). Huneck (*388*) gave a review on the significance of lichens and their metabolites.

Like many other plants lichens are vanishing in many parts of the world. The distribution of lichen-rich environments across the globe, their conservation status in regional, national and international terms and the threats from the environmental factors and human populations have been discussed by Wolseley (*736*).

2. Methods for Identification and Structure Elucidation of Lichen Substances

A comprehensive review on the identification of lichen substances has been presented by Huneck and Yoshimura (*415*). This book contains chapters on extraction, purification, colour reactions, UV, IR, NMR, MS, ORD, CD, microcrystallization, TLC, HPLC and derivatization of lichen substances. All known compounds are listed with their structure, melting point, spectroscopic properties and the corresponding reference lichen.

2.1. Spot Tests

Para-phenylenediamine is an excellent reagent for detection of an aldehyde group in a lichen substance. Its disadvantages are instability and suspected carcinogenic activity. To overcome the rapid oxidation of an alcoholic solution of *p*-phenylenediamine a method for stabilizing the solution for at least one year by adding disodium edetate (ethylenediamine tetraacetic disodium salt) has been described (*441*).

2.2. Thin Layer Chromatography (TLC)

Thin layer chromatography is still the method of choice for rapidly identifying lichen substances and is widely used by lichenologists and lichen chemists. White and James (*721*) have published a guide to microchemical techniques (including TLC) for identifying lichen substances. A standard method for TLC of lichen substances was reported by Culberson *et al.* (*89*, *102*) who published standardized R_f values of numerous lichen substances using a modified solvent system in which diethyl ether is substituted by methyl *tert.*-butyl ether (MTBE) (*n*-hexane:MTBE:formic acid = 140:72:18). Wright (*739*) described a simplified TLC method and Arup *et al.* (*21*) developed a method using high performance thin layer chromatography (HPTLC) which has the advantage of being more sensitive, of permitting the analysis of more samples in a shorter period of time, and greatly reducing the amount of solvent. A useful new solvent system (ethyl acetate:methanol:ammonia = 75:20:5, V/V/V) for TLC of orcinol *p*-depsides using 3-methyl-2-benzothiazolone hydrazone hydrochloride as spray reagent has been described (*16*) and according to Ramaut *et al.* (*607*) the best solvent system for the separation of atranorin and chloroatranorin is benzene:dioxane:acetic acid in a ratio of 90:25:4 (V/V/V, Pastuska mixture). R_f values of 39 lichen xanthones using ethyl acetate:cyclohexane 1:3 (V/V) and ethyl acetate:cyclohexane 1:1 (V/V) mixtures as solvent systems have been listed (*201*). A catalogue of standardized TLC data has been published (*206*).

2.3. High Performance Liquid Chromatography (HPLC)

High performance liquid chromatography is an ideal supplement of or replacement for identification of lichen substances because it is more sensitive and can be interpreted quantitatively. Applications of HPLC in lichen chemistry are summarized in Table 1.

Culberson *et al.* (*92*) found that retention times of a homologous series of orcinol depsides and phenolic units vary linearly with the number of carbon atoms. When two or more homologs are known, retention times of new members of the same series can be predicted accurately. Combination of this method with other microchemical techniques allows structural elucidation of depsides in small fragments of lichens taken from herbarium vouchers.

Table 1. HPLC of lichen substances

Analyzed compounds, groups of compounds or lichen	Ref.
Atranorin, vulpinic acid	(*673*)
Depsides, depsidones	(*510*)
Rhizocarpon superficiale	(*322*)
Different lichen substances	(*321*)
Aromatic compounds from the genera *Cladina* and *Cladonia*	(*417*)
Phenolic compounds from *Cladina stellaris* and *C. rangiferina*	(*416*)
Aromatic compounds from the genus *Cladina*	(*419*)
Aromatic compounds from the genus *Cladonia*, section Uncialis	(*420*)
Aromatic compounds	(*418*)
Aromatic compounds from the genus *Cladonia*, section Perviae	(*421*)
Aromatic compounds from the genus *Cladonia*, section Cocciferae	(*422*)
Aromatic compounds from the genus *Cladonia*, section Helopodium and subsection Foliosae	(*423*)
Extracts from *Evernia prunastri*	(*641*)
Lasallia brigantium, L. hispanica, L. pustulata	(*581, 583*)
Aromatic compounds from the genus *Cladonia*, section Cladonia, group Furcatae	(*424*)
Aromatic compounds from the genera *Cladina* and *Cladonia*	(*426*)
Sugars and Polyols	(*710*)
Usnic acid	(*707*)
Aromatic compounds from *Ramalina siliquosa*	(*98*)
Aromatic compounds from *Usena diffracta*	(*755*)
Phenol carboxylic acids and depsides	(*479*)
Aromatic lichen substances	(*291, 752*)
Aliphatic acids	(*393*)
Depsides and depsidones from lichens of the family Umbilicariaceae	(*556*)
(+)- and (−)-Usnic and isousnic acids	(*456*)
Protolichesterinic acid, fumarprotocetraric acid	(*338*)
Pulvinic acid derivatives	(*68*)

2.4. Droplet Counter-Current Chromatography

Vulpinic acid has been separated from an extract of *Letharia vulpina* by droplet counter-current chromatography using methanol:acetone:*n*-heptane 4:1:5 (V/V/V) (*178*).

2.5. NMR Spectroscopy

^{1}H- and ^{13}C-NMR spectroscopy and the corresponding special techniques are of unique importance in structure elucidation of natural products and are widely used in lichen chemistry. For example Gagnaire *et al.* (*309*) and Gagnaire and Vincendon (*310*) have discussed the NMR

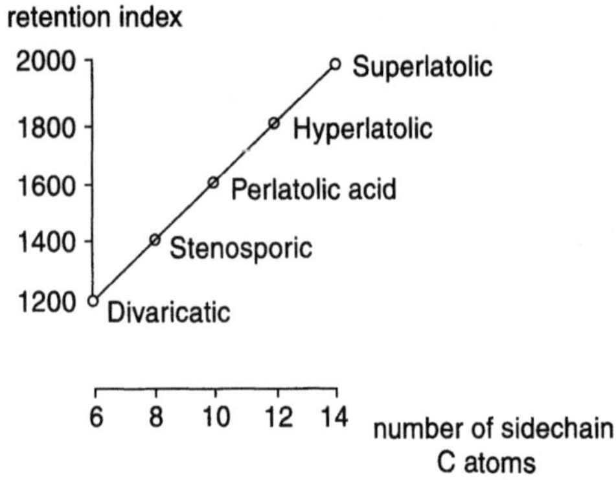

Fig. 1. Linear relationship of retention index and number of sidechain carbons in depsides

spectra of lichenin, β(1 → 4), β(1 → 3) polyglucane from lichens while Chapman (*66*) studied the water soluble compounds of some Antarctic lichen by ^{13}C-NMR spectroscopy. Also 3-acyltetronic acids have been shown by ^{13}C-NMR spectroscopy (*440*) to exist in two monoenolic forms A and B shown in Scheme 1.

According to Steglich *et al.* (*668*) the NMR spectra of pulvinic acids show large differences in the chemical shifts of the 2,6 and 2′,6′-aromatic protons. This allows unambiguous differentiation between the two possible isomers (A, B) of unsymmetrically substituted pulvinic acids shown in Fig. 2.

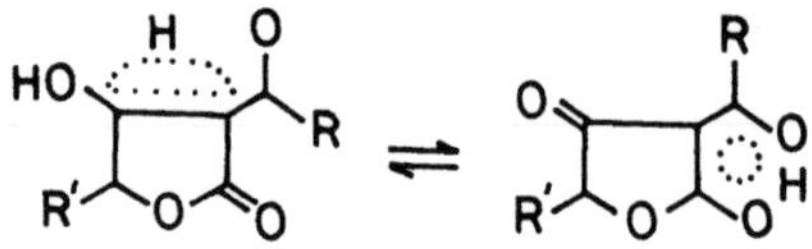

A B

Scheme 1. Monoenolic forms of 3-acetyltetronic acids

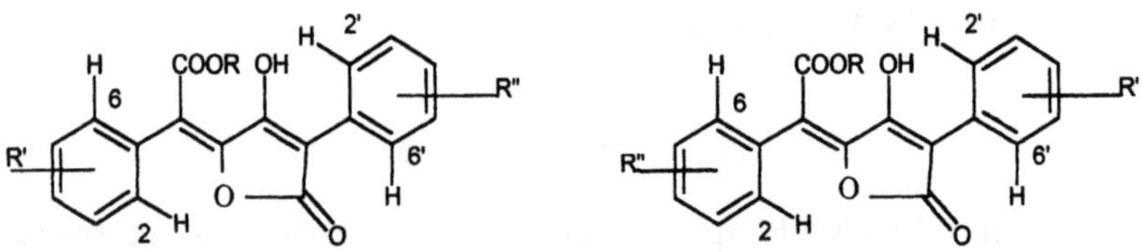

A B

Fig. 2. NMR differentiation of protons of pulvinic acids

Jakupovic and Huneck (*442*) correlated the ^{1}H-NMR spectra of 19 depsides, depsidones, depsones and dibenzofurans by NOE difference spectroscopy while Malterud and Anthonsen (*523*) recorded the ^{13}C-NMR spectra of 15 phenol derivatives. Spin–lattice relaxation times, T_1, were measured for the quarternary carbon atoms of these compounds and signals were assigned to the aromatic carbon atoms bound to hydroxyl and methoxy groups. T_1 values for C–OH are smaller than those for C–OMe within the same molecule. Furthermore, C–OMe groups with substituents in both ortho-positions have larger T_1 values than those with at least one unsubstituted ortho-position. Narui *et al.* (*557*) assigned the ^{1}H and ^{13}C-NMR spectra of the following depsides and tridepsides found in Umbilicariaceae: lecanoric acid, methyl lecanorate, evernic acid, methyl evernate, gyrophoric acid, hiascic acid, ovoic acid, umbilicaric acid, crustinic acid, lasallic acid, methyl gyrophorate, 4-*O*-methylgyrophoric acid, tenuiorin, papulosic acid, deliseic acid, and methyl 4-*O*-methyldeliseate, while ^{13}C-NMR spectra of 14 naphthoquinones and of 31 naphtho- and anthraquinones have been recorded by McDonald *et al.* (*539*) and Höfle (*376*) respectively. Numerous papers have appeared on ^{1}H and ^{13}C-NMR spectra of lichen triterpenoids, *e.g.* lupanes and hopanes (*10*, *719*, *651*, *731*), lupanes and taraxeranes (*59*), friedelanes and taraxasteranes (*572*), migrated hopanes and 21α-H-hopanes (*571*), fernenes (*729*) and compounds of the pyxinol group (*730*). The ^{13}C-NMR spectra of pentacyclic triterpenoids have been reviewed (*522*).

2.6. Mass Spectrometry (MS)

Addison (*2*) has applied pyrolysis-chemical ionization mass spectrometry with protonated molecular ions to the MS of the thermally labile lichen substances gyrophoric acid and tenuiorin. Tabacchi *et al.* (*683*, *686*) used tandem mass spectrometry (TMS) for analysis of *Pseudevernia furfuracea* and *Lobaria pulmonaria* and some lichen triterpenes [for the technique of TMS see Stach *et al.* (*664*)]. Holzmann and Leuckert (*375*) identified depsides, depsidones and usnic acid by their specific fragmentation patterns without prior isolation or purification. Mass spectra of ten di- and tri-oxygenated stictane triterpenoids and their trimethylsilyl derivatives have been studied (*374*). The major fragmentations involve ring C, but specific cleavages in ring A, B and E are also observed which allow the substitution pattern of the skeleton to be readily determined: see Scheme 2. The following classes of lichen substances have been analyzed by MS: depsides and depsidones (*177*,

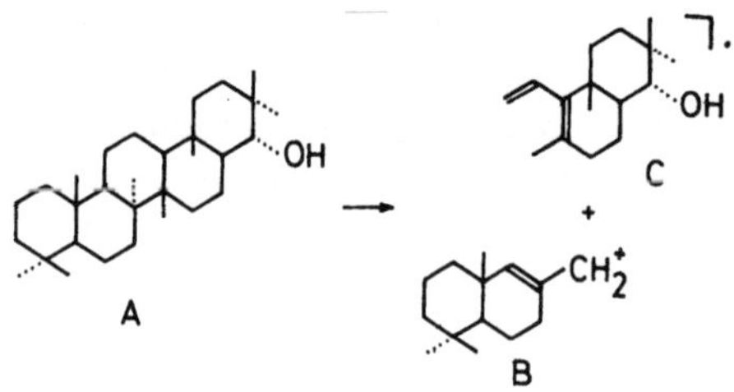

Scheme 2. MS fragmentation of 22α-hydroxystictane

425, *473*, *527*), xanthones (*623*), polyporic acid and derivatives (*337*) and usnic acid and derivatives (*404*, *639*).

Allemand (*9*) used MS-MS for identification of triterpenes and steroids in *Evernia prunastri* while Shojima *et al.* (*650*) reported the MS of about 100 natural and unsaturated pentacyclic triterpenoids, including hopanes and lupanes.

2.7. X-ray Analysis

X-ray analyses of lichen substances since the last review are summarized in Table 2.

2.8. Cathodoluminescence and X-ray Microanalysis

Mathey and Hoder (*530*) have applied the technique of cathodoluminescence and X-ray microanalysis to the identification of vinetorin, 3-*O*-methyl-2,5,7-trichloronorlichexanthone and of quinones in some crustose lichens.

2.9. Laser Microprobe Mass Spectrometry (LAMMA)

The principle of laser microprobe mass spectrometry has been described by Vogt *et al.* (*713*). A laser beam focussed on the object ionizes the organic compound(s) present in the sample; the ions are separated and recorded in a time-of-flight mass spectrometer. Mathey (*529*) discussed the prospects of LAMMA for lichenology, stating that this method allows not only to analyze minute amounts of compounds present in lichens (especially in tiny foliculous species) but also to correlate the data with certain structures. Thus lichexanthone could be identified in the tropical microlichen *Laurera benguelensis*. In a further

Table 2. X-ray analyses of lichen substances

Compound	Ref.
7-Acetyl-8,10-dihydroxy-3,9,10b-trimethylbenzofuro[2,3-g]-1,2-benzisoxazol-4(10bH)-one	(*81*)
8-Acetyl-5,7-dihydroxy-3,4a,6-trimethylbenzofuro[3,2-f]-1,2-benzisoxazol-4(4aH)-one	(*81*)
(+)-Aspicilin	(*597*)
Atranorin	(*40*)
Cercosporamide	(*679*)
Eriodermin	(*79*)
Graciliformin	(*7*)
Graphenone	(*549*)
Graphisquinone	(*549*)
22α-Hydroxy-3,4-seco-stictan-3-oic acid	(*82*)
3-Hydroxy-4,4,6-trimethoxy-2,5-di-(*p*-methoxyphenyl)cyclohexa-2,5-dien-1-one (intermediate of butlerin B)	(*284*)
Hypostrepsilic acid	(*548*)
Lepranthin	(*579*)
Leprolomin	(*205*)
Methyl 2,3-dihydroxy-6-hydroxy-5′-methoxy-4-(3-oxopentyl)-3,3′-dioxo-1′β-pentylspiro[benzofuran-2,2′-oxabicyclo [4.1.0]-hept-5-ene]-6′β-carboxylate (intermediate of the synthesis of epiphorellic acid 2)	(*77*)
Methyl everninate	(*35*)
Methyl isorhizonate	(*34*)
Pentaacetylnorstictic acid	(Göhrt, unpublished)
1-*n*-Pentyl-2-methoxycarbonyl-3,6-dimethoxy-8-methoxalyl-dibenzofuran (degradation product of didymic acid)	(*647*)
Phlebic acid D	(*24*)
(−)-Placodiolic acid	(*78*)
Retigeranic acids A and B	(*678*)
Roccellaric acid	(*552*)
Scabrosin dibutyrate (ambewelamide A)	(*283*)
Sekikaic acid	(*452*)
Solorinine	(*534*)
Stictane 3β,22α-diol	(*82*)
Tetraacetylsimonyellin	(*210*)
Thiomelin	(*218*)
Tri-*O*-acetyldemethylchodatin	(*254*)
Triacetylrhodocladonic acid	(*540*)
(+)-Usnic acid sodium salt 2½ hydrate	(*620*)
Vulpinic acid	(*41*)
Wrightiin	(*518*)

paper Mathey *et al.* (*533*) applied the method to other lichens and identified 1,3,6,8-tetrahydroxyanthraquinone in *Phaeographina chrysocarpa*, russulone, norlichexanthone and parietin in *Laurera sanguinaria* and elsinochrome A in a pathogenic lichen from Brazil.

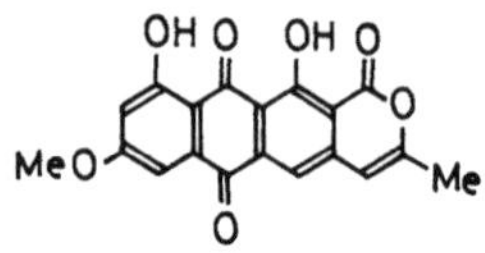

Russulone

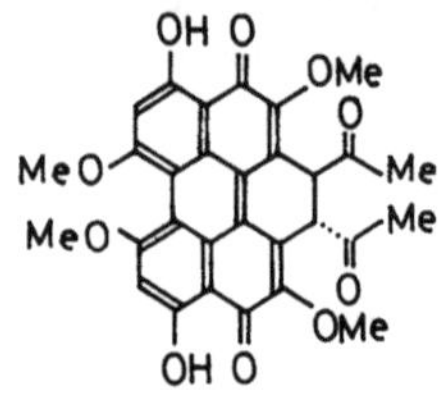

Elsinochrome A

LAMMA, in combination with IR and NMR data furnishes in most cases enough information for complete structure elucidation of lichen pigments (*705*, *531*).

2.10. Fluorescence Microscopy

Kauppi and Verseghy-Patay (*449*) studied the distribution of lichen substances (especially depsides and depsidones) in lichen thalli by fluorescence microscopy.

2.11. Computer Program for Identification of Lichen Substances

Mietzsch *et al.* (*544*) have developed a computer program (Wintabolites) for identification of lichen substances based on spot tests, TLC R_f values and HPLC R_I values. The third edition of this data base contains data for over 700 lichen substances (*545*).

3. General Methods for Derivatization and Regeneration of Lichen Substances

3.1. Carboxyl groups

3.1.1. Methylation

A very useful reagent for methylation of carboxylic acids is diazomethane. At 0°C and within a few seconds only the carboxylic group of phenolic carboxylic acids is esterified. Alkaline saponification regenerates the carboxylic acid, but in compounds with a free hydroxyl group in the *para* position decarboxylation occurs. Another reagent for regeneration of the acid from a methyl ester is lithium propylmercaptide. α-Methylene-γ-lactonecarboxylic acids like protolichesterinic acid react with diazomethane to form methyl esters of pyrazolines. In order to prepare the methyl ester the carboxylic acid is first converted into the silver salt which on reaction with methyl iodide furnishes the corresponding methyl ester (see Scheme 3). Carboxyl groups can also be protected by formation of an oxazoline (see Scheme 4).

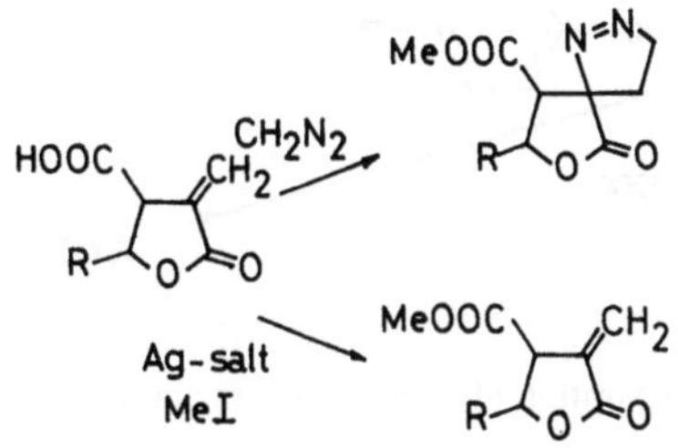

Scheme 3. Reaction of α-methylene-γ-lactonecarboxylic acids with diazomethane and of their silver salts with methyl iodide

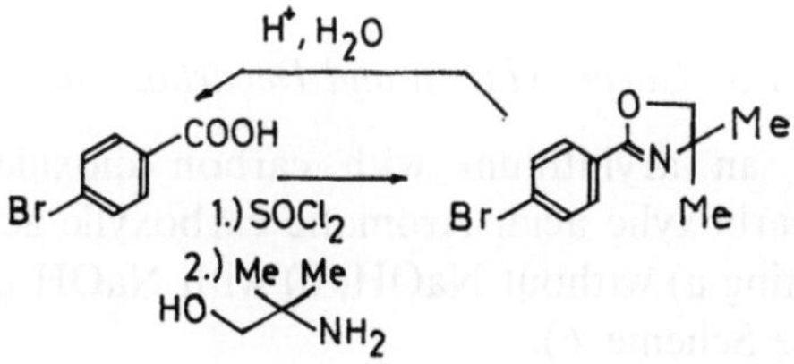

Scheme 4. Transformation of a carboxylic acid into an oxazoline

Phenacyl esters are prepared by heating the potassium salt with phenacylbromide in the presence of 18-crown-6 in acetonitrile. Such esters exhibit strong UV absorption near 240 nm and are useful derivatives for HPLC (*393*). *Para*-nitrophenyl esters are prepared from the acid and *p*-nitrophenol in the presence of *N*,*N*′-dicyclohexylcarbodiimide and absorb at about 270 nm (see Scheme 5).

3.1.2. Benzylation and Debenzylation

Benzylation of a carboxylic acid is a key step in the synthesis of depsides and is carried out using a) benzylbromide (BnBr) and

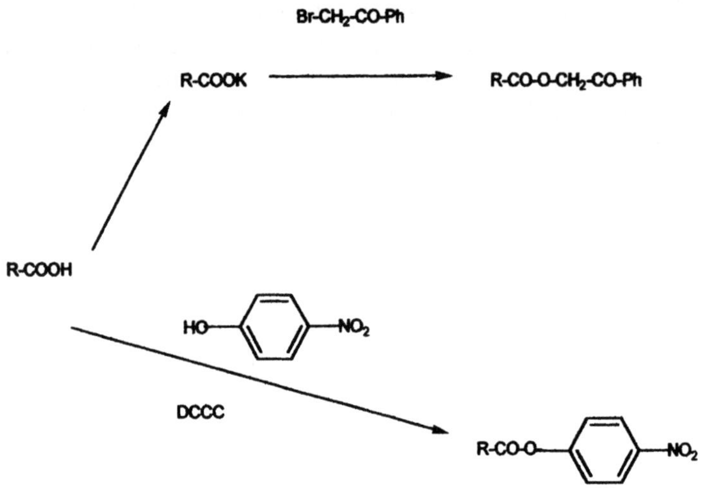

Scheme 5. Preparation of phenacyl esters and *p*-nitrophenyl esters

potassium hydrogen carbonate or b) phenyldiazomethane. Hydrogenation of the benzyl ester with PD–C regenerates the free carboxylic acid.

3.1.3. Carboxylation and Decarboxylation

Reaction of an aryllithium with carbon dioxide leads to the corresponding carboxylic acid. Aromatic carboxylic acids are decarboxylated by heating a) without NaOH, b) with NaOH or c) with Cu_2O in dipyridyle (see Scheme 6).

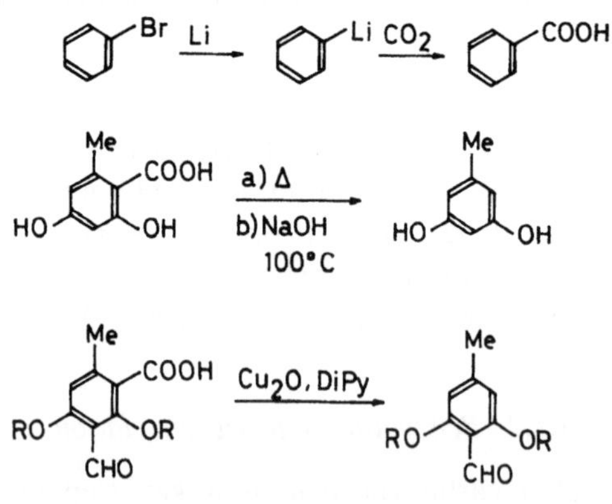

Scheme 6. Carboxylation and decarboxylation of aromatic carboxylic acids

3.2. Alcoholic and Phenolic groups

3.2.1. Acetylation

Primary and secondary aliphatic and phenolic hydroxyls can be acetylated either by acetanhydride–pyridine (Ac_2O–Py) or by acetanhydride–sulfuric acid (Ac_2O–H_2SO_4). Aromatic aldehydes react with Ac_2O–H_2SO_4 to form the corresponding diacetates (see Scheme 7).

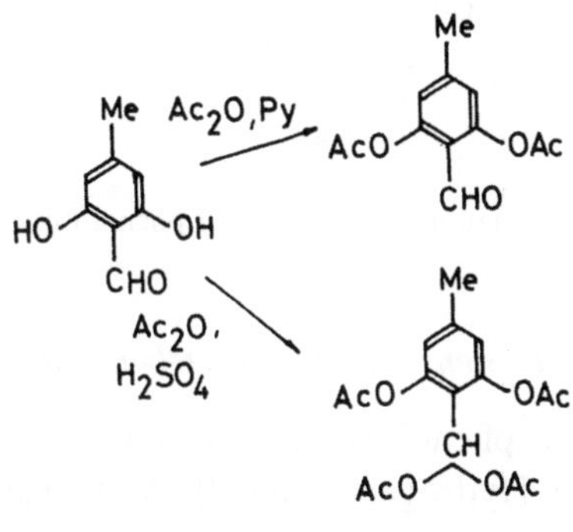

Scheme 7. Reaction of aromatic aldehydes with Ac_2O–H_2SO_4 to diacetates

3.2.2. Methylation and Demethylation

Methylating agents for phenolic groups are a) diazomethane (the reaction time varies from minutes to 24 hours), b) dimethyl sulphate (DMS) – potassium carbonate in acetone or dimethylformamide (DMF) and c) methyl iodide – potassium carbonate in acetone. For demethylation of a methyl ether group ortho to a methyl ester group boron tribromide is used, while demethylation of a methyl ether para to a methyl ester group is carried out with boron trichloride (see Scheme 8).

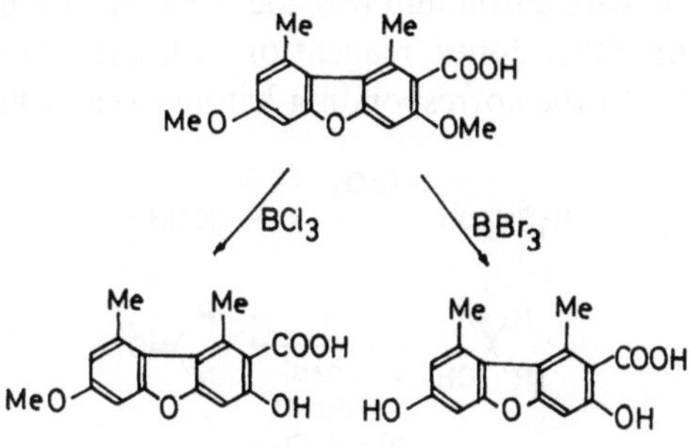

Scheme 8. Demethylation of methyl ethers with boron tribromide and boron trichloride

3.2.3. *Isopropylation and Deisopropylation*

Phenols are isopropylated with isopropyl bromide – potassium carbonate. The products are cleaved with either BCl_3 or with $TiCl_4$ in methylenechloride at 0°C (see Scheme 9).

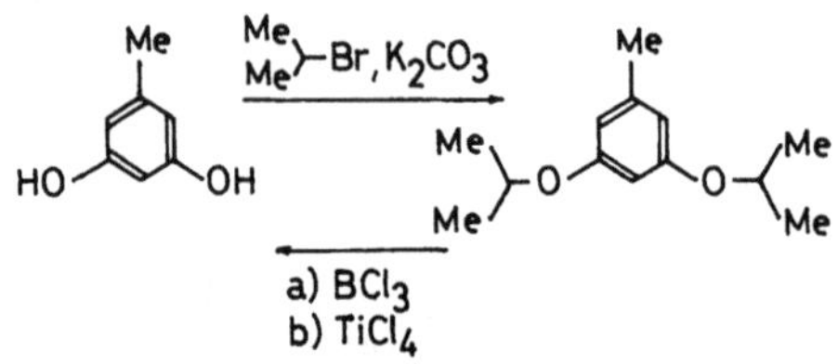

Scheme 9. Formation and cleavage of phenol-isopropyl ethers

3.2.4. *Benzylation and Debenzylation*

Benzylation of a phenol is carried out using benzylbromide – potassium carbonate. Hydrogenation with Pd–C regenerates the phenol (see Scheme 10).

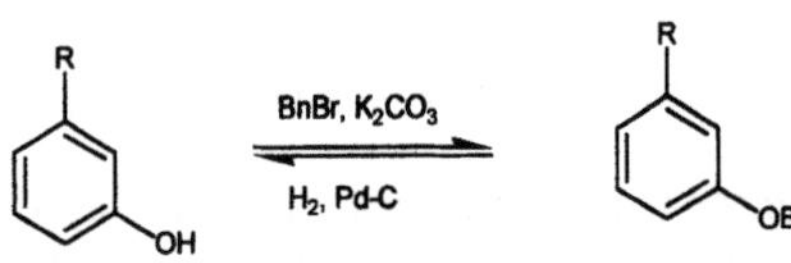

Scheme 10. Formation of a phenol-benzyl ether and its cleavage

3.2.5. *Oxidation of Alcohols*

Primary aliphatic alcohols produce the corresponding carboxylic acid on oxidation with chromium trioxide. Secondary aliphatic alcohols are oxidized with either Jones' reagent or with pyridinium chlorochromate ($PyClCrO_4$) to the corresponding ketones (see Scheme 11).

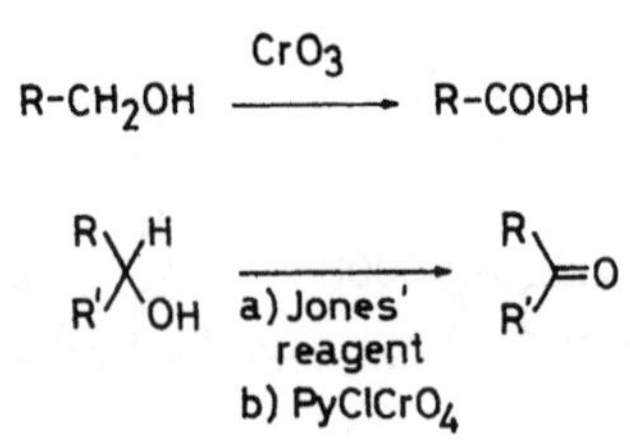

Scheme 11. Oxidation of aliphatic alcohols

3.3. Substitution at Aromatic Rings

3.3.1. Halogenation and Dehalogenation

Halogenation and dehalogenation of aromatic compounds are summarized in Scheme 12.

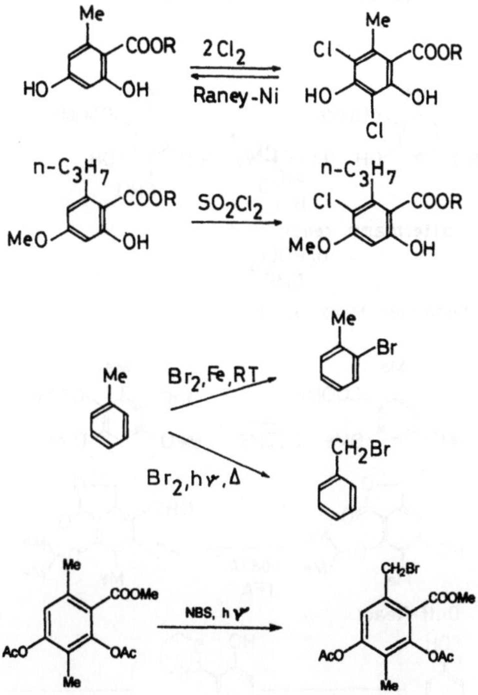

Scheme 12. Halogenation and dehalogenation of aromatic compounds

3.3.2. Hydroxylation

Ethyl everninate is hydroxylated to ethyl 5-hydroxyeverninate by oxidation with potassium peroxodisulphate (see Scheme 13).

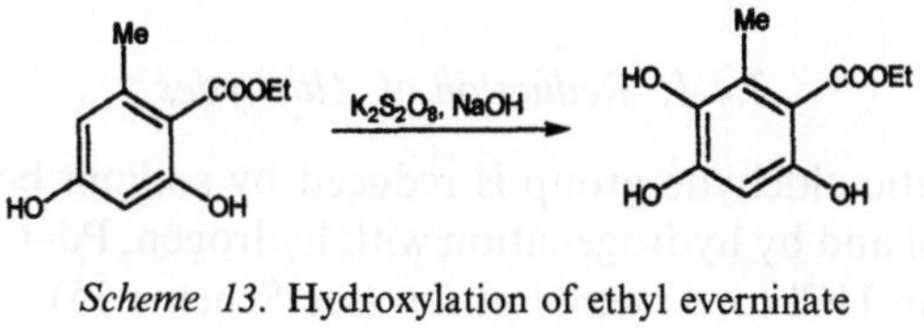

Scheme 13. Hydroxylation of ethyl everninate

3.4. Synthesis of Aldehydes

Common syntheses of aromatic aldehydes employ a) the Gattermann reaction, b) the Vilsmeier–Haack reaction, c) formylation with dichlorodimethyl ether (DCDME), d) the Duff reaction, and e) the Gassman–Amick reaction (see Scheme 14). Aromatic aldehydes can be protected either as diacetates or Schiff's base.

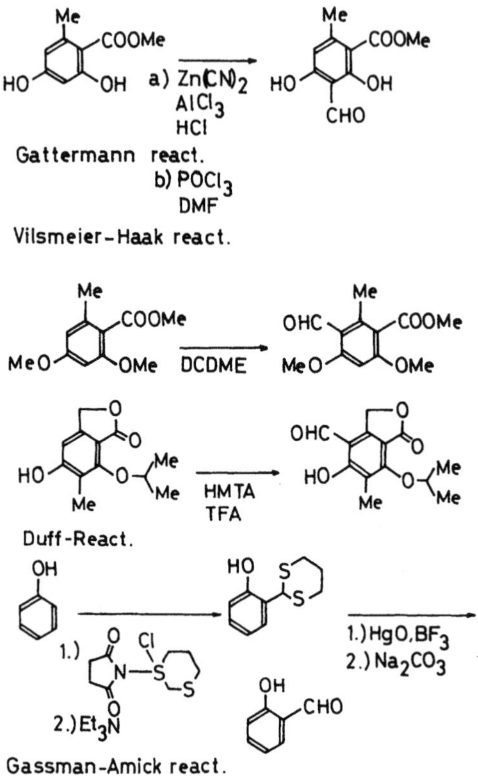

Scheme 14. Syntheses of aromatic aldehydes

3.4.1. Reduction of Aldehydes

An aromatic aldehyde group is reduced by sodium borohydride to benzyl alcohol and by hydrogenation with hydrogen, Pd–C and reaction with Zn–Hg in HCl to a methyl group (see Scheme 15).

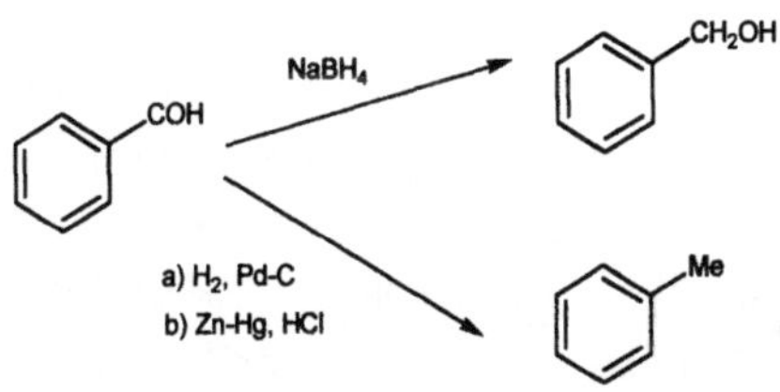

Scheme 15. Reduction of aromatic aldehydes

3.4.2. Oxidation of Aldehydes

Aromatic aldehydes are oxidized a) with sodium chlorite ($NaClO_2$) in the presence of sulfamic acid to the corresponding carboxylic acid (Lindgren–Nilson reaction) and b) with hydrogen peroxide or *m*-chloroperbenzoic acid to the corresponding phenols (see Scheme 16).

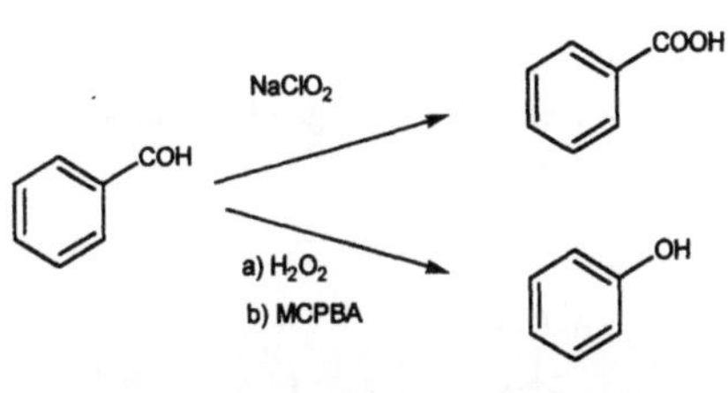

Scheme 16. Oxidation of aromatic aldehydes to carboxylic acids and phenols

3.5. Cleavage and Formation of the Depside Ester Bond

Depsides may be cleaved with either conc. sulphuric acid at 0°C or with KOH–H_2O or KOH–MeOH. Prolonged boiling (5–20 hours) of a depside with *tert.*-butanol (TBA) yields the *tert.*-butyl ester of the S-part of the depside and the phenolic unit (*23*, *384*, *542*) (see Scheme 17). Depsides with a keto group in the β-position of the side chain in the S-part of the molecule give with KOH at room temperature a γ,δ-unsaturated δ-lactone, *e.g.* confluentic acid → 4-*O*-methylolivetonide (see Scheme 18).

Trifluoroacetic anhydride (TFAA) and *N*,*N'*-dicyclohexylcarbodiimide (DCCD) are modern reagents for the synthesis of depsides (see Scheme 19). A mild conversion of carboxylic acids to esters using

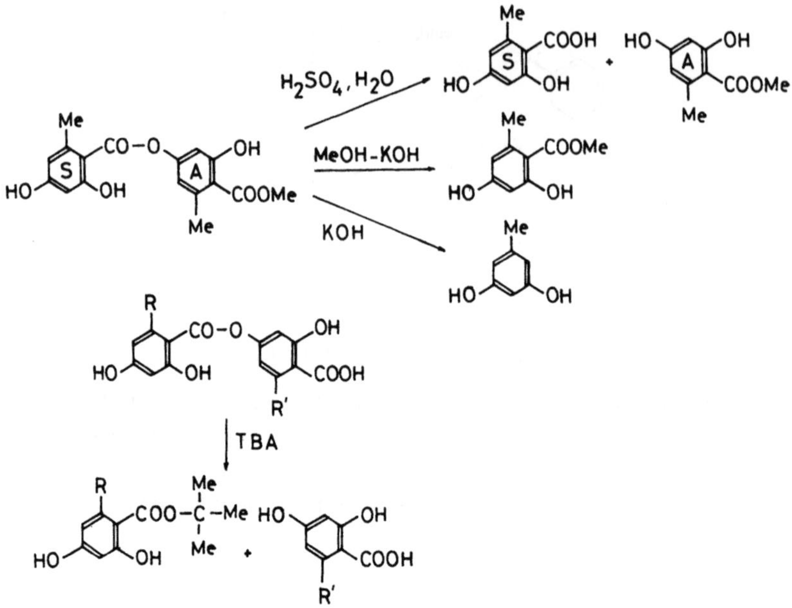

Scheme 17. Cleavage of depsides

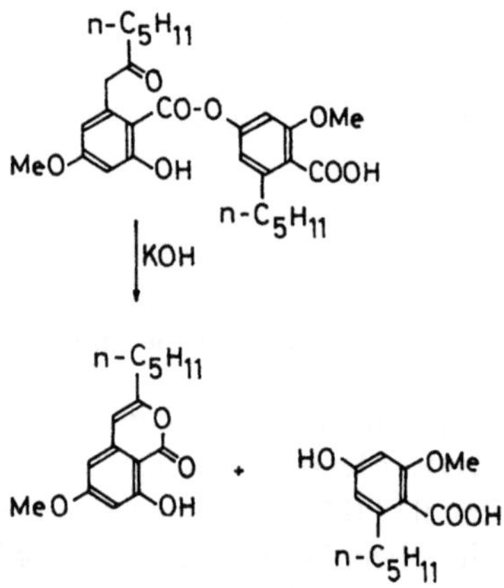

Scheme 18. Cleavage of confluentic acid with KOH

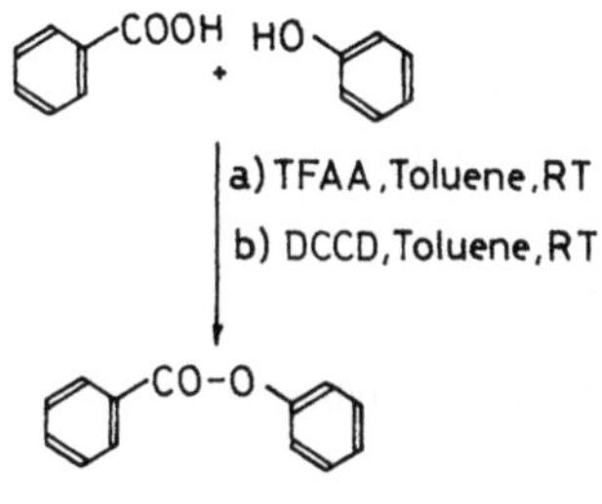

Scheme 19. Synthesis of depsides

transfer catalysis was described by Puntambekar *et al.* (*592*) and may be useful for the synthesis of lichen depsides.

4. Structure Elucidation and Synthesis of Lichen Substances

4.1. Nitrogen Containing Compounds

Fujikawa *et al.* (*304–306*) analyzed the free amino acids in Japanese lichens. Solberg (*657*) isolated from *Anaptychia fusca* a mixture of an aminopeptidol ($C_5H_{13}NO_4$) and an aminocyclitetrol ($C_6H_{15}NO_4$) and from *Stereocaulon paschale* an aminohexitol ($C_6H_{15}NO_5$); the structures of these constituents are unknown. Solorinin (**1**) is a new amino acid derivative from *Solorina crocea* whose structure was determined by X-ray diffraction (*451*, *534*). Glutathione (**2**) and glutathione disulfide (**3**) have been found in *Pseudevernia furfuracea* (*472*). The structure of picroroccellin (**4**) proposed initially (*301*) has been revised to **5** by Marcuccio and Elix (*524*). Later on these workers also synthesized methylanhydropicroroccellin (**6**) and dimethylpicroroccellin (**7**) and established their trans-stereochemistry (*525*). The properties of dimethylpicroroccellin were identical with those recorded for this compound prepared from the natural product from *Roccella fuciformis* (*670*) (see Scheme 20).

Shin *et al.* (*649*) synthesized picroroccellin diastereomers and their regiomers (see Scheme 21).

Huneck *et al.* (*400*) established the structure of arthonin (**8**), the dibenzoate of new amino acid amino alcohol from *Arthonia endlicheri* as (−)-*N*-benzoyl-L-valinyl-*N*′-benzoyl-L-isoleucinate. The synthesis of arthonin is shown in Scheme 22; the isomeric *iso*-arthonin (**9**) was also synthesized.

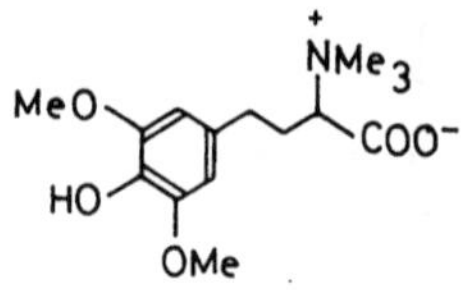

1: Solorinin

mp 149-150, $[\alpha]_D + 9$ (MeOH)

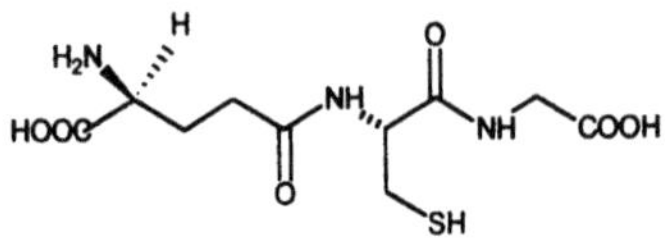

2: Glutathione

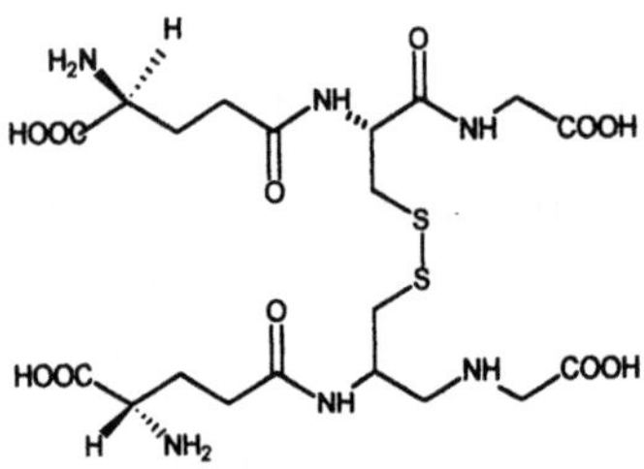

3: Glutathionedisulfide

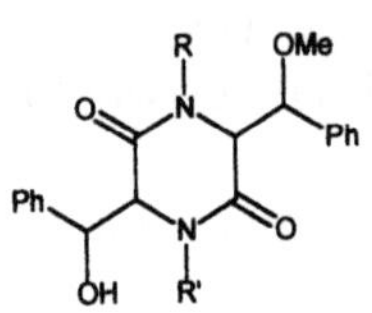

4: R = H, R′= Me or R = Me, R′= H

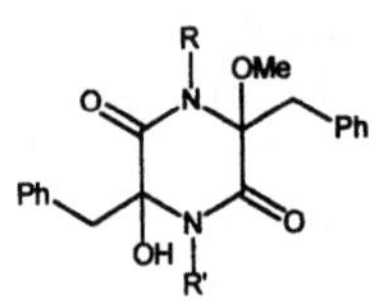

5: Picroroccellin: R = H,R′ = Me or R = Me, R′ = H

Another dibenzoate of an aminoacid amino alcohol is hypothallin, (−)-*N*-benzoyl-L-phenyl-alaninyl-*N*′-benzoyl-L-leucinate (**10**) from the lichen *Schismatomma hypothallinum* (*402*). It was synthesized by condensation of (−)-*N*-benzoxycarbonyl-L-leucine and (−)-*N*′-benzoyl-L-phenylalaninol in the presence of *N*, *N*′-carbonyldiimidazole and subsequent debenzoxycarbonylation and benzoylation (see Scheme 23).

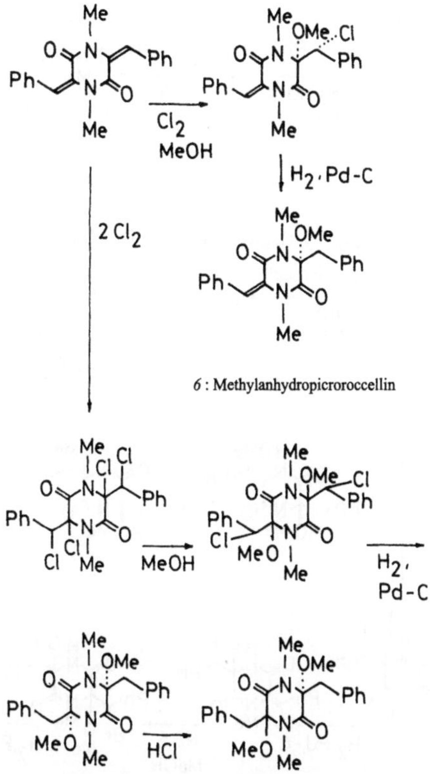

6 : Methylanhydropicroroccellin

7: Dimethylpicroroccellin

Scheme 20. Synthesis of methylanhydropicroroccellin and dimethylpicroroccellin

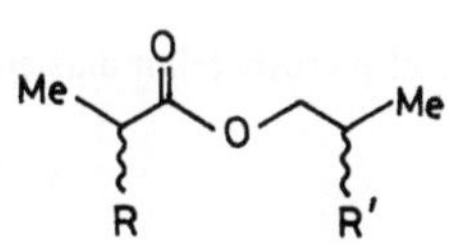

11	R	R'
a:	◀ NH-CO-Ph	◀ NH-CO-Ph
b:	····· NH-CO-Ph	····· NH-CO-Ph
c:	◀ NH-CO-Ph	····· NH-CO-Ph
d:	····· NH-CO-Ph	◀ NH-CO-Ph

Huneck and Porzel (*399*) synthesized all stereoisomers esters of L-and D-*N*-benzoylalanine and L- and D-benzoylalaninol (**11a–11d**) and recorded their CD spectra; in this way the stereochemistry of new esters can be correlated easily.

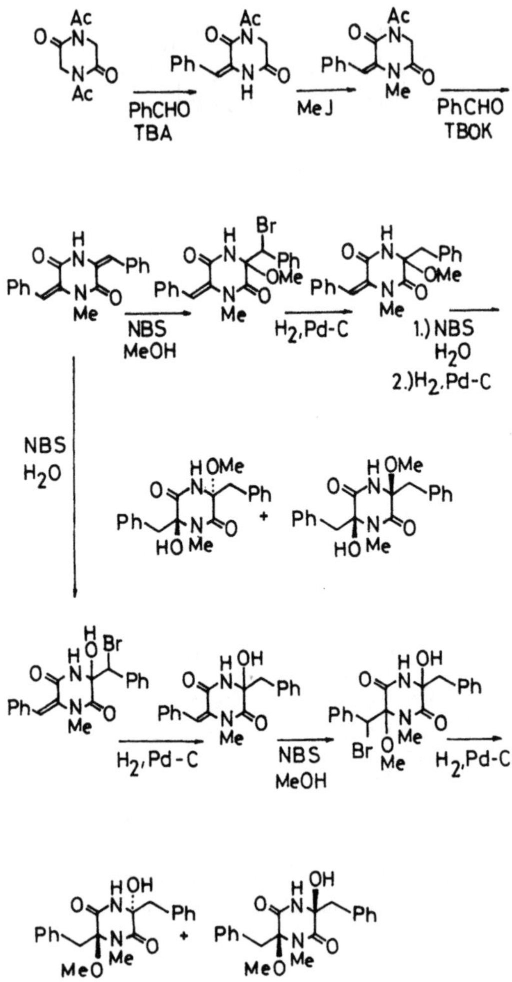

Scheme 21. Synthesis of picroroccellin diastereomers and their regiomers

Huneck and Follmann (*394*) isolated from *Arthothelium galapagoense* a neutral compound which they named arthogalin. Later on Huneck *et al.* (*395*) elucidated the structure of arthogalin as **12** by hydrolysis to S-phenylalanine, S-valine and two molecules of R-α-hydroxyisovalerianic acid. The connectivities of these units were fixed by proton-detected multiple bond ^{1}H, ^{13}C correlation (HMBC).

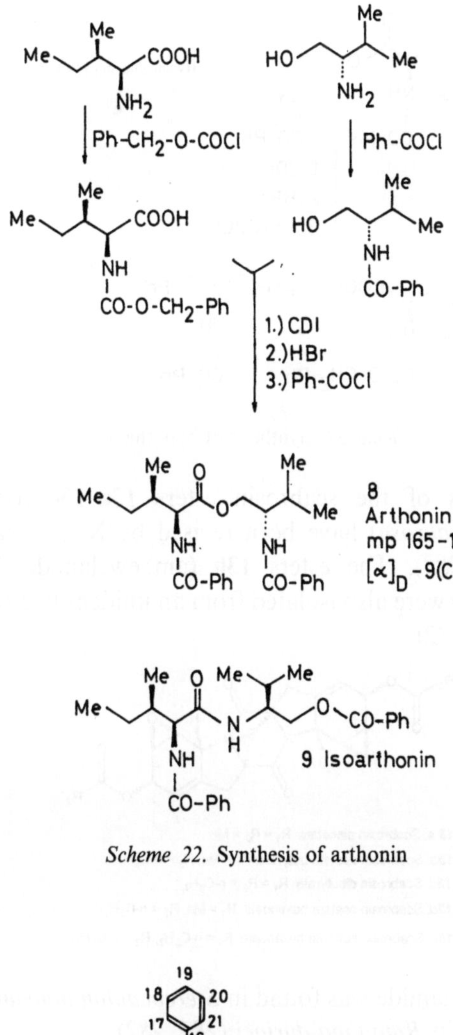

Scheme 22. Synthesis of arthonin

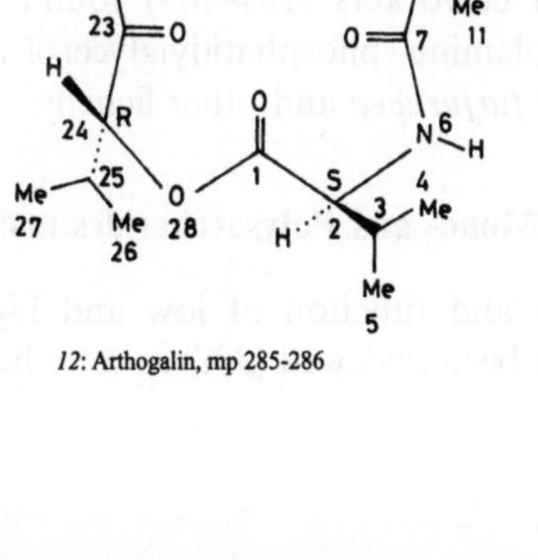

12: Arthogalin, mp 285-286

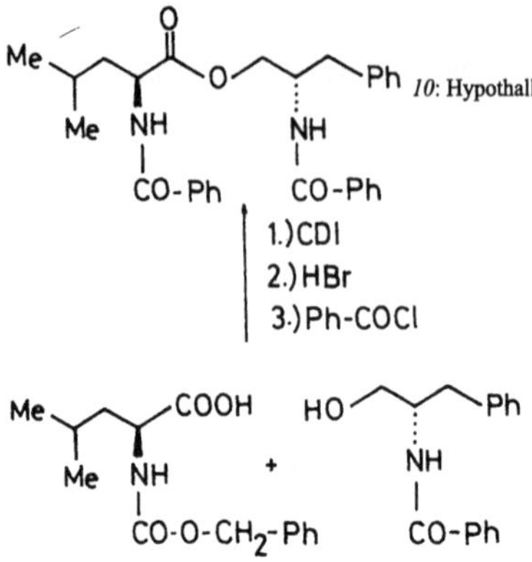

10: Hypothallin, mp 176-177

Scheme 23. Synthesis of hypothallin

The structures of the scabrosin esters **13a–13e** (isolated from *Xanthoparmelia scabrosa*) have been revised by X-ray analysis of the dibutyrate **13c** (*283*). The esters **13b** (ambewelamide A) and **13d** (ambewelamide B) were also isolated from an unidentified *Usnea* species from Sri Lanka (*732*).

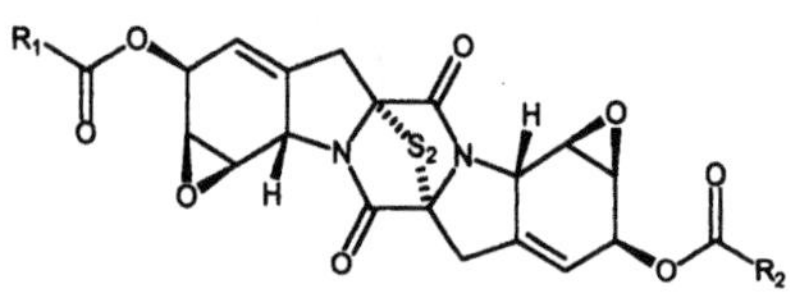

13a: Scabrosin diacetate: $R_1 = R_2 = Me$
13b: Scabrosin acetate butyrate: $R_1 = Me$, $R_2 = n\text{-}C_3H_7$
13c: Scabrosin dibutyrate: $R_1 = R_2 = n\text{-}C_3H_7$
13d: Scabrosin acetate hexanoate: $R_1 = Me$, $R_2 = n\text{-}C_5H_{11}$
13e: Scabrosin butyrate hexanoate: $R_1 = n\text{-}C_3H_7$, $R_2 = n\text{-}C_5H_{11}$

9-*cis*-Octadecenamide was found in *Stereocaulon alpinum* (*436a*) and indolylacetic acid in *Ramalina duriaei* (*281*, *282*).

4.2. Phosphorus Containing Compounds

Dembitsky and coworkers (*164–169*) found phosphatidylcholine, phosphatidylethanolamine, phosphatidylglycerol and phosphatidylserine in *Pseudevernia furfuracea* and other lichens.

4.3. Polyols, Mono- and Polysaccharides and Carbohydrates

The occurrence and function of low and high molecular weight carbohydrates have been reviewed (*332*) as were lichen glycosides (*587*).

According to Pueyo (*586*) the following diholosides and heterosides were found in lichens: saccharose, trehalose, galactosyl-arabitol, galactosyl-pentitol, galactosyl-ribitol, glucosyl-mannitol, galactosyl-mannitol, and mannosyl-mannitol, while lactose has been isolated from the lichen *Sticta damaecornis* (*159*). Syntheses of tri- and tetra-saccharides related to fine structures of lichenan and cereal β-D-glucanes were based on the silver tirfluoromethanesulfonate-promoted Königs-Knorr condensation (*692*). The ^{1}H-NMR spectrum of lichenin confirmed its structure (14) as a polysaccharide containing glucose units linked by β(1 → 4) and β(1 → 3) glycosidic linkages in the ratio 2:1 (*309*). The crystalline conformation of lichenin was investigated by Tvaroska *et al.* (*700*).

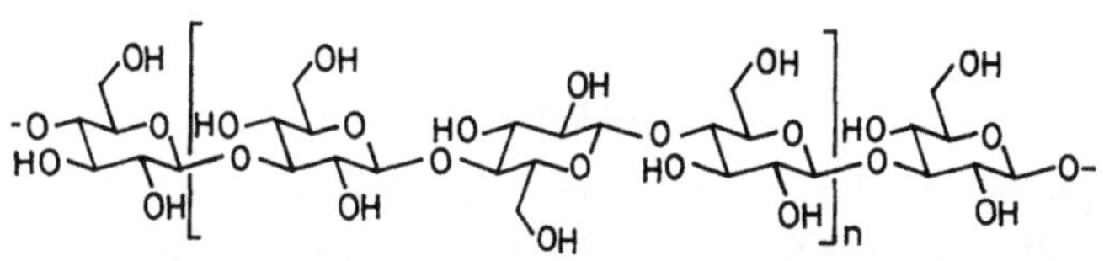

14: Lichenin

The ^{13}C-NMR spectra of lichenin (*310*) and pustulan (*674*) have been reported. Water-soluble polysaccharides from five *Stereocaulon* and two *Umbilicaria* species have been analyzed by Takahashi *et al.* (*688*) and Kjølberg and Kvernheim (*462*) respectively. *Umbilicaria* species contain partly *O*-acetylated pustulan. D-Galacto-D-mannans were isolated from six Canadian lichens (*333*) and from *Umbilicaria pustulata* and *U. spodochroa* (*463*). Iacomini and coworkers analyzed the carbohydrates of *Cladonia alpestris*, *C. confusa*, *C. amaurocraea* (*431*), *Cora pavonia* (*432*), *Stereocaulon ramulosum* (*26*, *27*), *Letharia vulpina*, *Actinogyra muehlenbergii*, *Usnea* spec. (*430*), *Ramalina ecklonii* (*543*), and *Evernia prunastri* (*695*). The polysaccharides from *Neuropogon aurantiaco-ater* consist of lichenan, isolichenan and a galactomannan (*28*), while an unusual glucomannan from *Tornabenia intricata* has been isolated (*694*) which has a (1 → 6)-linked α-D-Man p main chain substituted at O-2 mainly with side chains of α-DF-Man p. Three aldoses, six alditols, eight oligosaccharides and a glycopeptide were found in *Sticta* species from Brazil (*160*). The glycosphingolipid from *Ramalina celastri* was characterized as an *O*-β-D-galactopyranosyl-(1 → 1′)-ceramide; its main lipid components are (4E)-sphingenine, sphinganine and eicosa-sphinganin, esterified with palmitic, oleic and 2-hydroxypalmitic acids (*520*).

4.4. Aliphatic and Cycloaliphatic Compounds

4.4.1. Hydrocarbons

The plant growth hormone ethylene was detected in the lichen *Ramalina duriaei* by Epstein (*281*, *282*). Ott and Zwoch (*569*) analyzed *Cetraria islandica*, *Cladonia rangiferina*, *C. arbuscula*, *C. sulfurina*, *C. cornuta*, *C. pyxidata*, *Peltigera lactucifolia* and *Usnea aurantiaco-ater* for ethylene and found it in amounts of 0.8–5.5 μl/g dry weight/hour. The presence of hydrocarbons with 17–33 carbon atoms has been reported (*317*) in *Cetraria nivalis*, *C. crispa* and *Siphula ceratites*; in other lichens C-27, C-29 and C-31 were the main hydrocarbons (*765*). The hydrocarbons of the odoriferous fraction of *Evernia prunastri* have been examined (*357*).

4.4.2. Aliphatic Alcohols

Aliphatic alcohols have been identified in *Cetraria islandica* ($C_{22}H_{41}OH$, $C_{24}H_{45}OH$, $C_{26}H_{49}OH$) (*658*), *Cetraria delisei*, *Lobaria pulmonaria*, *Stereocaulon tomentosum* and *Usnea hirta* (*659*). 1-(*O*-α-D-Glucopyranosyl)-3S,25R-hexacosanediol (**15**) is a higher alcohol from *Solorina crocea* (*451*, *453*). The absolute configurations at C-2 and C-25 were determined by FT NMR (Mosher's ^{1}H method) of the MTPA-derivatives (see Scheme 24).

4.4.3. Aliphatic Acids and Esters

The fatty acid composition of lichens has been discussed extensively by Solberg (*658*) and Dembitsky *et al.* (*163*, *164*, *167–170*). Řezanka and Dembitsky (*619*) isolated from *Acarospora gobiensis* (from saline soils near the lake Ysyk Köl, Kyrgyzstan, Central Asia) the first bromine- containing lichen substances, (5E,17E)-18-bromo-octa-5,17-dien-15-yonic acid and 18-bromo-octadeca-5,7,17-triyonic acid. Solberg (*658*, *659*) has identified oxalic, malonic, fumaric, succinic, glutaric, adipic, pimelic, octa-1,8-dienedioic, suberic, azelaeic and sebaic acids in *Cetraria islandica*, *C. delisei*, *Usnea hirta*, *Lobaria pulmonaria* and *Stereocaulon tomentosum* but already earlier Schade (*635*) had found in numerous *Usnea* species calcium oxalate which may be deposited in the form of two different hydrates: the tetragonale weddellite, $(COO)_2Ca \cdot 2H_2O$, and the monocline whewellite, $(COO)_2Ca \cdot H_2O$, as shown by X-ray diffraction and scanning microscopy (SEM) (*715*). The glistening pruina of 25 investigated Pyxinaceae species consisted of whewellite, whereas the spotted pruina consisted of weddellite (*15*). Manganese

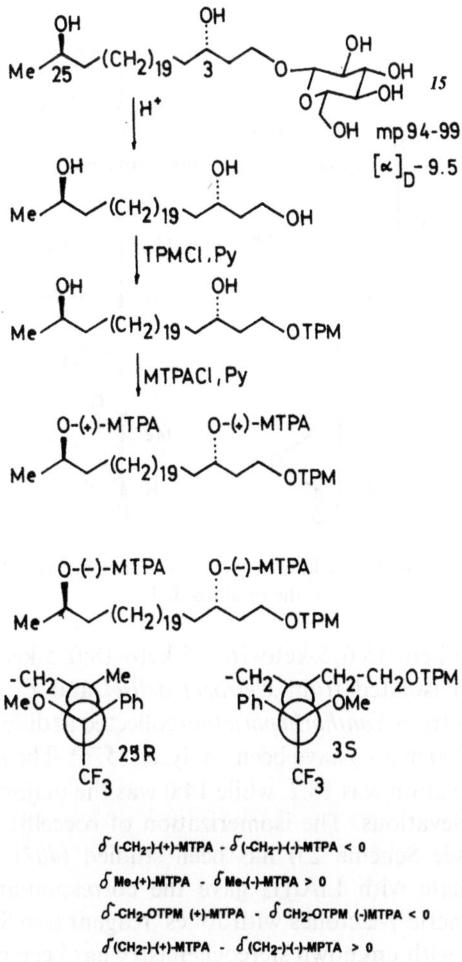

Scheme 24. Determination of the absolute configuration at C-3 and C-25 of 1(*O*-α-D-glucopyranosyl)-3S,25R-hexacosanediol

oxalate has been detected in *Pertusaria corallina* (*733*) and fumaric acid was isolated from a *Stereocaulon* species from Venezuela (*541*). Small amounts of branched aliphatic acids were found in *Aspicilia transbaicalica*, *Evernia mesomorpha* and *E. prunastri* (*168*) and a mixture of tetrahydroxy fatty acids (**16**) has been isolated from *Alectoria fremontii* (*655*).

$Me\text{-}(CH_2)_n\text{-}CHOH\text{-}CHOH\text{-}CH_2\text{-}CHOH\text{-}CHOH\text{-}(CH_2)_7\text{-}COOH$

16: n = 4, 6, 7, 8, 9

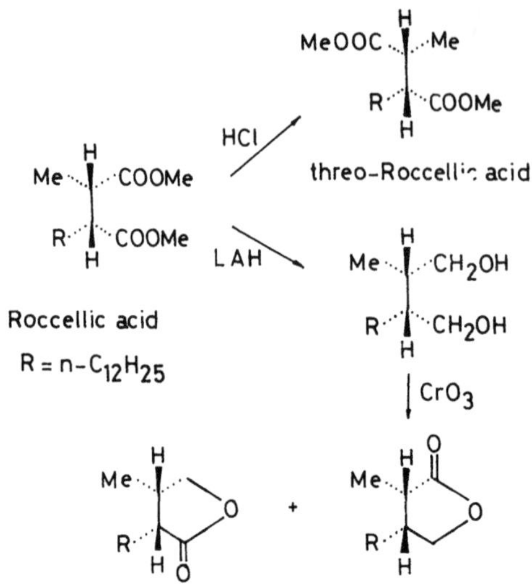

Scheme 25. Isomerization of roccellic acid, reduction of dimethyl roccellate and oxidation of the resulting diol

Keto acids (5-keto-15:0, 5-keto-16:0, 5-keto-18:0, 5-keto-21:0, 5-keto-23:0) have been isolated from *Cetraria delisei* and *C. islandica* (*659*, *658*). Fatty acids from *Xanthoria parietina* collected at different elevations (400, 800 and 1300 m a.s.) have been analyzed (*576*). The main fatty acid at the lowest elevation was 18:2, while 14:0 was the major component at the other two elevations. The isomerization of roccellic acid to threo-roccellic acid (see Scheme 25) has been studied (*407*). Reduction of dimethyl roccellate with $LiAlH_4$ gave the corresponding diol which yielded two isomeric γ-lactones with Jones' reagent (see Scheme 25).

Structure **17** with unknown stereochemistry has been proposed (*502*) for angardianic acid from *Leproloma angardianum*. If its biosynthesis is similar to that of roccellic acid angardianic acid should have stereostructure **17a**.

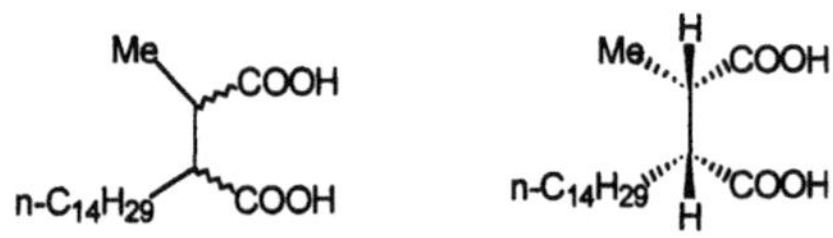

17 **17 a**

Angardianic acid

An enantiospecific synthesis (see Scheme 26) of (+)-bourgeanic acid (**18**) (*37*) *via* (−)-hemibourgeanic defines the latter as 2S,3S,4R,6R-3-hydroxy-2,4,6-trimethyloctanoic acid (*722*, *723*). Conformational

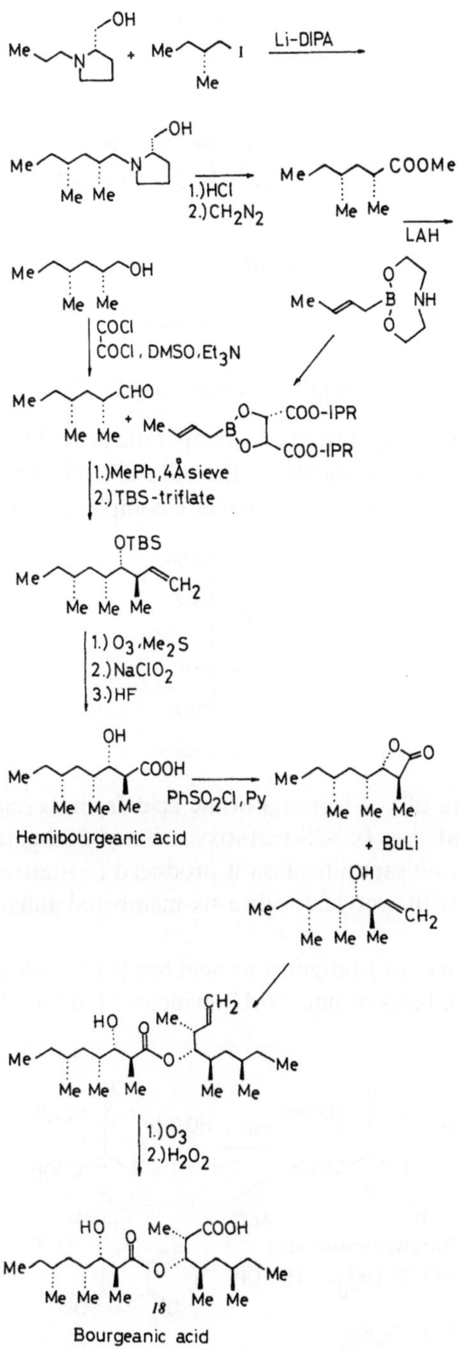

Scheme 26. Synthesis of (+)-bourgeanic acid

analysis of bourgeanic lactone (**19**) shows that the eight-membered ring adopts a crown conformation with C_2 symmetry and all substituents in equatorial orientation (*722*).

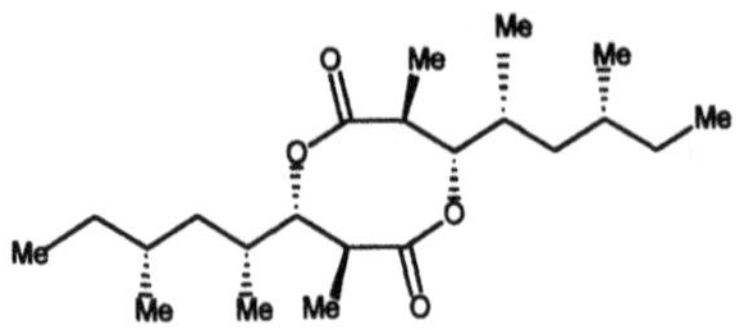

19 : Bourgeanic lactone

4.4.4. Aliphatic Tricarboxylic Acids

The absolute configuration of norcaperatic acid (**20**) was determined as 2S,3S by CD measurements on the natural acid and optically active synthetic compounds as their molybdate complexes (*38*).

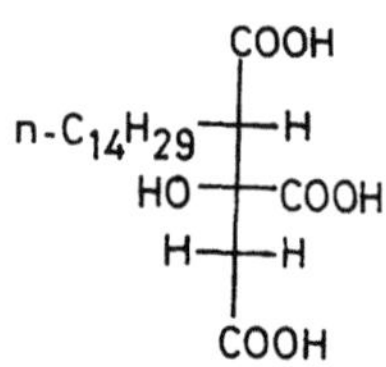

20 : Norcaperatic acic

The structure of (+)-isorangiformic acid from *Lecanora stenotropa* has been elucidated as (+)-2S-methoxycarbonyl-3S-heptadecanedicarboxylic acid (**21**); on saponification it produced (+)-norrangiformic acid and on heating with acetylchloride a six-membered anhydride (*381*) (see Scheme 27).

The structure of (+)-rangiformic acid has been revised from **22** (*411*) to **23** (*31*) on the basis of new NMR evidence. Kümmerling *et al.* (*488*)

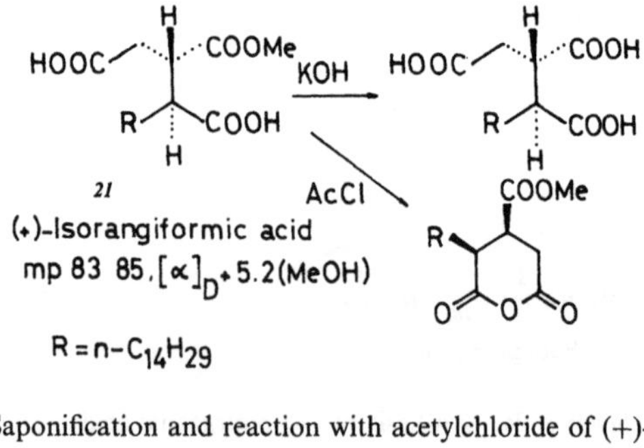

Scheme 27. Saponification and reaction with acetylchloride of (+)-isorangiformic acid

proposed structure **24** for jackinic acid from *Lepraria jackinii*; the close relationship with rangiformic acid makes stereochemistry **24a** likely.

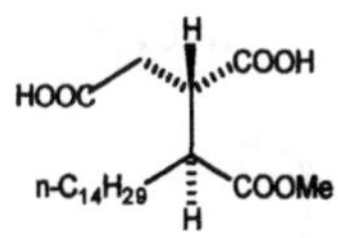

22

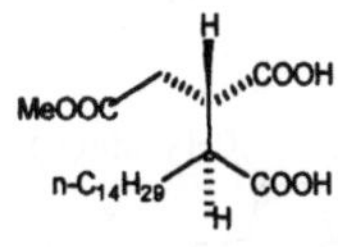

23: Rangiformic acid

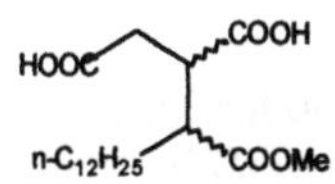

24

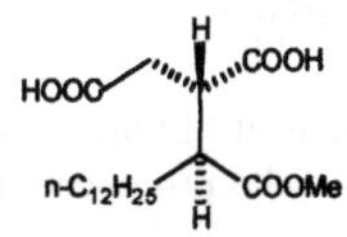

24 a: Jackinic acid

4.4.5. γ-Lactonic Acids

Two γ-lactonic acids of unknown stereochemistry were found in *Lecanora rupicola* (*174*) resp. *Cladonia impexa* (*173*). **25** from the former and **26** from the latter species.

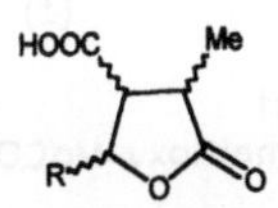

25: R = $C_{11}H_{23}$
26: R = $C_{15}H_{31}$

Numerous papers have been published dealing with the synthesis of stereochemistry of protolichesterinic acid and related compounds. According to Löffler *et al.* (*512*) β′-bromodimethylmethylmesaconate gave in a Reformatsky reaction with myristic aldehyde and zinc a mixture of racemic methyl protolichesterinate (**27**) and methyl allopro-tolichesterinate (**28**) (see Scheme 28). Another strategy was followed by Martin *et al.* (*526*): the key step was the carboxylation of γ-butyrolac-tones with methyl methoxymagnesium carbonate (see Scheme 29). Damon and Schlessinger (*158*) published a stereoselective synthesis of racemic protolichesterinic acid based on the conjugate addition of (*tris*-(methylthio)methyl) lithium to γ-tridecylbutenolide and *in situ* trap-ping with formaldehyde, followed by hydrolysis and methylene formation (see Scheme 30). Carlson and Oyler (*58*) used itaconic acid as starting material for the synthesis of racemic protolichesterinic,

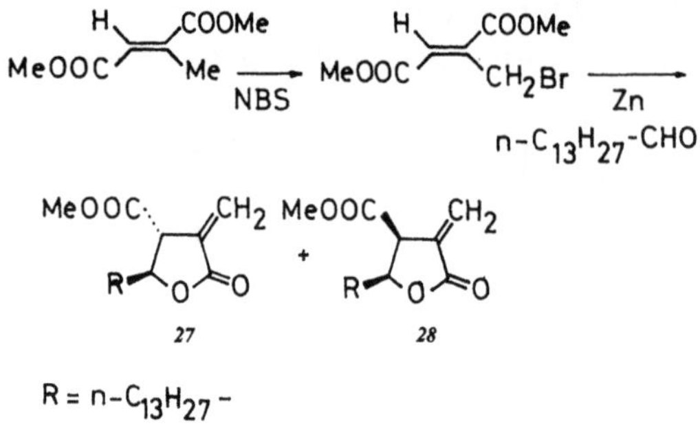

Scheme 28. Synthesis of (±)-methyl protolichesterinate and (±)-methyl allo-protoliche-sterinate (*512*)

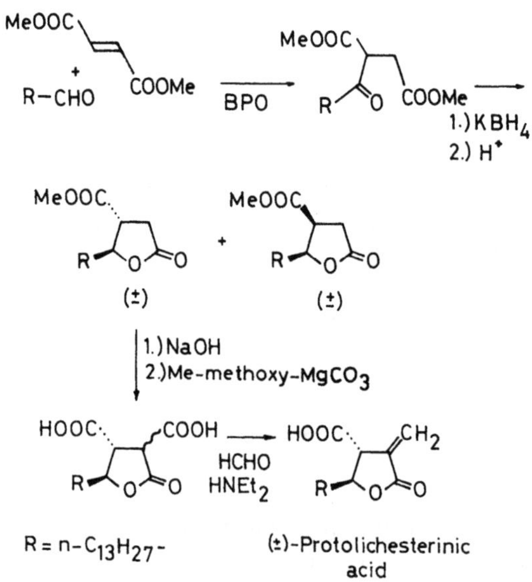

Scheme 29. Synthesis of racemic protolichesterinic acid by carboxylation of γ-butyrolactones with methyl methoxymagnesium carbonate

allo-protolichesterinic and nephrosterinic acids (see Scheme 31). Finally Murta *et al.* (*555*) described the first synthesis of (−)-protolichesterinic acid *via* a facially selective 2+2 cycloaddition reaction of dichloroketene with an enantiopure *O*-alkylenol ether (see Scheme 32).

19-Acetoxylichesterinic acid (neuropogolic acid) (**29**), 19-acetoxy-protolichesterinic acid (**30**), (−)-allopertusaric acid (**31**) and (−)-dihydropertusaric acid (**32**) are γ-lactone carboxylic acids with an oxygen function in the side chain from *Neuropogon trachycarpus* (*323*)

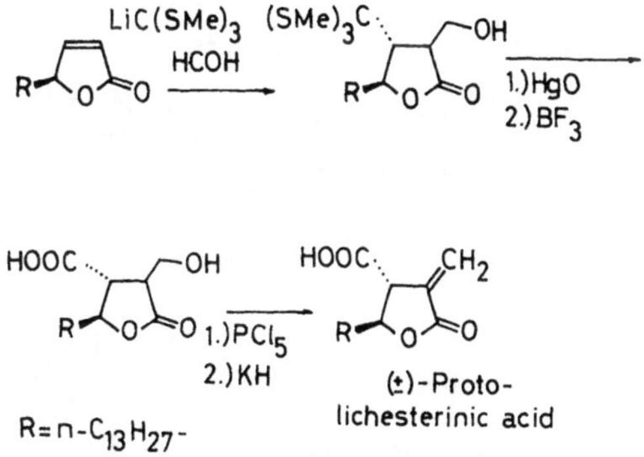

Scheme 30. Synthesis of racemic protolichesterinic acid (*158*)

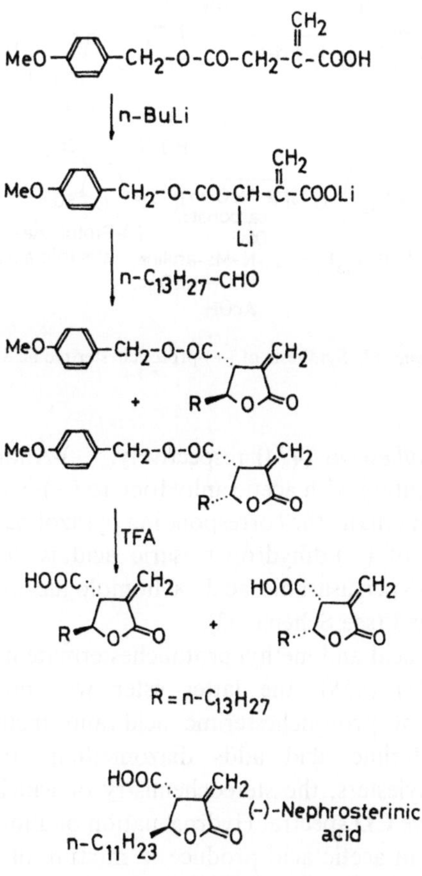

Scheme 31. Synthesis of racemic protolichesterinic, allo-protolichesterinic and nephrosterinic acids (*58*)

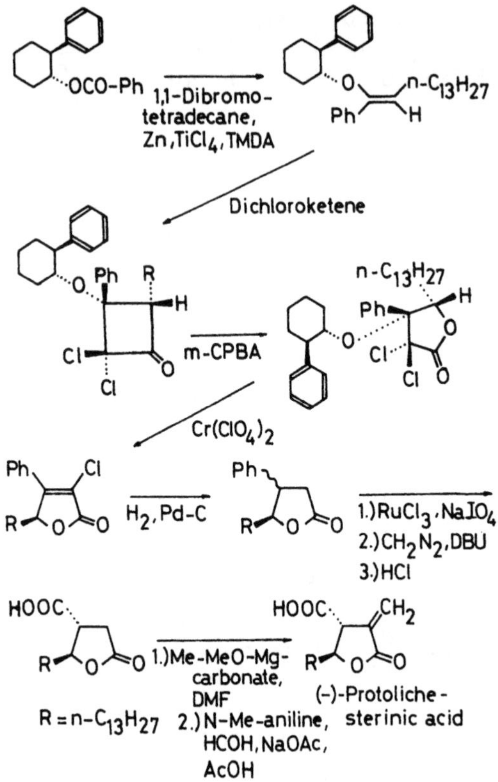

Scheme 32. Synthesis of (–)-protolichesterinic acid (*555*)

and *Pertusaria albescens* (*414*) respectively. (–)-Allopertusaric acid is isomerized on heating with acetic anhydride to (–)-isomuronic acid and gives with diazomethane the corresponding pyrazoline methyl ester. The dithiospiroketal of (–)-dihydropertusaric acid is desulphurized with Raney-Ni to (–)-pertusaranic acid, a homologue of (–)-dihydroprotolichesterinic acid (see Scheme 33).

Lichesterylic acid and methyl protolichesterinate have been found in *Cetraria islandica* (*658*); the latter ester was prepared from the potassium salt of protolichesterinic acid and methyl iodide (*413*). Allo-protolichesterinic acid adds diazomethane to form isomeric pyrazoline methylesters, the stereochemistry of which was established by their respective CD spectra. Hydrogenation of allo-protolichesterinic acid with Pd–C in acetic acid produces a mixture of (–)-dihydro-allo-protolichesterinic acid (**33**), (–)-nephromopsinic acid (**34**) and (–)-lichesterinic acid (**35**) (see Scheme 34).

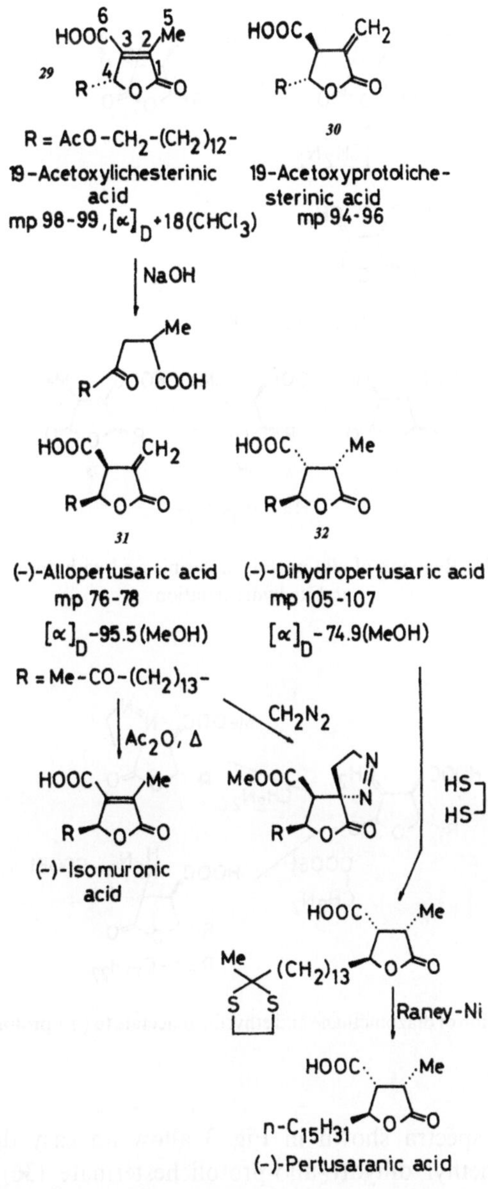

Scheme 33. Structure and chemistry of 19-acetoxylichesterinic, 19-acetoxyprotolichesterinic, (−)-allopertusaric and (−)-dihydropertusaric acids

(+)-Protolichesterinic acid adds diazomethane to give the 2R-methylesterpyrazoline and adds ethyl diazoacetate to give the corresponding 2R-adduct (see Scheme 35).

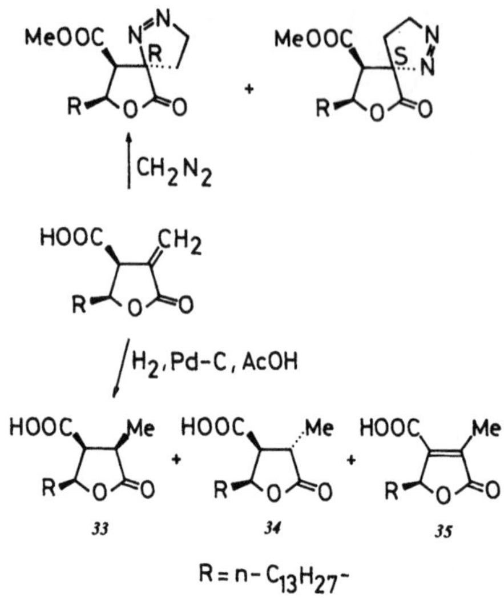

Scheme 34. Reaction of allo-protolichesterinic acid with diazomethane and its hydrogenation

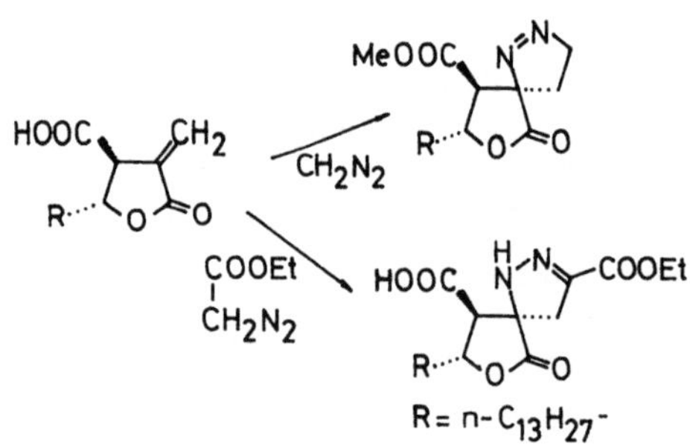

Scheme 35. Addition of diazomethane and ethyl diazoacetate to (+)-protolichesterinic acid

The ORD spectra shown in Fig. 3 allow an easy differentiation between (−)-methyl dihydro-allo-protolichesterinate (**36**), (−)-methyl nephromopsinate (**37**), (−)-methylroccellarate (**38**) and (−)-methyl dihydroprotolichesterinate (**39**) (see Fig. 3).

(+)-Praesorediosic acid (**40**) and (+)-protopraesorediosic acid (**41**) are γ-lactone carboxylic acids from *Parmotrema praesorediosum* (*162*).

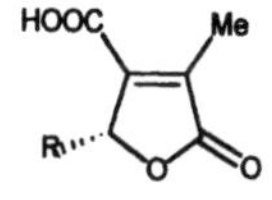

40

R = HOOC-$(CH_2)_{14}$-

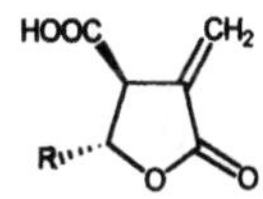

41

(+)-Praesorediosic acid, mp 139-140.5

(+)-Protopraesorediosic acid, mp 123-124

Mulzer *et al.* (*553*) synthesized (−)-nephromopsinic acid by the addition of a chiral allylic bromide to tetradecanal (see Scheme 36). The same group (*628*, *666*) also described total syntheses of (−)- and (+)-roccellaric acid and (−)-dihydroprotolichesterinic acid and secured the relative configuration of (+)-roccellaric acid by an X-ray analysis. One route started from (2S,3E)-1,2-isopropylidene-5-*O*-benzyl-3-pentene-1,2,5-triol (**42**) (see Scheme 37). The second route included the ozonisation of methyl-(2RS,3R,4E)-3-benzyloxy-5-hydroxy-oxolan-3-one and subsequent condensation with tridecyltriphenylphosponium bromide (see Scheme 38). Finally they synthesized (+)-roccellaric acid and (+)-dihydroprotolichesterinic acid from D-glucose as the chiral pool (*552*) (see Scheme 39).

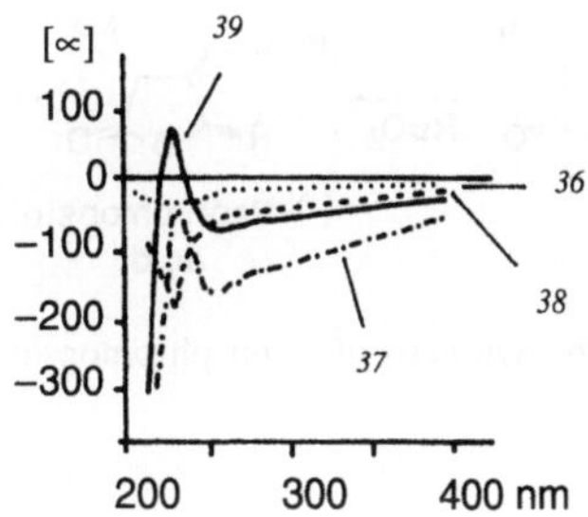

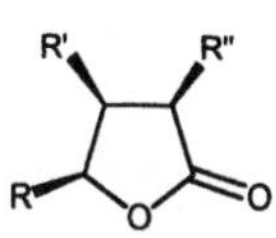

36

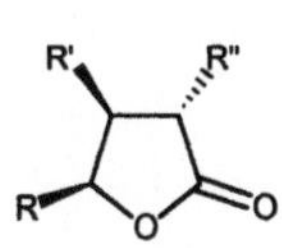

37

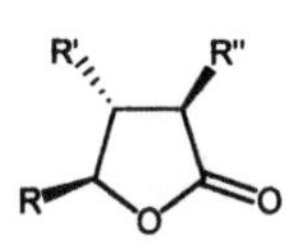

38

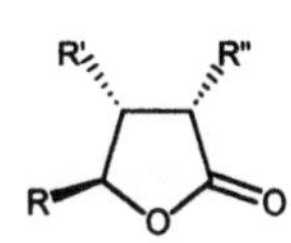

39

R = n-$C_{13}H_{27}$, R' = COOMe, R" = Me

Fig. 3. ORD spectra of (−)-methyl dihydro-allo-protolichesterinate (**36**), (−)-methyl nephromopsinate (**37**), (−)-methyl roccellarate (**38**), and (−)-methyl dihydroprotolichesterinate (**39**)

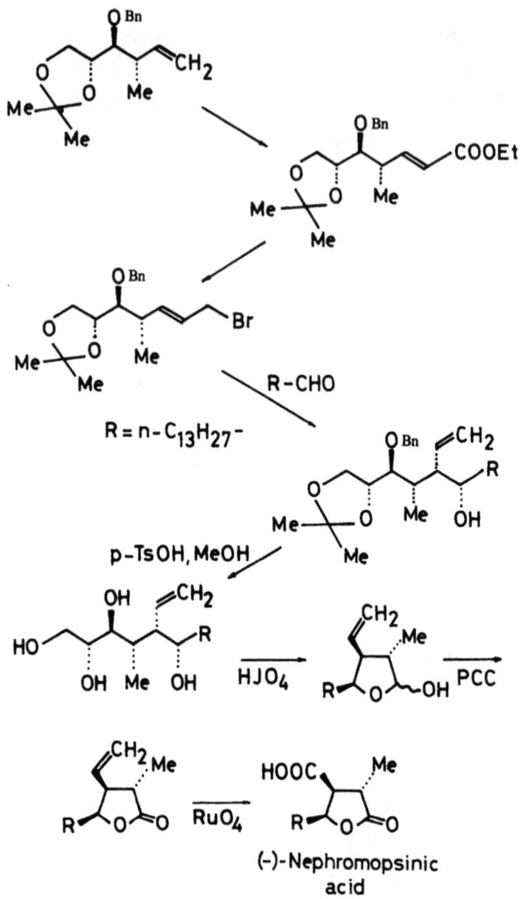

Scheme 36. Synthesis of (–)-nephromopsinic acid

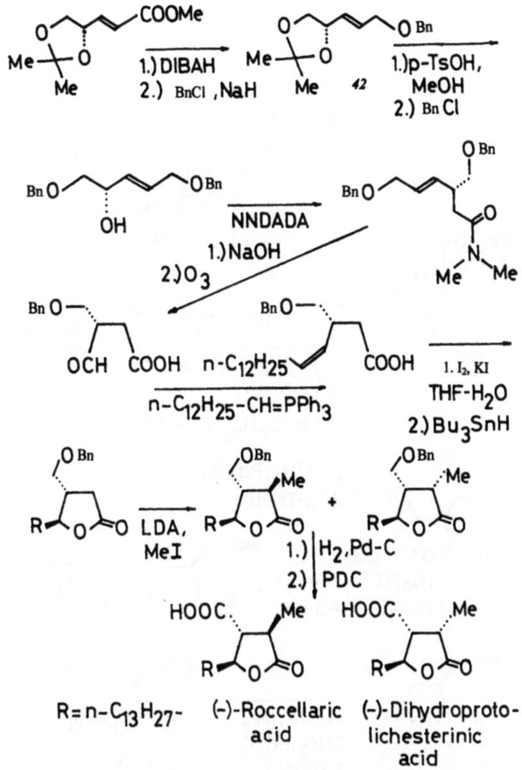

Scheme 37. Synthesis of (−)-roccellaric and (−)-dihydroprotolichesterinic acids starting from (2S,3E)-1,2-isopropylidene-5-*O*-benzyl-3-pentene-1,2,5-triol

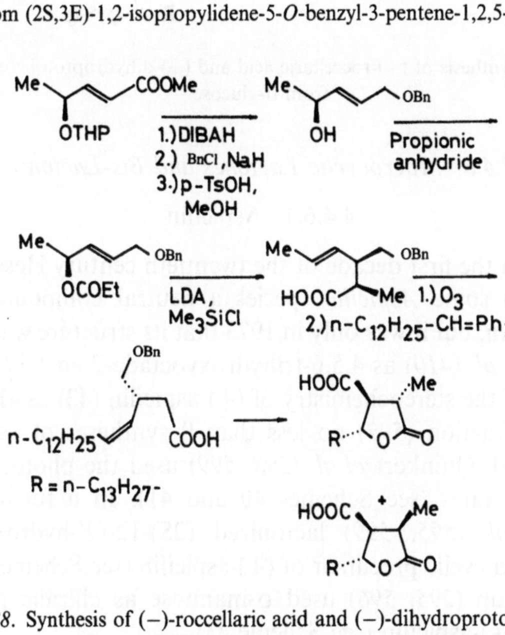

Scheme 38. Synthesis of (−)-roccellaric acid and (−)-dihydroprotolichesterinic acid *via* (4S,2E)-1-*O*-benzyl-2-pentene-1,4-diol

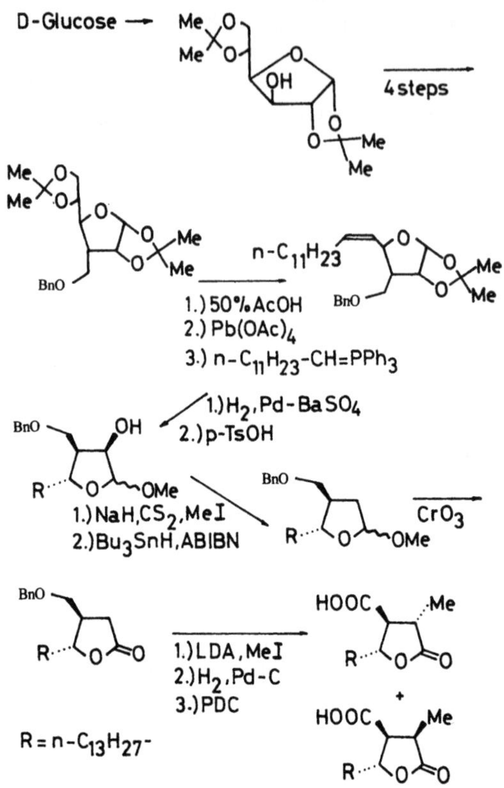

Scheme 39. Synthesis of (+)-roccellaric acid and (+)-dihydroprotolichesterinic acid from D-glucose

4.4.6. Macrocyclic Lactones and Bis-Lactones

4.4.6.1. Aspicilin

Already in the first decade of the twentieth century Hesse (*364*, *365*) isolated from some *Aspicilia* species a neutral compound which he named aspicilin, but it was only in 1973 that its structure was recognized by Huneck *et al.* (*410*) as 4,5,6-trihydroxyoctade-2-en-1:17-olide. After elucidation of the stereochemistry of (+)-aspicilin (**43**) as 4R,5S,6R,17S by X-ray diffraction (*597*) no less than 9 syntheses of aspicilin have been described. Quinkert *et al.* (*598*, *599*) used the photolactonization of *o*-quinolacetates (see Schemes 40 and 41). In a further synthesis Quinkert *et al.* (*595*, *599*) lactonized (2S)-12-(2′-hydroxyphenyl)-2-dodecanol to a cyclic precursor of (+)-aspicilin (see Scheme 42). Finally the same group (*295*, *596*) used D-mannose as chiralic pool for the synthesis of (+)-aspicilin (see Scheme 43).

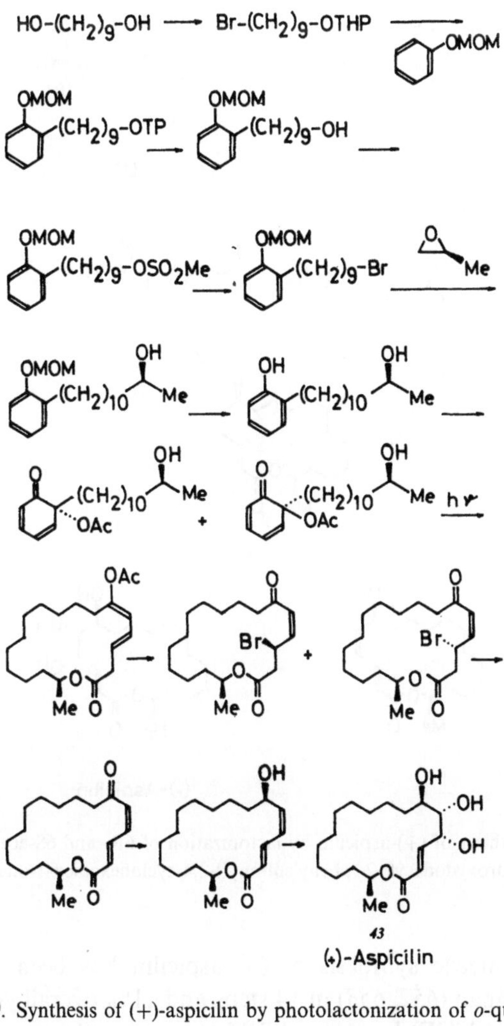

Scheme 40. Synthesis of (+)-aspicilin by photolactonization of *o*-quinolacetates

Waanders *et al.* (*714*) started the synthesis of (−)-aspicilin from 2R,3R-dihydroxysuccinic acid and 2-propyne-1-ol as shown in Scheme 44. Solladie *et al.* (*662*) synthesized (−)-aspicilin by condensation of triphenylphosponium iodide **44** with aldehyde **45** prepared from C-3/C-9 fragment **46** (see Scheme 45). The C-3/C-9 fragment was synthesized *via* two different routes (*661*, *68a*) (see Schemes 46a and 46b).

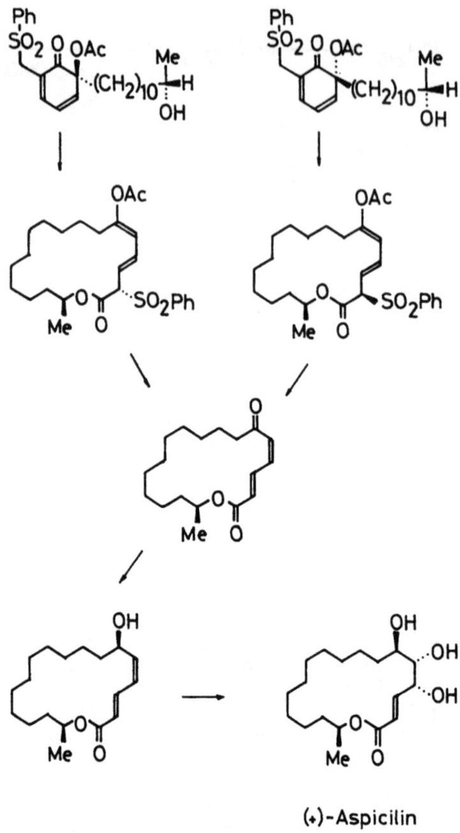

Scheme 41. Synthesis of (+)-aspicilin by lactonization of 6R- and 6S-acetoxy-(6-[(11′S)-hydroxydodecyl]-2-(phenylsulfonyl)-2,4-cyclohexadien-1-ones

The asymmetric synthesis of (+)-aspicilin has been achieved by Sinha and Keinan (*652*, *653*) in 14 steps and 11% overall yield, starting from the achiral (3E)-hexadeca-1,3,15-triene (see Scheme 47). Enders and Prokopenko (*279*) realized the synthesis of (+)-aspicilin in 19 steps and with high stereoselectivity. Three of the four stereogenic centres were generated by employing the SAMP/RAMP hydrazone method (see Scheme 48a). Finally Kobayashi *et al.* (*467a*) described the chiral synthesis of (+)-aspicilin by using a furyl group as the masked γ-oxo-α,β-unsaturated carboxylic acid (see Scheme 48b).

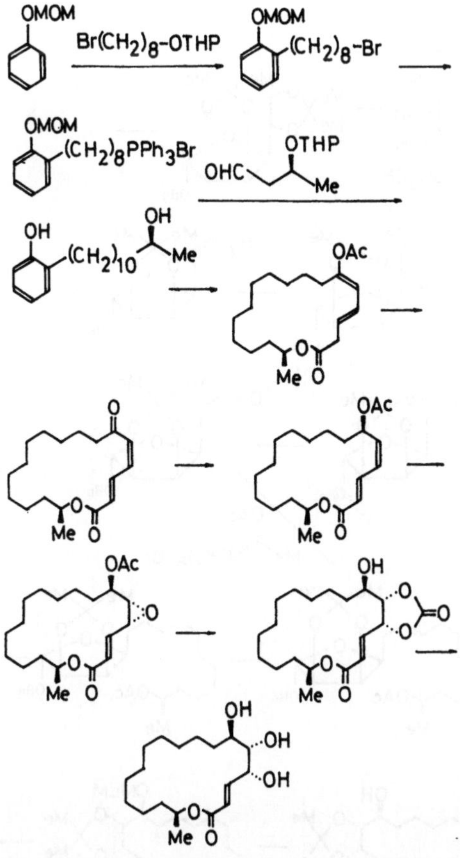

Scheme 42. Synthesis of (+)-aspicilin *via* 6-acetoxyoctadeca-3,5-diene-17-olide

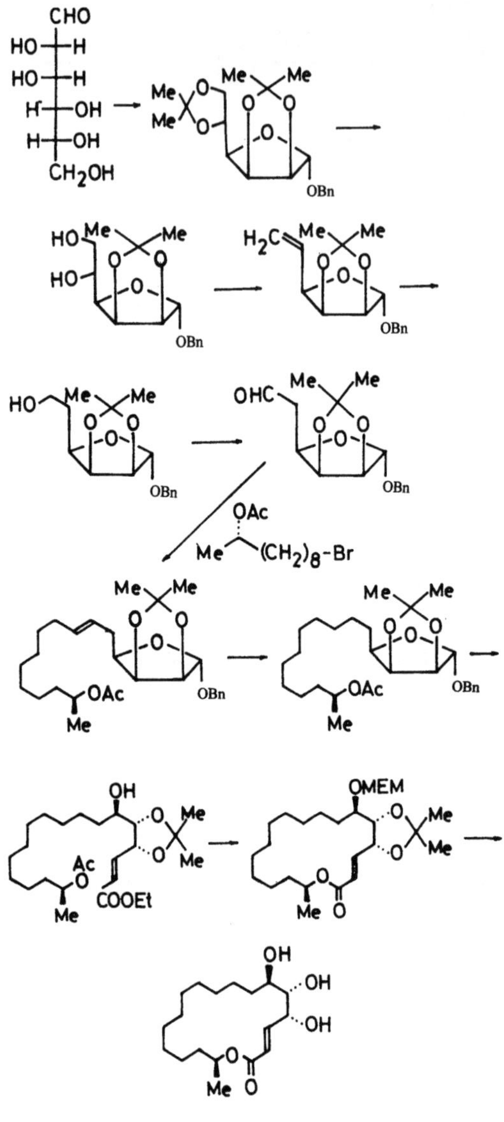

Scheme 43. Synthesis of (+)-aspicilin from D-mannose

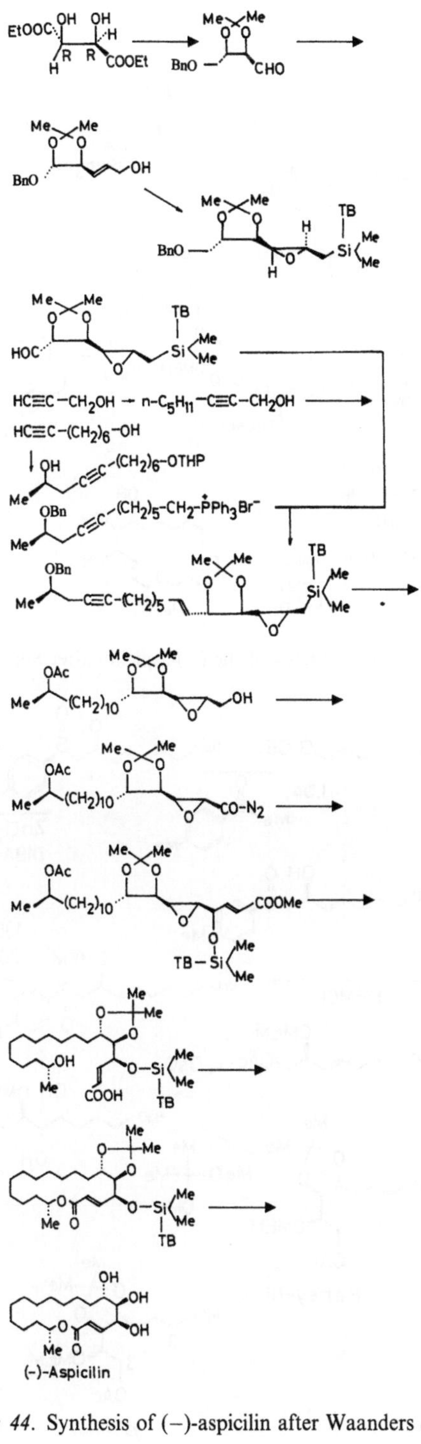

Scheme 44. Synthesis of (−)-aspicilin after Waanders *et al.*

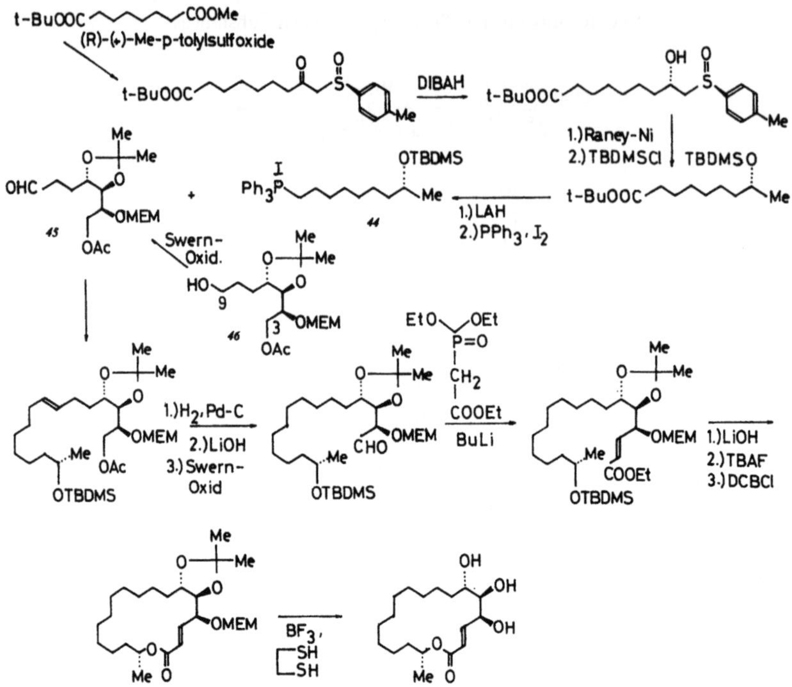

Scheme 45. Synthesis of the (−)-aspicilin after Solladie *et al.*

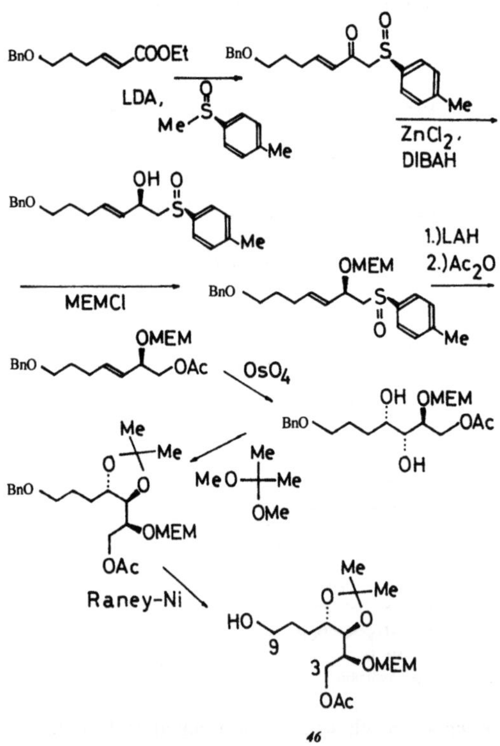

Scheme 46a. Synthesis of the C-3/C-9-fragment of (−)-aspicilin after Solladie *et al.*

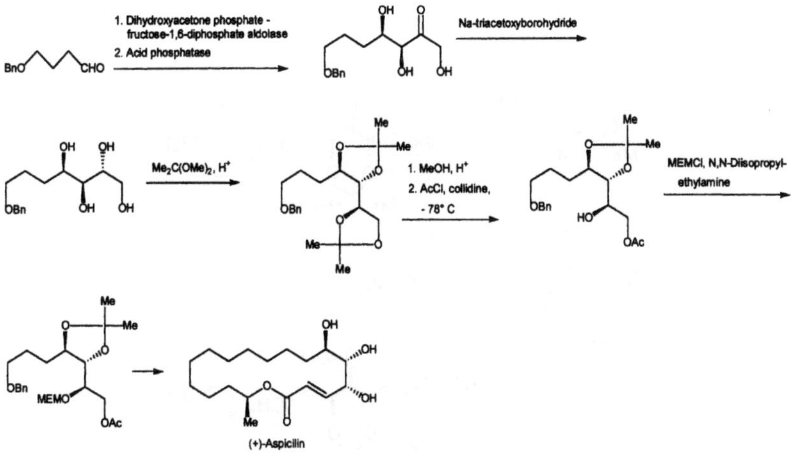

Scheme 46b. Synthesis of the C-3/C-9-fragment of (+)-aspicilin after Chenevert *et al.*

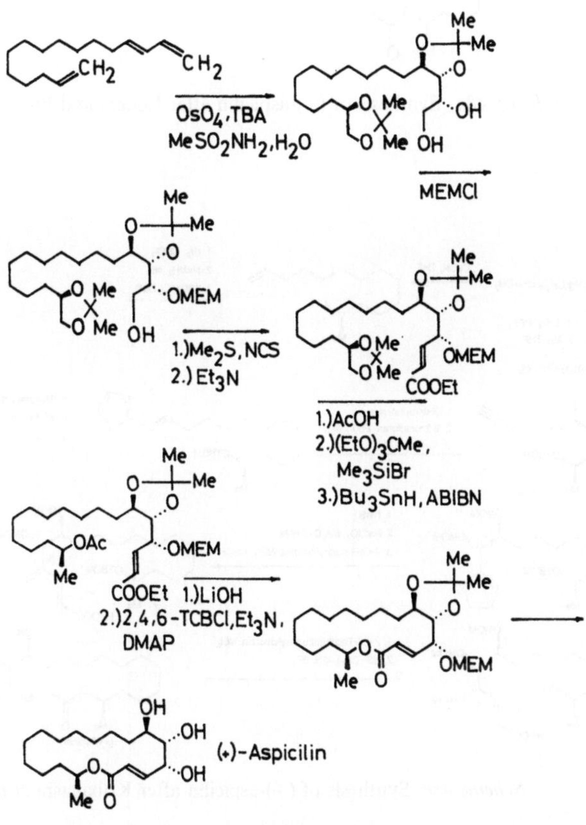

Scheme 47. Synthesis of (+)-aspicilin from (3E)-hexadeca-1,3,15-triene after Sinha and Keinan

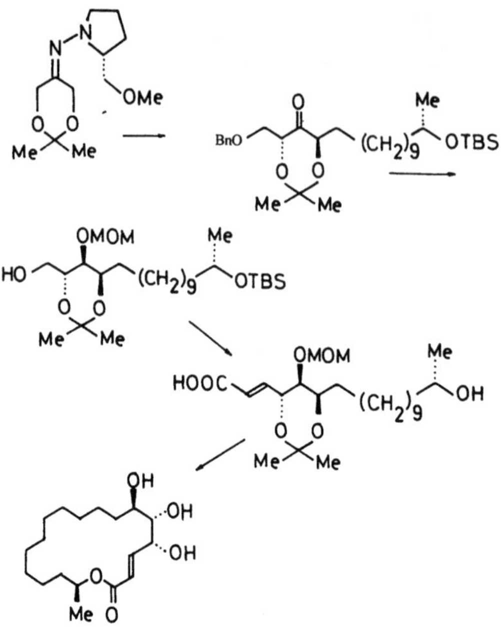

Scheme 48a. Synthesis of (+)-aspicilin after Enders and Prokopenko

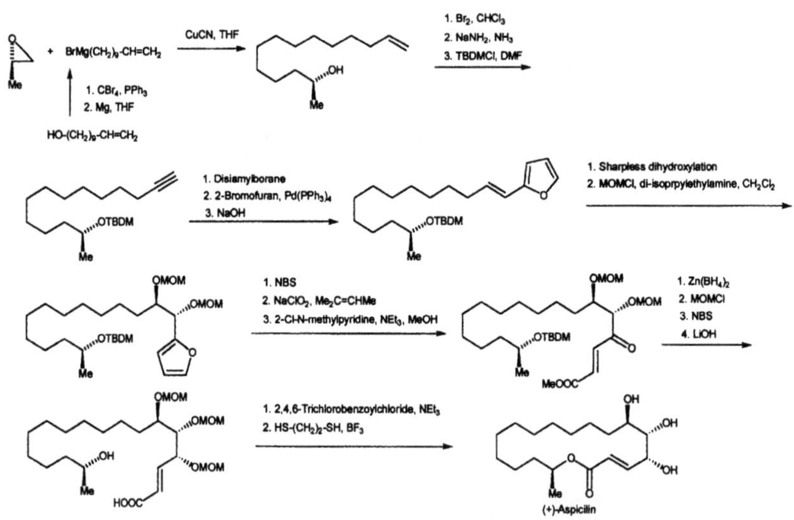

Scheme 48b. Synthesis of (+)-aspicilin after Kobayashi *et al.*

4.4.6.2. Lepranthin

More than 90 years ago Zopf (*763*) described a new neutral compound lepranthin from *Arthonia impolita* (syn. *Leprantha impolita*). Much later, in 1995, Polborn *et al.* (*579*) elucidated the structure of lepranthin as the *bis*-lactone **47** by NMR spectroscopy and X-ray analysis.

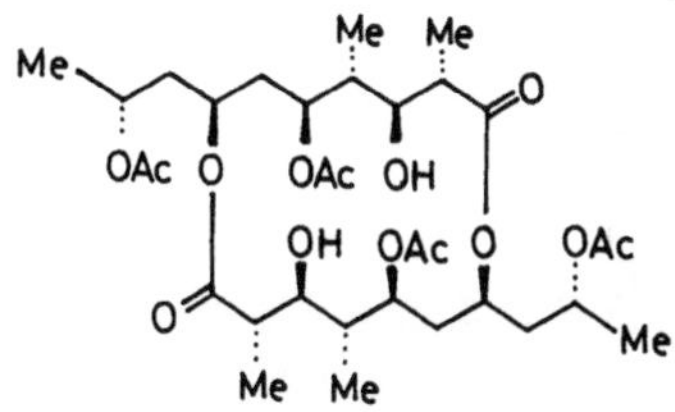

47: Lepranthin, mp 185
CD (EtOH) Δε -4.17 (216 nm)

4.4.7. Anhydrides

Graphenone (**48**) is a dicarboxylic anhydride from the mycobiont of *Graphis scripta*; its structure was elucidated by an X-ray analysis (*549*). The *bis*-anhydride homoheveadride (**49**) was isolated by Archer and Taylor (*17*) from *Cladonia polycarpoides*.

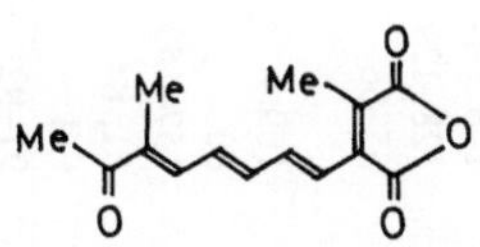

48: Graphenone, orange plates, mp 163-165

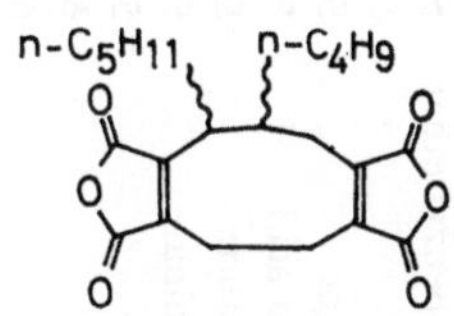

49: Homoheveadride, $[\alpha]_D$ + 118 ($CHCl_3$)

4.5. Aromatic Compounds

4.5.1. Aromatic Units

Sources, structure elucidation and syntheses of aromatic units are summarized in Table 3. It is of interest that some orsellinic acid

Table 3. Aromatic units from lichens

Compound	Formula No.	M.p. (°C)	Lichen	Synth. Scheme	Ref.
2′-Acetoxyolivetol	**50**	138–139		49	(*181*)
3′-Acetoxyolivetol	**51**	Oil		49	(*181*)
Atranol	**52**	124	*Evernia prunastri*, *Stereocaulon vesuvianum*		(*51*)
				From orcinol with DCDMe–$TiCl_4$	(*85*)
3-Chloroorsellinic acid di-*O*-benzyl ether	**53**	146.5–147		50	(*681*)
Dichloroisoeverninic acid	**54**	128–129		51, 52	(*179*, *180*)
3,5-Dihydroxy-*n*-heptylbenzene	**55**	55–56		53	(*19*, *297*)
3,5-Dihydroxy-4-methylacetic acid	**56**			54	(*605*)
5,7-Dihydroxy-6-methylphthalide	**57**	169.4	*Alectoria nigricans*		(*655*, *391*)
1,1-Dimethylnoneicosyl everninate	**58**	60	*Evernia prunastri*		(*561*)
Divaric acid	**59**	157–159, 179	*Protusnea malacea*	55, 56	(*182*, *64*, *183*)
Divarinol	**60**	51, 84			(*452*)
Divarinolmonomethyl ether	**61**	Oil			(*452*)
Ethyl 3,5-dimethylorsellinate (2,7-di-^{14}C)	**62**	91–92		57	(*29*)
Ethyl 3,5-dimethylorsellinate (1-^{13}C,7-^{13}C,-$CO^{18}OEt$)	**63**			58	(*642*)
Ethyl divaricatinate	**64**	41–42	*Protusnea malacea*		(*64*)
Ethyl haematommate	**65**			59	(*521*)
Ethyl 4-*O*-methylolivetolcarboxylate	**66**				(*329*)

Ethyl 5-hydroxyevernninate	**67**	106–107		By hydroxylation of ethyl evernninate with $K_2S_2O_8$-NaOH	(*3*)
4-Formyl-5,7-dihydroxy-6-methylphthalide	**68**			60	(*588*)
4-Formyl-5-hydroxy-7-iso-propyloxyphthalide	**69**			60	(*588*)
Glomellin	**70**	85		61	(*546*)
Grammitic acid	**71**				(*735*)
6-*n*-Heptyl-2,4-dihydroxybenzoic acid	**72**			55	(*183*)
Haematommic acid	**73**	172–173			(*521*, *672*)
Hierridin	**74**	70–72	*Ramalina hierrensis*	62	(*330*, *328*)
5-Hydroxy-7-acetoxyphthalide	**75**				(*353*)
5-Hydroxyevernninic acid	**76**	172–173		By hydroxylation of evernninic acid with $K_2S_2O_8$–NaOH	(*3*)
3-Hydroxy-5-methyl-4-6-dimethoxy-2-(tricos-9′-enyl)-benzene	**77**				(*685*)
5-Hydroxyorsellinic acid	**78**	132		63	(*669*)
Insignin	**79**	64	*Porpidia glaucophaea*	64	(*265*)
3-Iodoorsellinic acid	**80**	182–183		50	(*681*)
5-Iodoorsellinic acid	**81**	178–179		50	(*681*)
Methyl 3-bromoevernninate	**82**	118–118		65	(*85*)
Methyl 5-bromoevernninate	**83**	143–144			(*85*, *54*)
Methyl 3-chloroorsellinate	**84**	133–134			(*287*)
5-Methyl-2,4-di-*O*-methyl-orsellinic acid	**85**	208–210		66	(*55*)
Methyl haematommate	**86**	144	*Cetraria islandica*		(*658*)
Methyl 5-hydroxyevernninate	**87**	153–154			(*3*)

Table 3 (continued)

Compound	Formula No.	M.p. (°C)	Lichen	Synth. Scheme	Ref.
Methyl 3-hydroxyorsellinate	**88**	154–155		67	(*3*, *308*)
Methyl-5-hydroxyorsellinate	**89**			68	(*308*)
Methyl 4-*O*-methylhaematommate	**90**			65	(*85*)
Methyl olivetolcarboxylate (5,6-di-^{13}C,1-^{14}C)	**91**			69	(*580*)
Methyl olivetolcarboxylate (1′-^{13}C)	**92**			69	(*580*)
Methyl 2-*O*-methyl-β-orcinol-carboxylate	**93**	156–157		X-ray analysis	(*34*)
				70	(*339*)
Methyl β-orcinolcarboxylate	**94**	143–144		70	(*51*)
4-*O*-Methylolivetolcarboxylic acid	**95**	123			(*429*)
Methyl orsellinate		208–210		From methyl acetoacetate and *n*-BuLi with NaH in THF and subsequent acidification	(*70*)
Olivetol	**96**	39–41	*Cladina macaronesica*	53	(*329*)
				71	(*297*)
				72	(*474*)
				73	(*470*)
				74	(*443*, *19*)
Olivetol (5′-$^{2}H_3$)	**97**			75	(*326*)
Olivetolcarboxylic acid	**98**		*Cladina macaronesica*		(*329*)
Olivetolcarboxylic acid (5,6-di-^{13}C,1-^{14}C)	**99**			69	(*580*)
Olivetoldimethyl ether	**100**			76	(*637*)
Olivetolmonomethyl ether	**101**		*Cladina macaronesica*		(*329*)
Orcinol	**102**		*Evernia prunastri*		(*19*, *320*)

β-Orcinol	**103**		*Evernia prunastri*		(*320*)
				By reduction of atranol with Zn–Hg/HCl	(*720*)
β-Orcinolcarboxylic acid	**104**		*Pseudevernia furfuracea*		(*339*)
Orsellinic acid	**105**			55	(*183*)
				56	(*182*)
				57	(*29*)
				64	(*669*)
				77	(*355*)
				78	(*701*)
				79	(*631*)
				80	(*318*)
					(*448*, *336*)
				81	(*71*)
				82	(*70*)
				83	(*356*)
					(*448*, *478*, *335*, *336*)
Orsellinic acid (2-^{14}C)	**106**			84	(*702*)
Phthalic acid			*Evernia prunastri*		(*320*)
Rhizonic acid	**107**	196, 204	*Myrothecium* spec.		(*699*)

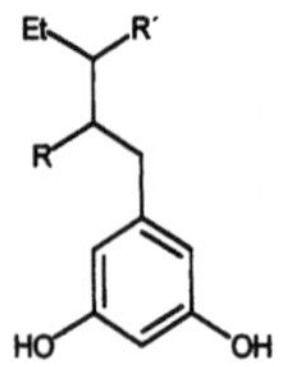

50: R = OAc, R′= H

51: R = H, R′= OAc

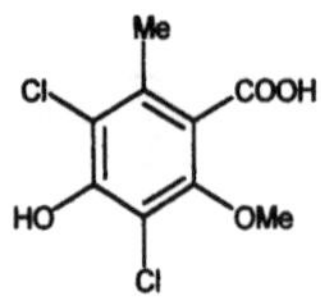

54

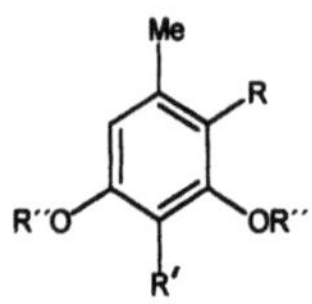

52: R = R′′ = H, R′= CHO

53: R = COOH, R′= Cl, R′′ = Bn

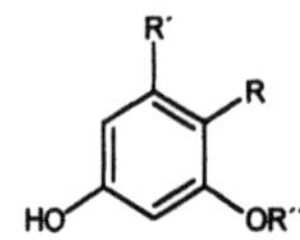

55: R = R′′ = H, R′= n-C_7H_{15}

59: R = COOH, R′= n-C_3H_7, R′′ = H

60: R = R′′ = H, R′= n-$_3H_7$

61: R = H, R' = n-C_3H_7, R" = Me

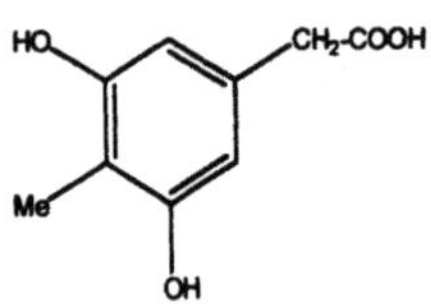

56

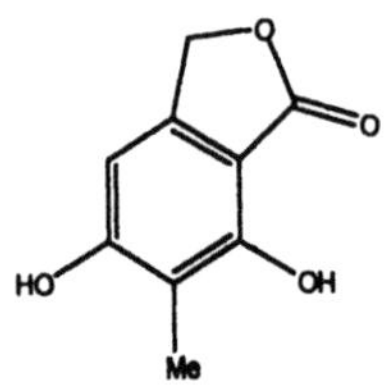

57

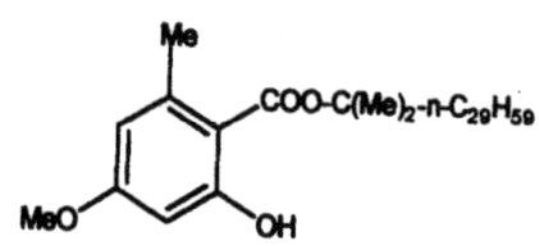

58

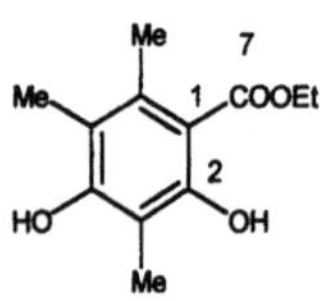

62: ^{14}C-2, ^{14}C-7

63: ^{14}C-2, ^{14}C-7, -C^{18}OOEt

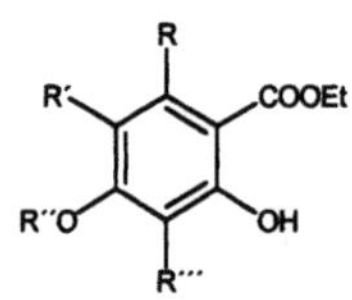

64: R = n-C_3H_7, R′= R′′ = R′′′= H

65: R = Me, R′= R′′ = H, R′′′= CHO

66: R = n-C_5H_{11}, R′= R′′′= H, R′′ = Me

67: R = Me, R′= OH, R′′ = Me, R′′′= H

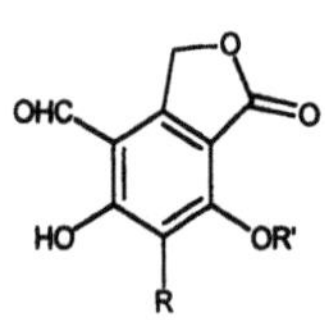

68: R = Me, R' = H

69: R = H, R' = CH$(Me)_2$

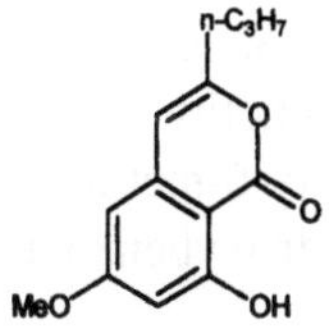

70

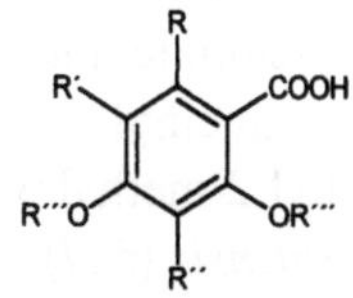

71 : R = n-$C_{11}H_{23}$, R´= R´´´ = H, R´´ = Me

72 : R = n-C_7H_{15}, R´= R´´ = R´´´= H

73 : R = Me, R´= R´´´= H, R´´ = CHO

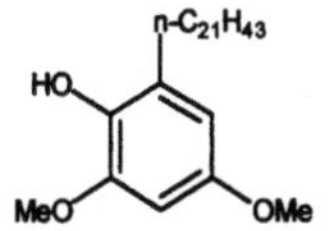

74

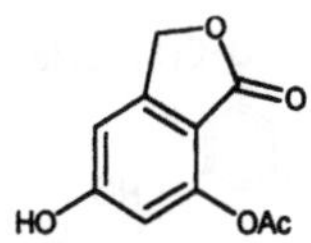

75

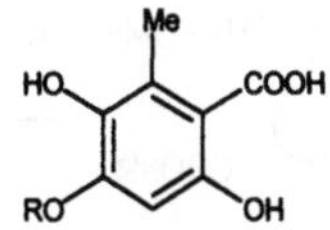

76: R = Me

78: R = H

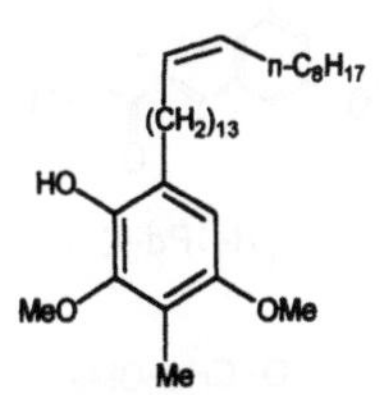

77

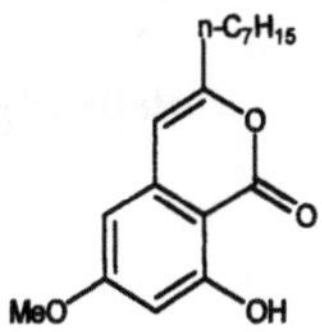

79

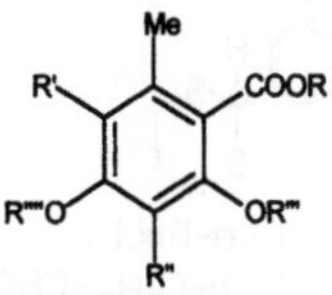

80: R = R' = R''' = R'''' = H, R'' = I

81: R = R'' = R''' = R'''' = H, R' = I

82: R = R'''' = Me, R' = R''' = H, R'' = Br

83: R = R'''' = Me, R' = Br, R'' = R''' = H

84: R = Me, R' = R''' = R'''' = H, R'' = Cl

85: R = R´´ = H, R´= Me, R´´´= R´´´´ = Me

86: R = Me, R´= R´´´= R´´´´ = H, R´´ = COH

87: R = R´´´´ = Me, R´= OH, R´´ = R´´´= H

88: R = Me, R´= R´´´= R´´´´ = H, R´´ = OH

89: R = Me, R´= OH, R´´ = R´´´= R´´´´ = H

90: R = R´´´´ = Me, R´= R´´´= H, R´´ = COH

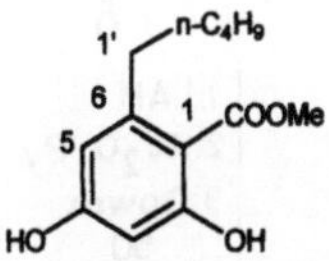

91: ^{13}C-5, ^{13}C-6, ^{14}C-1

92: . ^{13}C-1'

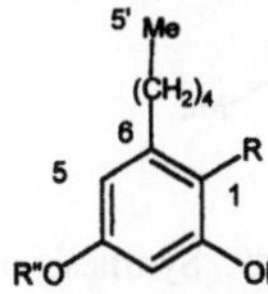

95: R = COOH, R' = H, R" = Me

96: R = R' = R" = H

97: $^2H_3$-5'

98: R = COOH, R' = R" = H

99: R = COOH, R' = R" = H, ^{13}C-5, ^{13}C-6, ^{14}C-1

100: R = H, R' = R" = Me

101: R = R' = H, R" = Me

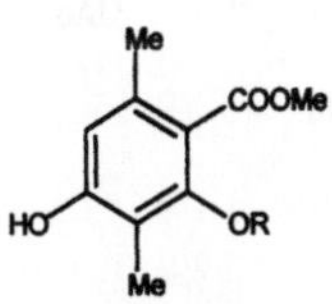

93: R = Me

94: *R = H*

102: R = R' = H

103: R = H, R' = Me

104: R = COOH, R´= Me

105: R = COOH, R' = H

106: R = COOH, R' = H, ^{14}C-2

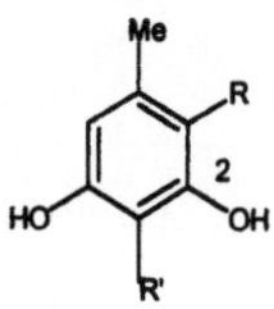

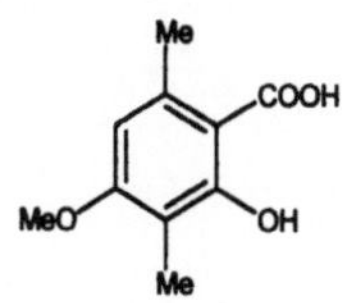

107

derivatives were found in liverworts: methyl everninate in *Blasia pusilla* (*751*) and methyl orsellinate, methyl everninate, methyl 2-methyl-3,4,6,-tri-*O*-methylbenzoate,methyl 2-methyl-3,4-methylenedioxy-6-methoxybenzoate, and methyl 2-methyl-3,4-methylenedioxy-6-hydroxybenzoate in *Plagiochila killarniensis* (*624*).

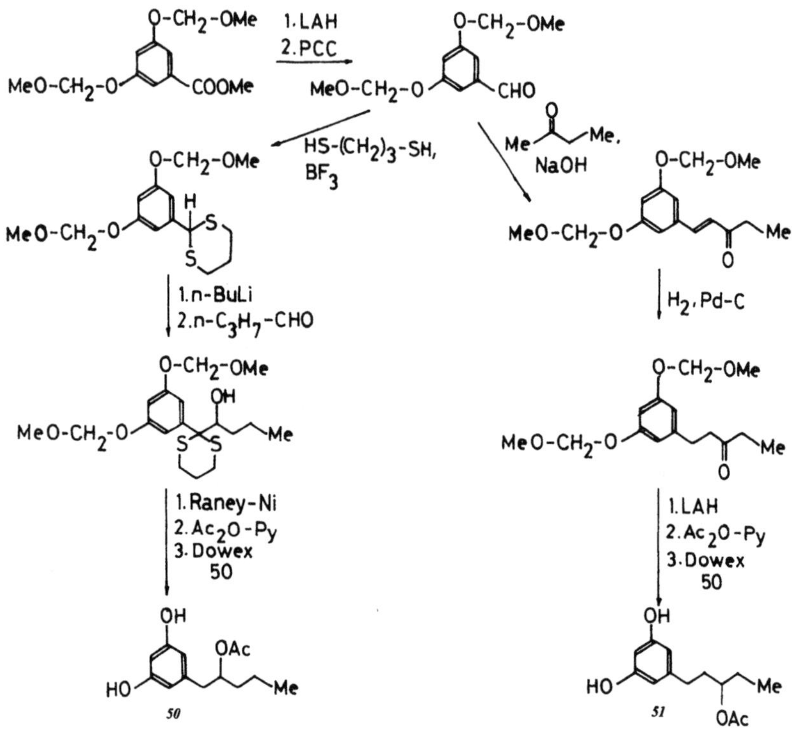

50 *51*

Scheme 49. Synthesis of 2′- and 3′-acetoxyolivetol

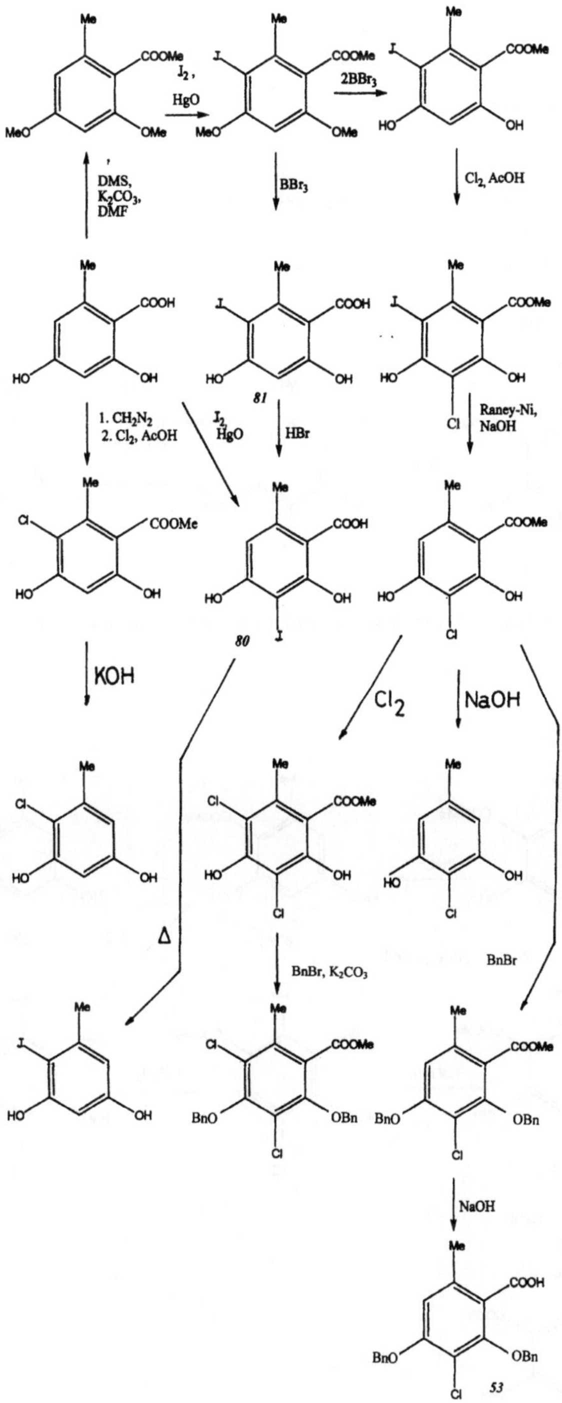

Scheme 50. Synthesis of 3-chloroorsellinic acid-di-*O*-benzylether

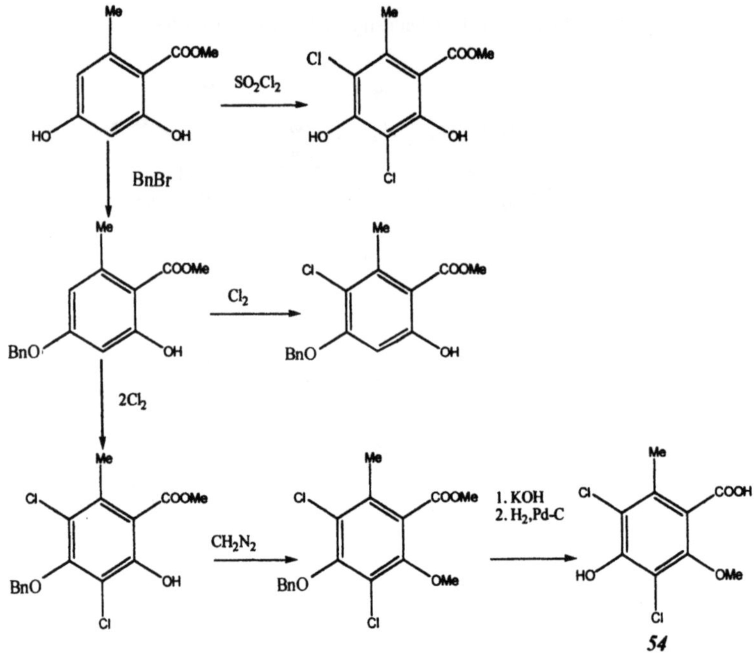

Scheme 51. Synthesis of dichloroisoeverninic acid after (*179*)

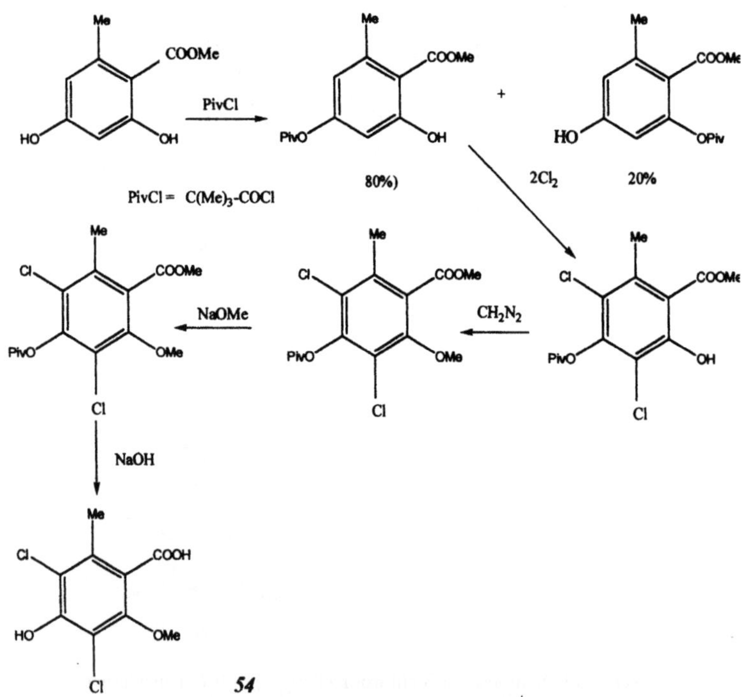

Scheme 52. Synthesis of dichloroisoeverninic acid after (*180*)

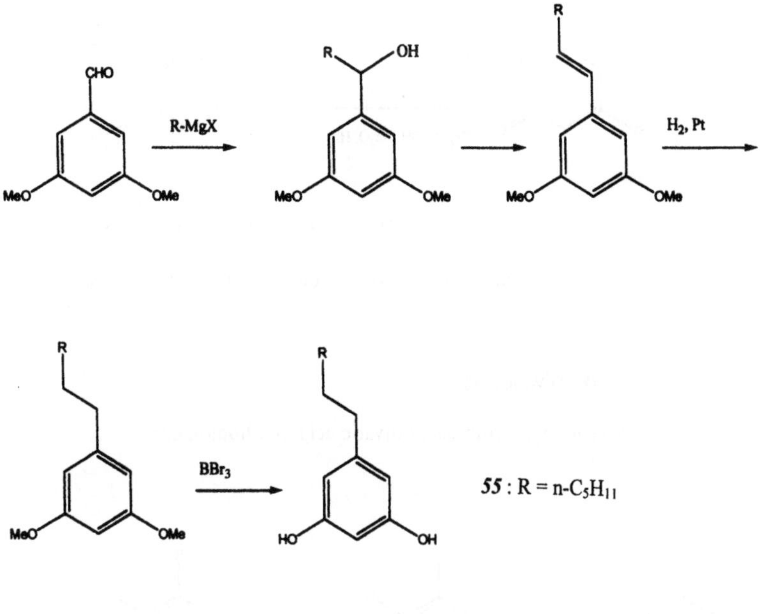

55 : R = n-C_5H_{11}

R = n-C_3H_7, n-C_4H_9, n-C_5H_{11}, n-C_6H_{13}

Scheme 53. Synthesis of 3,5-dihydroxy-*n*-pentylbenzene and homologues

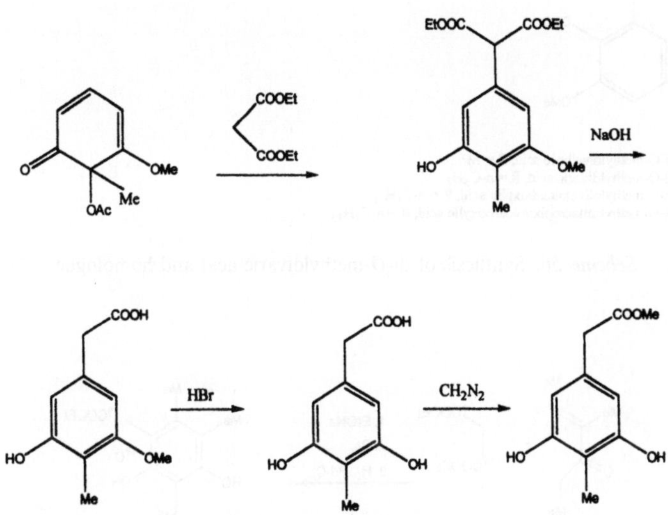

56

Scheme 54. Synthesis of 3,5-dihydroxy-4-methyl-acetic acid

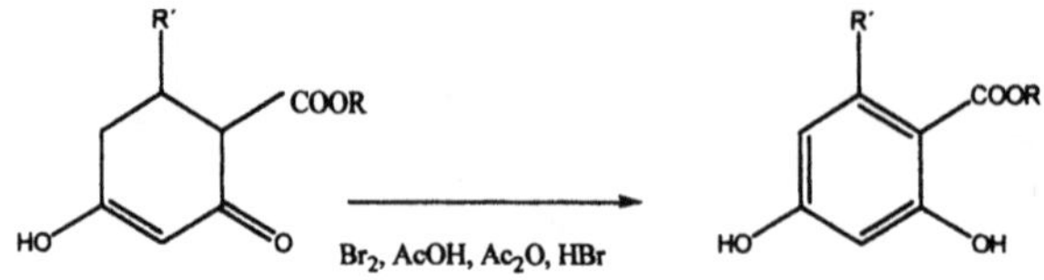

R:	H	H	H	H	Et	Et	Et	Et	Me
R':	Me	n-C_3H_7	n-C_5H_{11}	n-C_7H_{15}	Me	n-C_3H_7	n-C_5H_{11}	n-C_7H_{15}	Me

↓

59 : Divaric acid

Scheme 55. Synthesis of divaric acid and homologues

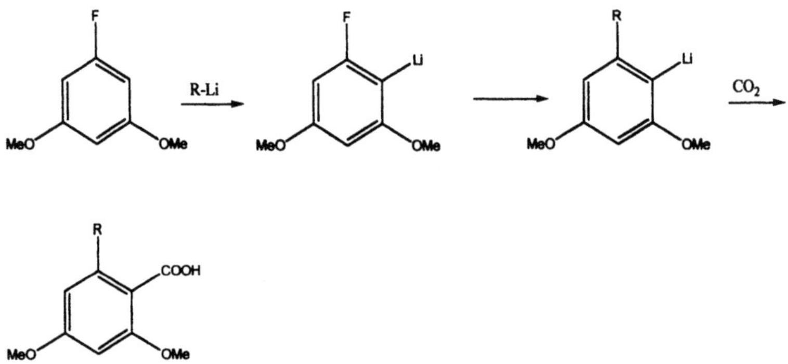

Di-O-methylorsellinic acid, R = Me
Di-O-methyldivaric acid, R = n-C_3H_7
Di-O-methylolivetolcarboxylic acid, R = n-C_5H_{11}
Di-O-methylsphaerophorolcarboxylic acid, R = n-C_7H_{15}

Scheme 56. Synthesis of di-*O*-methyldivaric acid and homologues

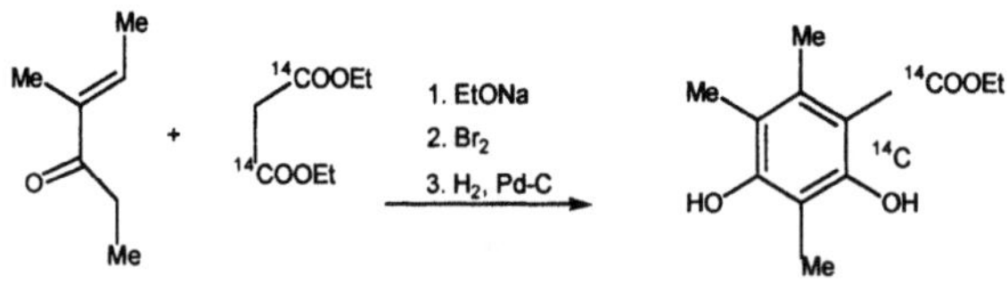

62

Scheme 57. Synthesis of ethyl 3,5-dimethylorsellinate (2,7-di-^{14}C)

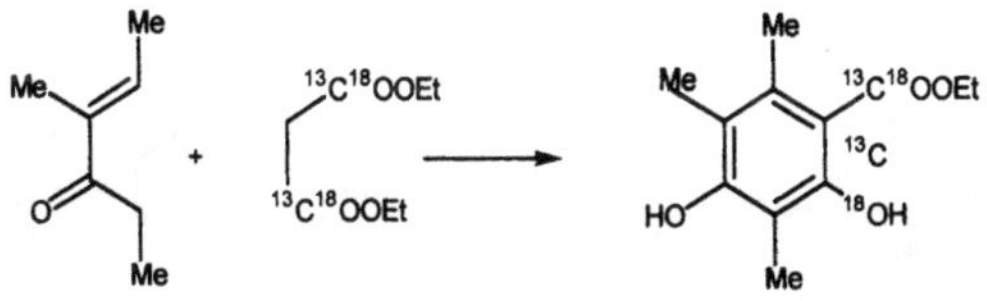

63

Scheme 58. Synthesis of ethyl 3,5-dimethoxyorsellinate (1-^{13}C,7-^{13}C,-CO^{18}OEt)

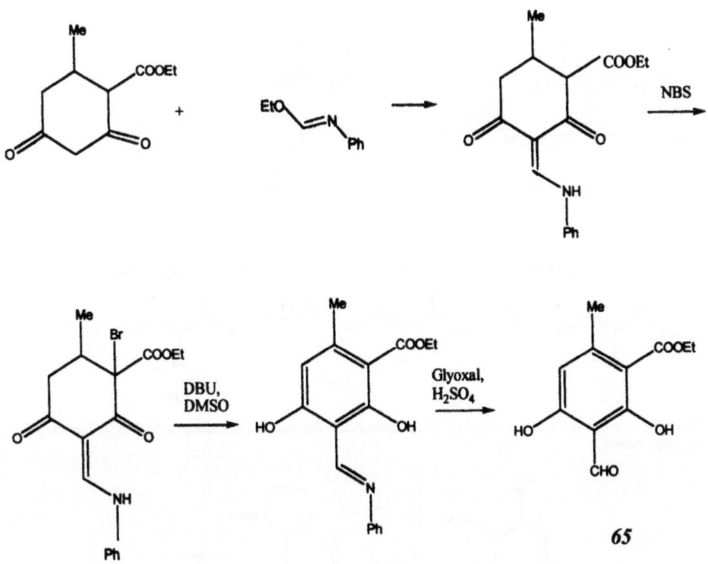

65

Scheme 59. Synthesis of ethyl haematommate

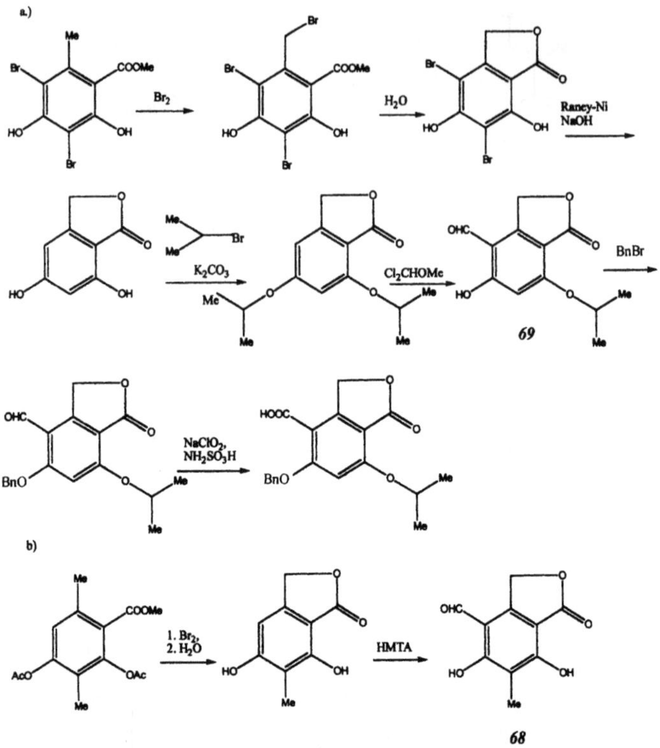

Neither the carboxylic acid I nor the benzylated β-orcinol II did react trifluoroacetic anhydride to the wanted benzophenones.

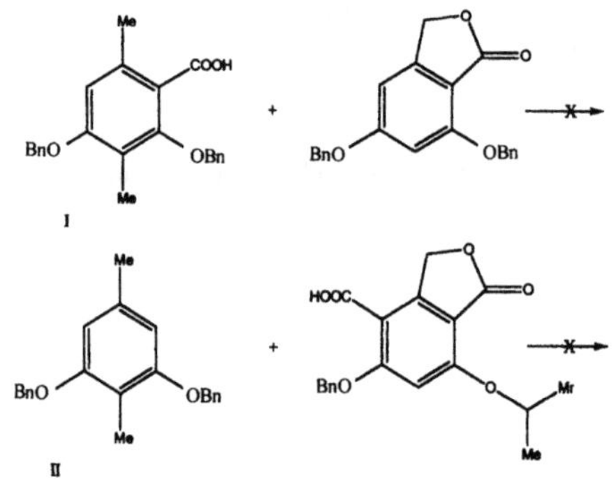

The Smiles rearrangement of the synthetic depside III also failed:

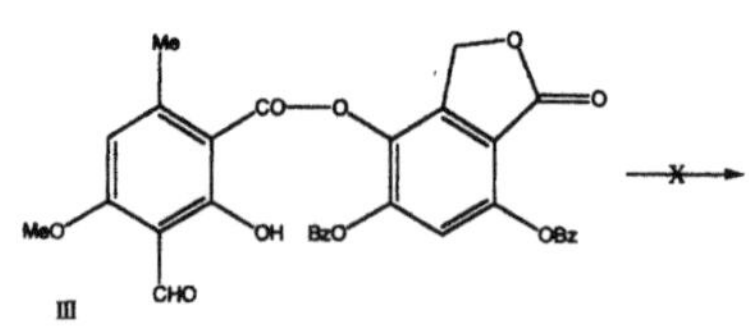

Scheme 60. Synthesis of 4-formyl-5,7-dihydroxy-6-methylphthalide and 4-formyl-5-hydroxy-7-isopropyloxyphthalide

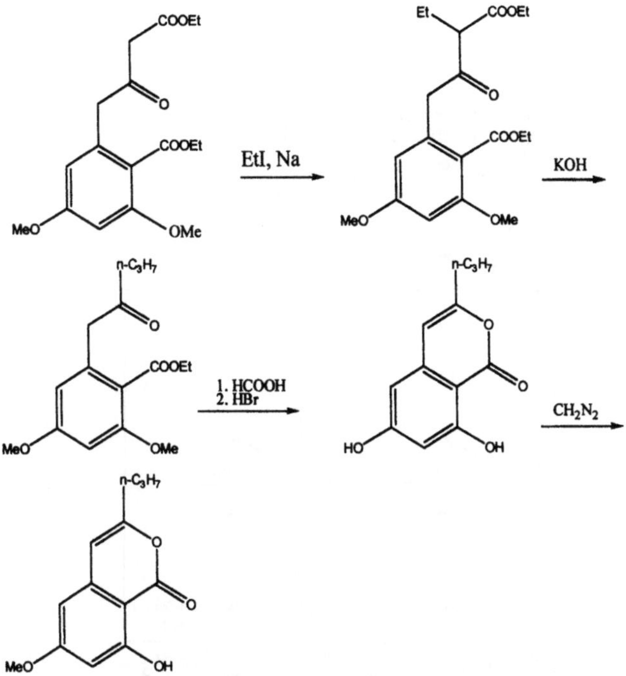

70

Scheme 61. Synthesis of glomellin

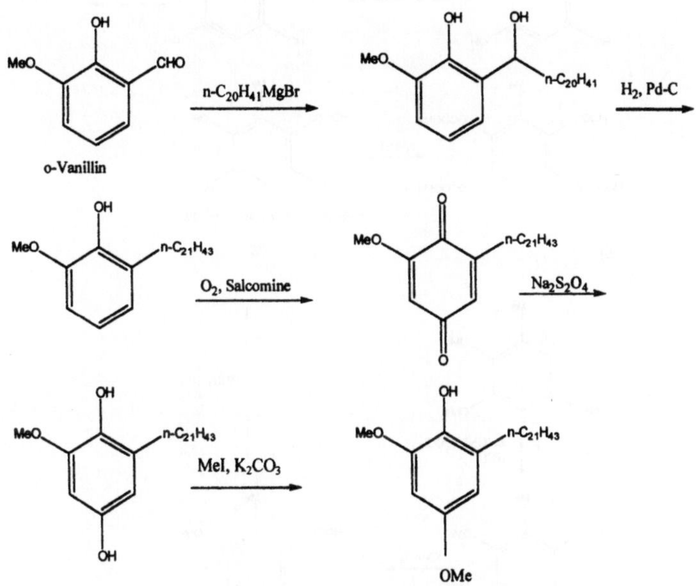

74 : Hierridin

Scheme 62. Synthesis of hierridin

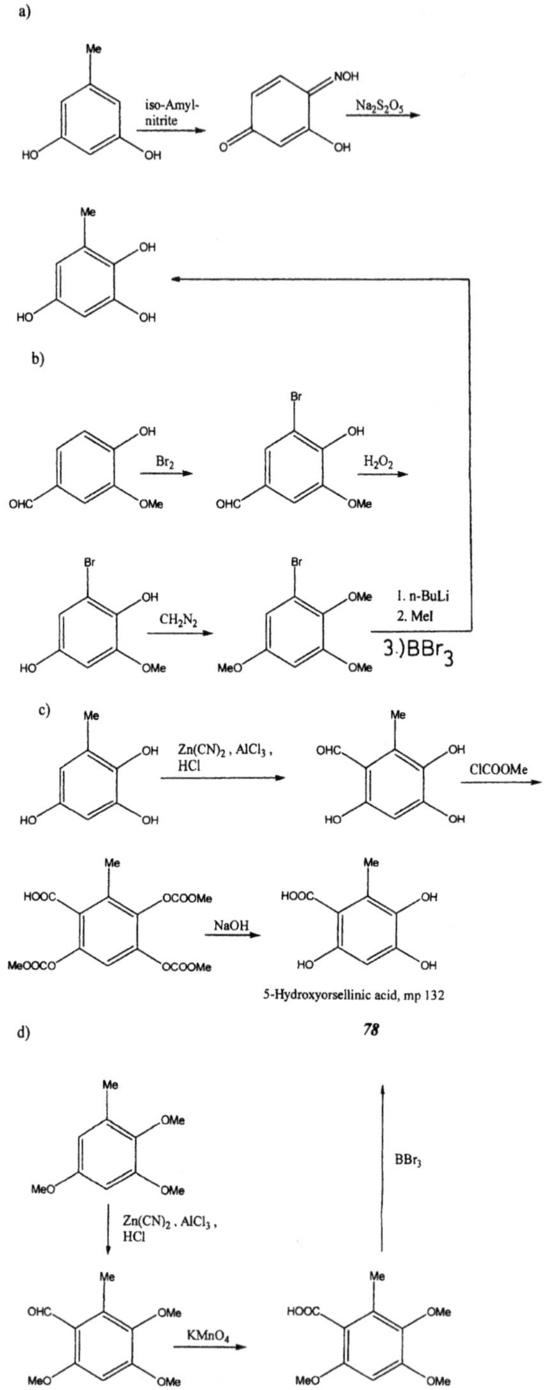

Scheme 63. Synthesis of 5-hydroxyorsellinic acid

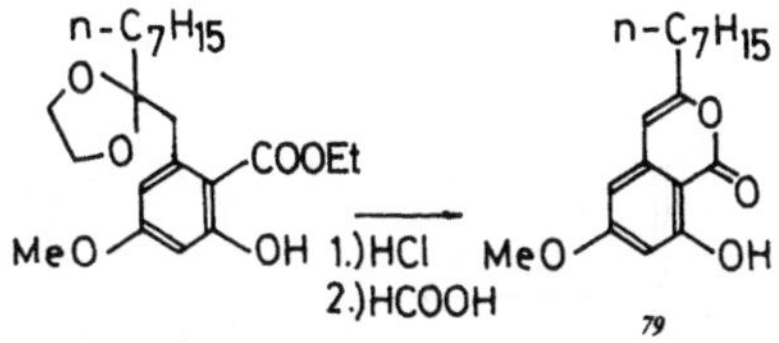

79

Scheme 64. Synthesis of insignin

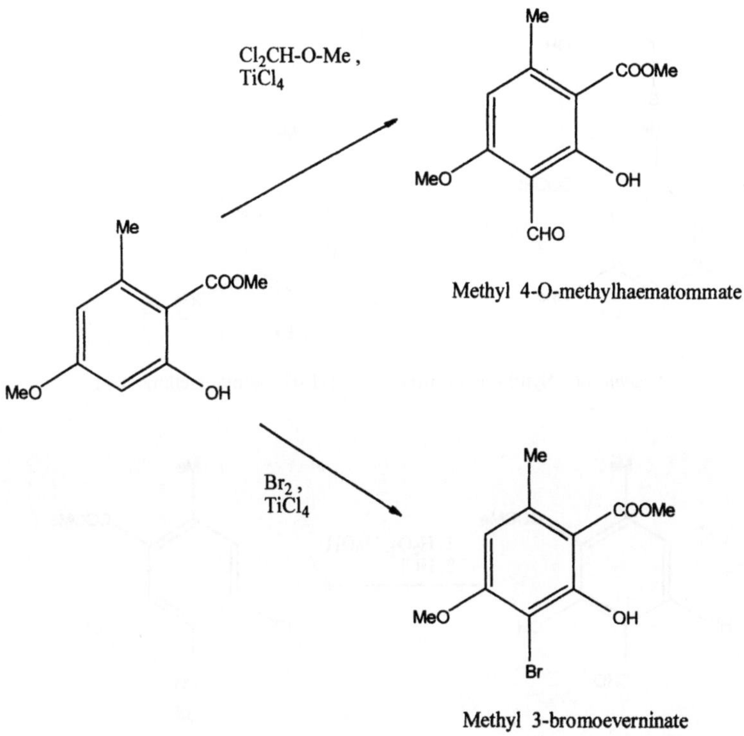

82

Scheme 65. Synthesis of methyl 3-bromoeverninate

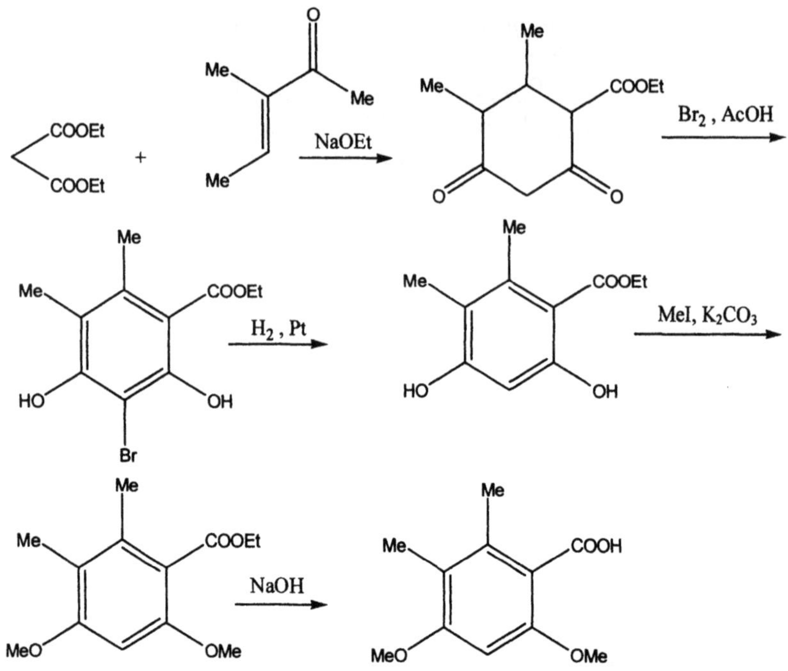

85

Scheme 66. Synthesis of 5-methyl-2,4-di-*O*-methylorsellinic acid

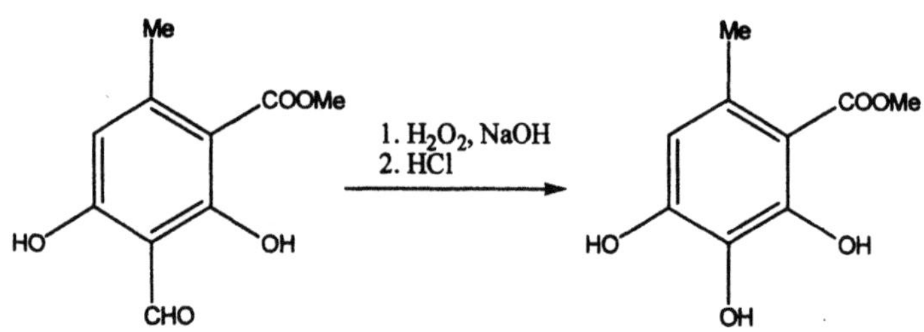

88

Scheme 67. Synthesis of methyl 3-hydroxyorsellinate

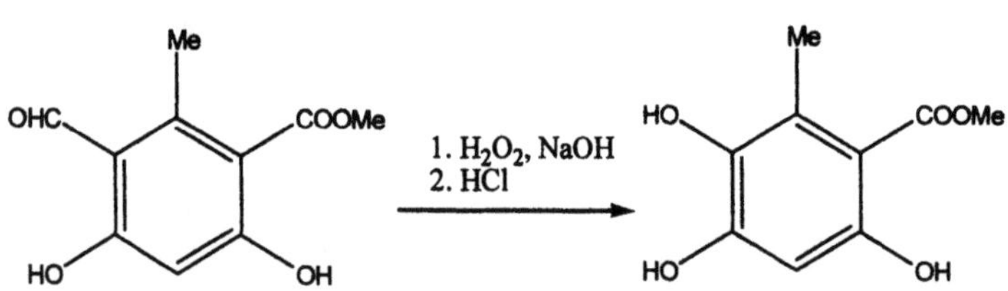

89

Scheme 68. Synthesis of methyl 5-hydroxyorsellinate

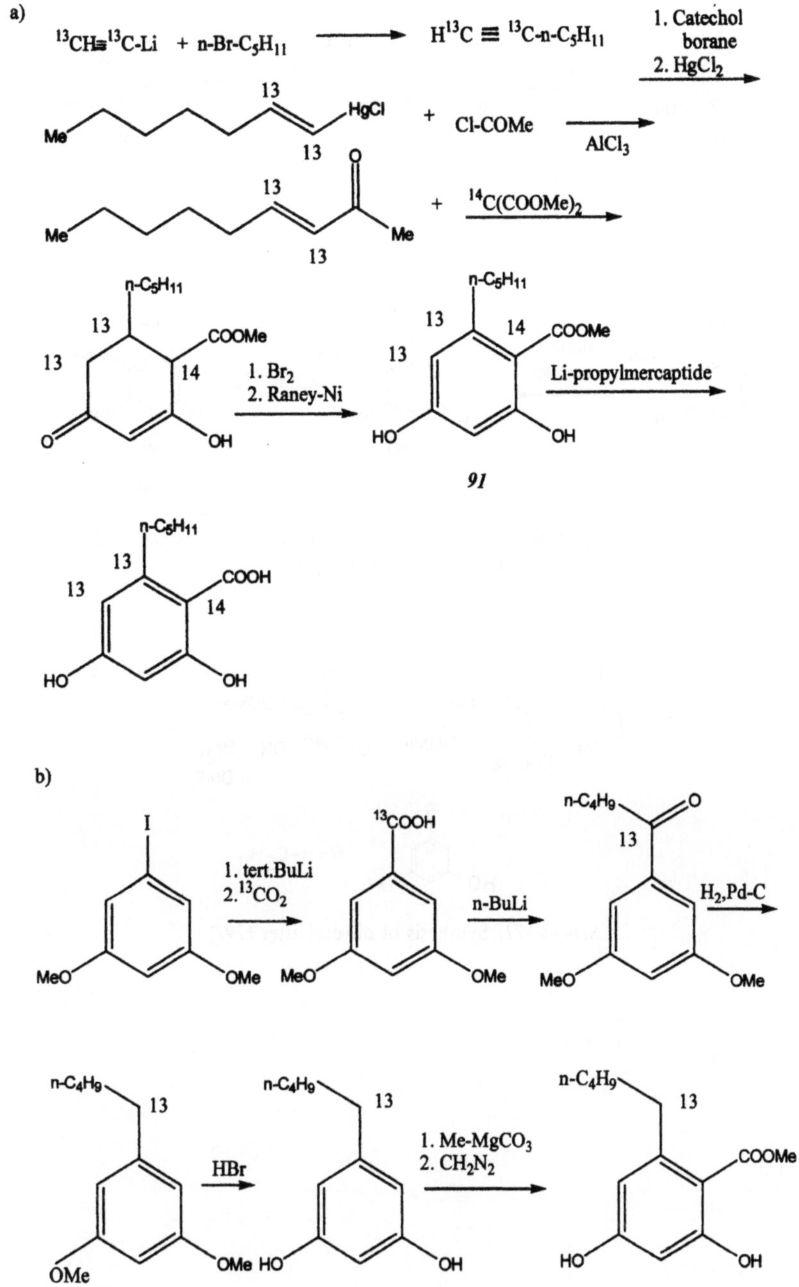

92

Scheme 69. Synthesis of methyl olivetolcarboxylate (5,6-di-^{13}C,1-^{14}C) and methyl olivetolcarboxylate (1′-^{13}C)

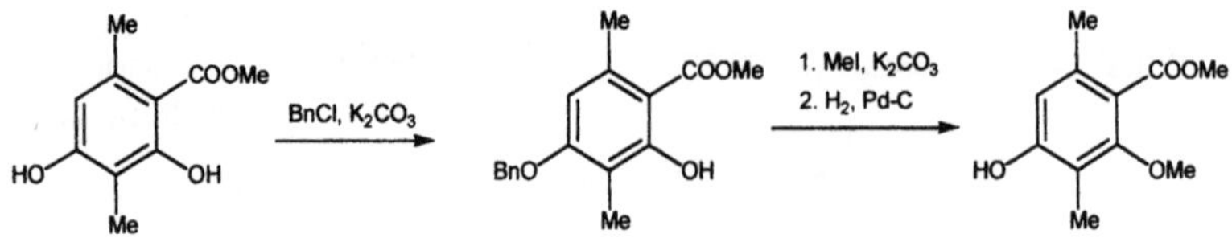

93

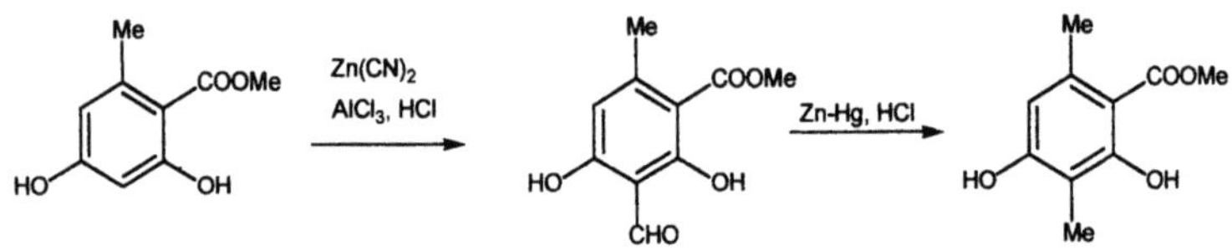

94

Scheme 70. Synthesis of methyl 3-*O*-methyl-β-orcinolcarboxylate and methyl β-orcinolcarboxylate

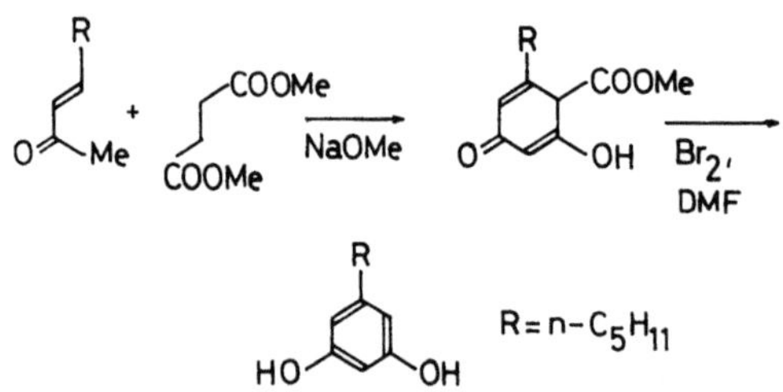

Scheme 71. Synthesis of olivetol after (*297*)

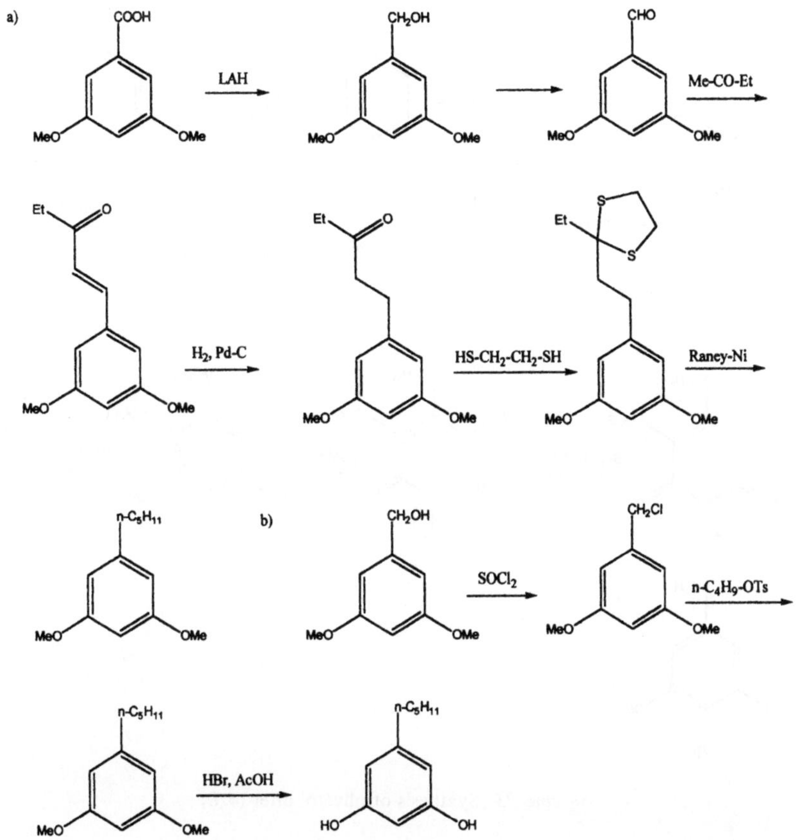

96

Scheme 72. Synthesis of olivetol after (*474*)

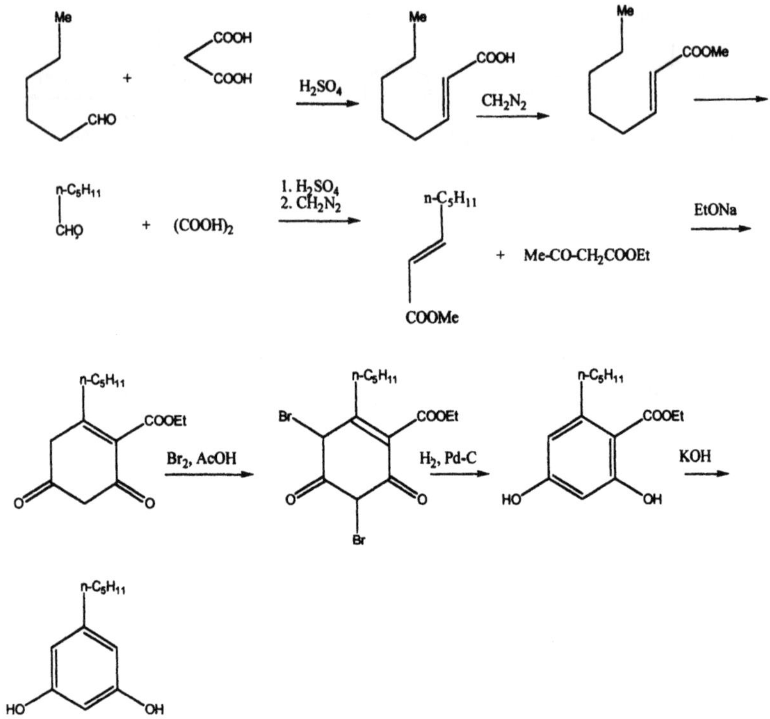

96

Scheme 73. Synthesis of olivetol after (*470*)

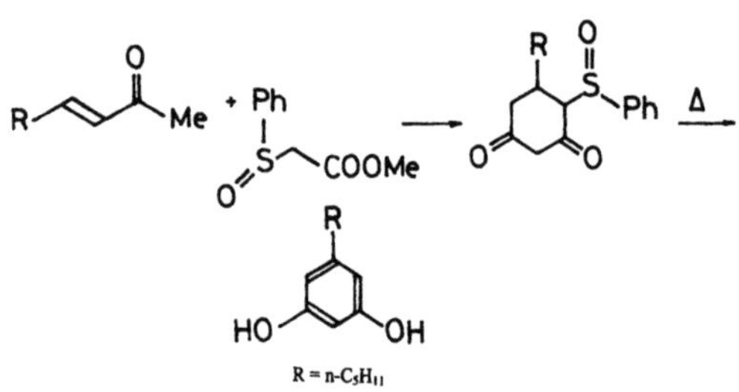

Scheme 74. Synthesis of olivetol after (*443*)

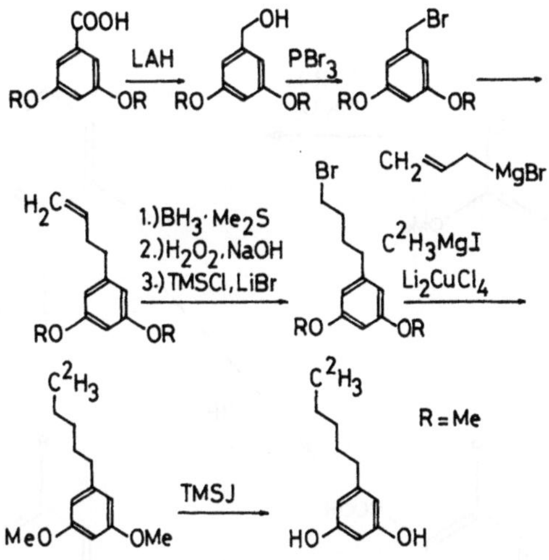

97

Scheme 75. Synthesis of olivetol (5′-$^2H_3$)

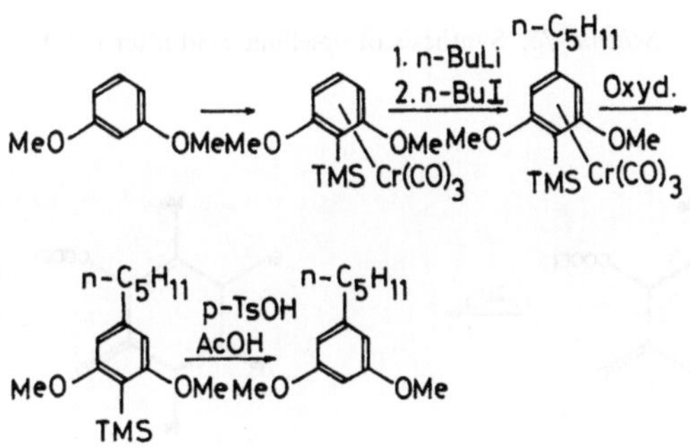

Scheme 76. Synthesis of olivetoldimethylether

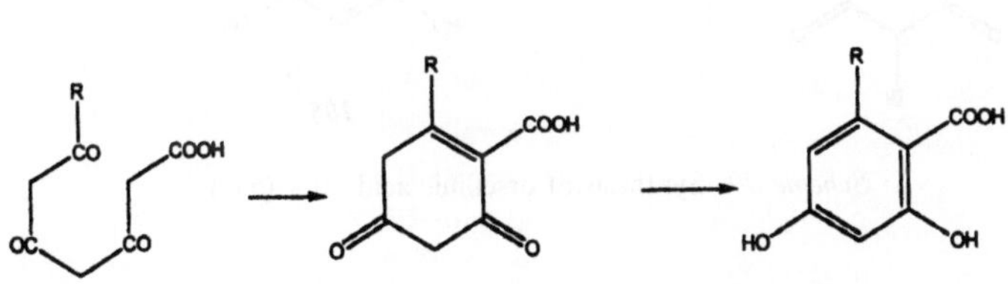

105 **: Orsellinic acid, R = Me**

Scheme 77. Synthesis of orsellinic acid after (*355*)

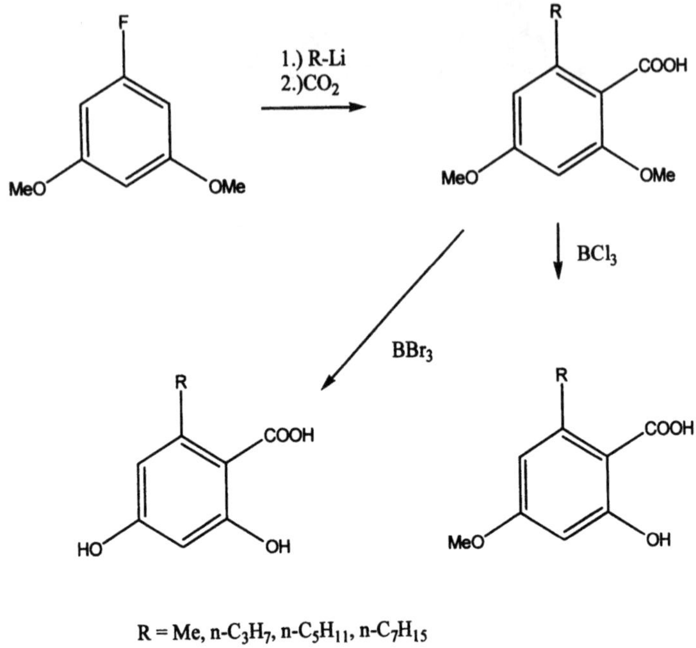

R = Me, n-C_3H_7, n-C_5H_{11}, n-C_7H_{15}

Scheme 78. Synthesis of orsellinic acid after (*701*)

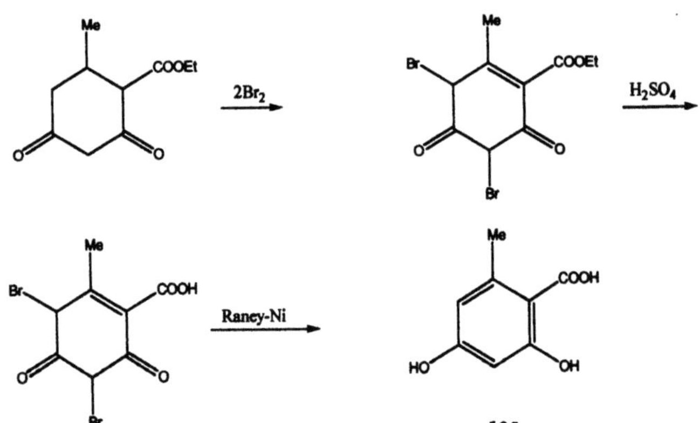

105

Scheme 79. Synthesis of orsellinic acid after (*631*)

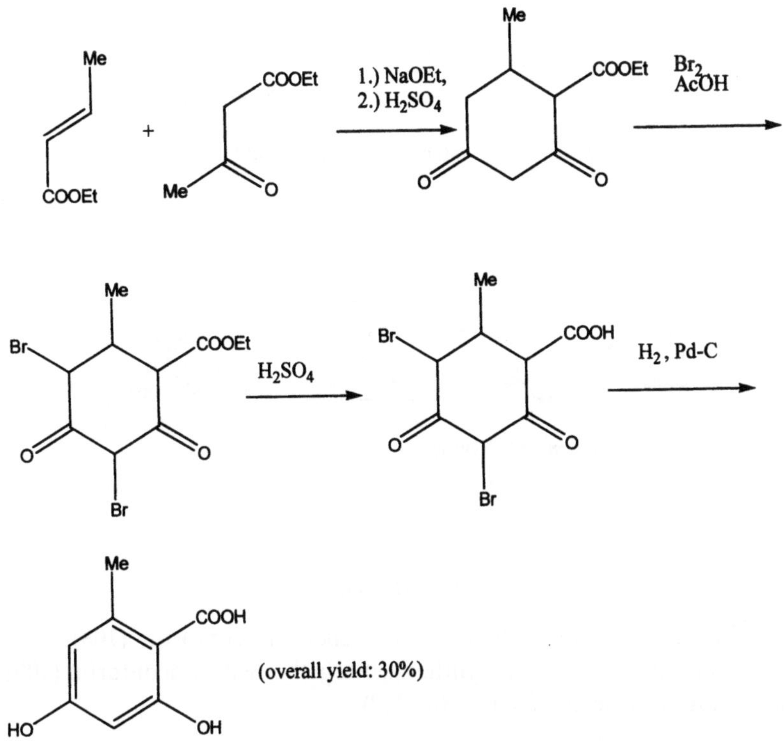

Scheme 80. Synthesis of orsellinic acid after (*318*)

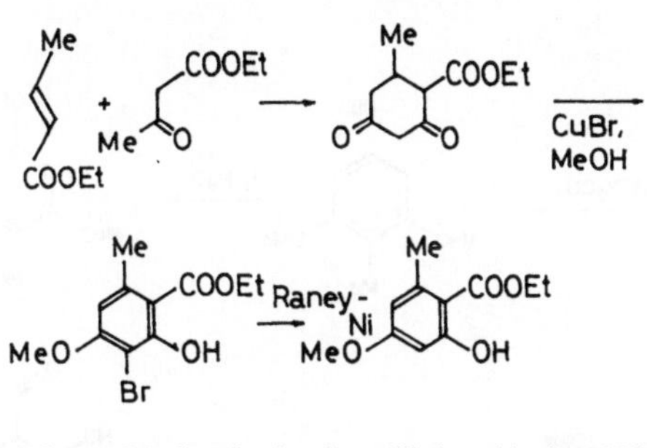

Scheme 81. Synthesis of orsellinic acid after (*71*)

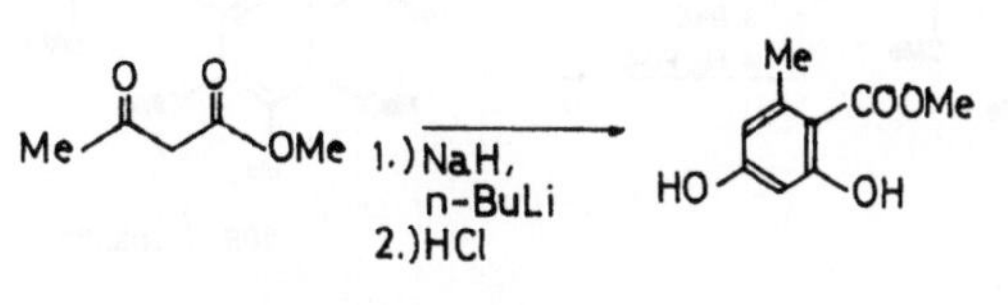

Scheme 82. Synthesis of orsellinic acid after (*70*)

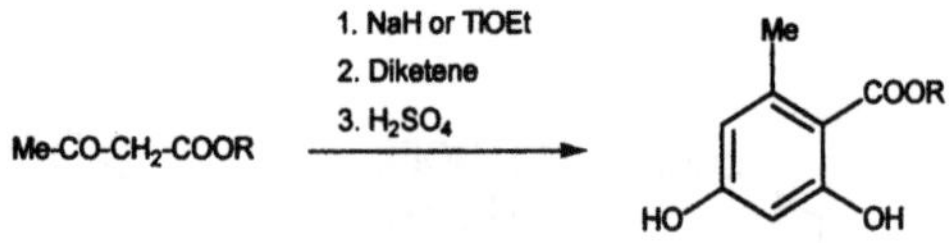

Scheme 83. Synthesis of orsellinic acid (*356*)

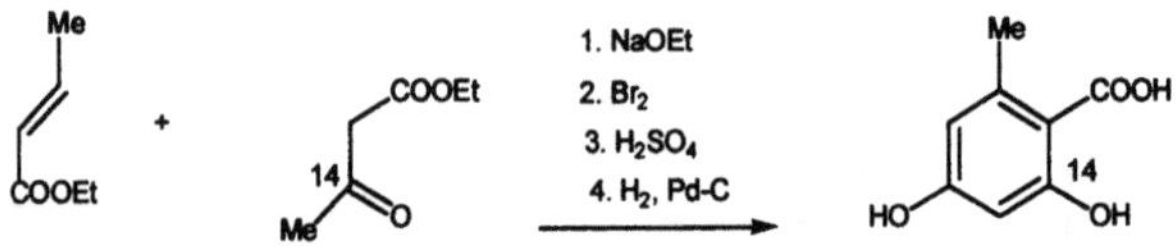

Scheme 84. Synthesis of orsellinic acid (2-^{14}C)

4.5.2. Biphenyls

The only known biphenyl from lichens is contortin (**108**) from *Psoroma contortum*. It was synthesized together with isocontortin (**109**) according to Schemes 85 and 86 (*222*).

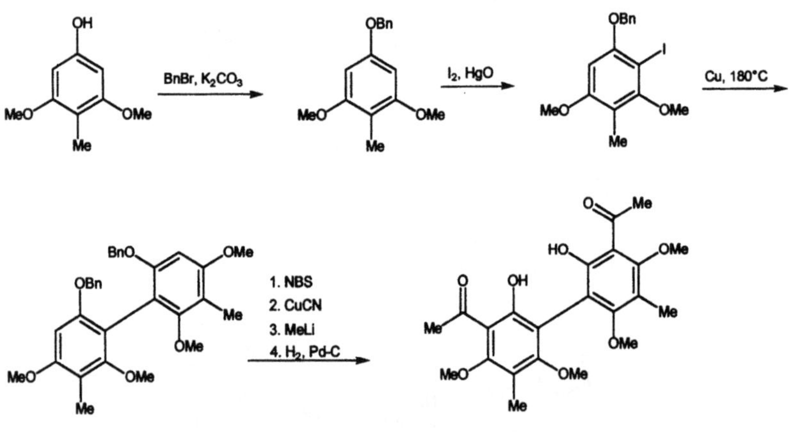

108 : Contortin

Scheme 85. Synthesis of contortin

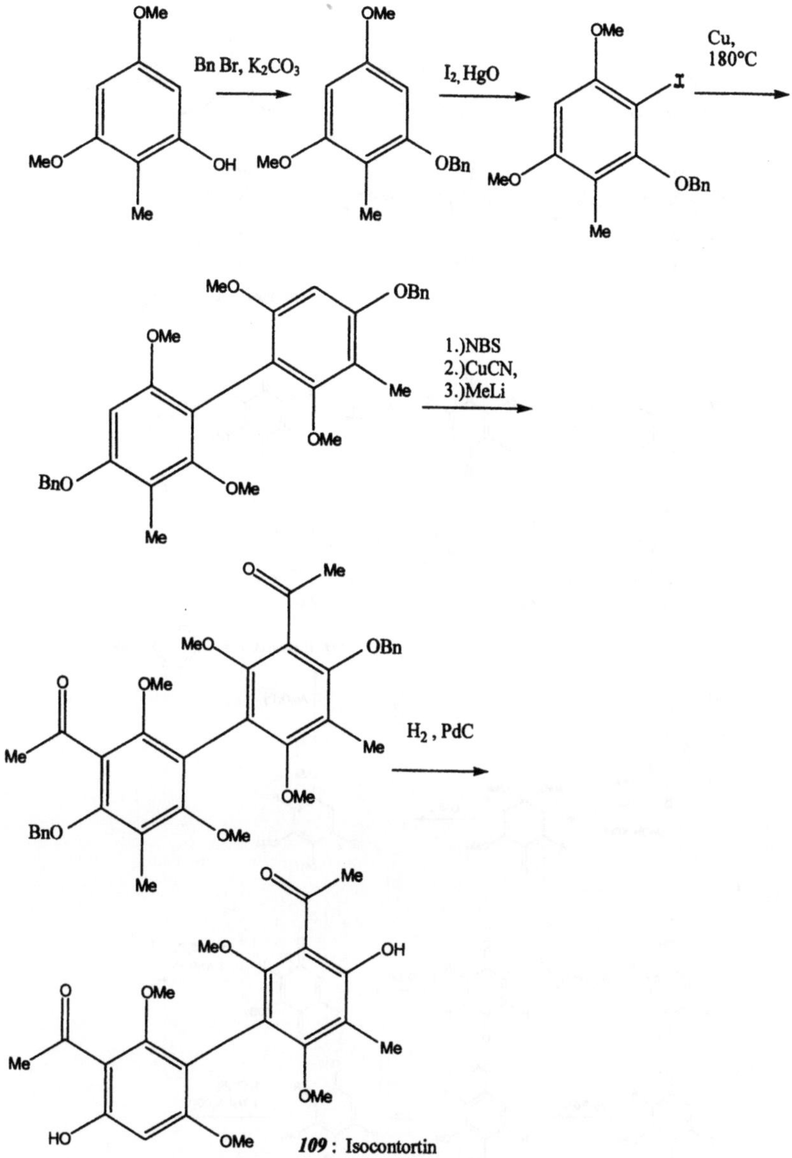

109 : Isocontortin

Scheme 86. Synthesis of isocontortin

4.5.3. Terphenyls

Butlerins A (**110**), B (**111**), C (**112**), D (**113**), E (**114**) and F (**115**) are *p*-terphenyls from *Parmelia butleri* (*212*) and *Relicina connivens* (*207*)

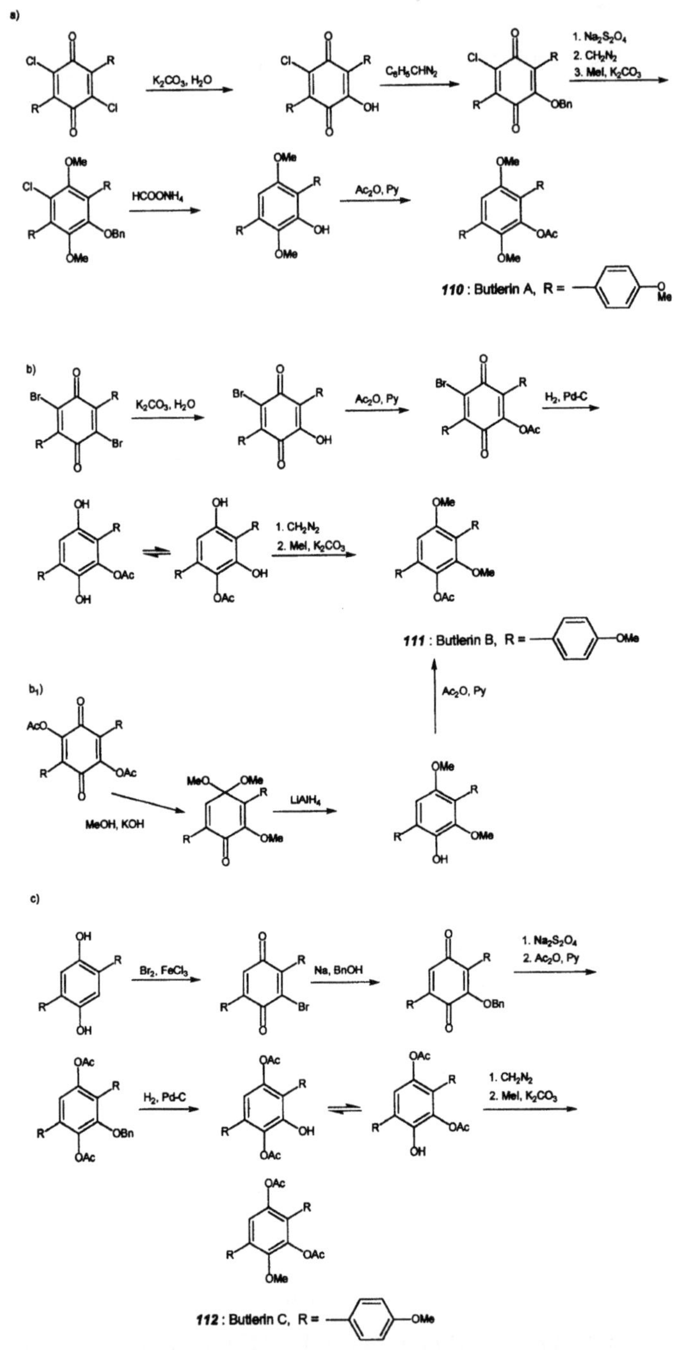

Scheme 87. Synthesis of the butlerins A, B and C

respectively. The structures of **110** and **111** were established by X-ray analysis; furthermore all butlerins have been synthesized (*207*, *284*) (see Schemes 87 and 88).

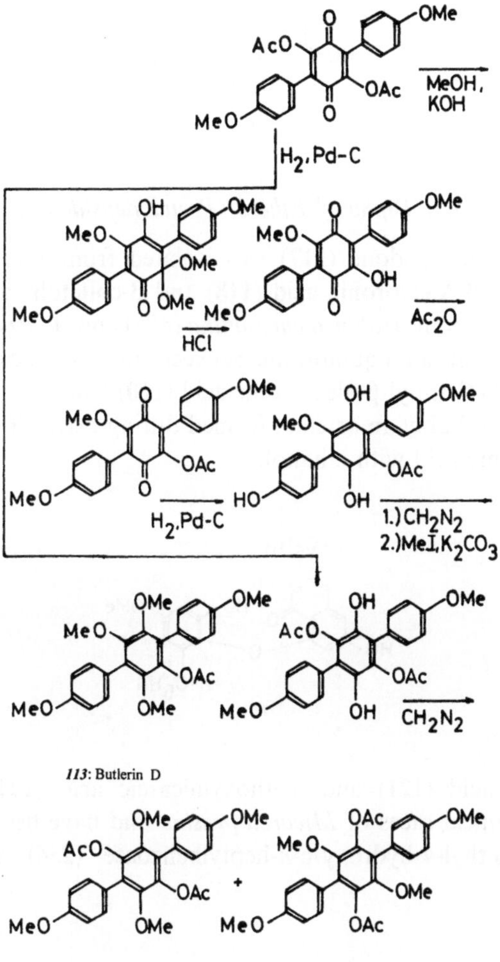

113: Butlerin D

114: Butlerin E

115: Butlerin F

Scheme 88. Synthesis of the butlerins D, E and F

4.5.4. Diphenylmethanes

2,4,2′,4′-Tetrahydroxy-6,6′-di-*n*-propyldiphenylmethane (**116**) was isolated from *Solorina crocea* (*453*).

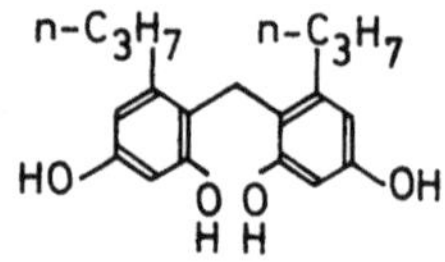

116

4.5.5. Diphenyl Ethers (Pseudodepsidones)

2′-*O*-Methylphysodone (**117**) was isolated from *Pseudevernia furfuracea* (*339*). β-Alectoronic acid (**118**) and β-collatolic acid (**119**) are diphenyl ethers from *Asahinea chrysantha* (*476*) and *A. scholanderi* (*476*, *672*) and exhibit an equilibrium between the keto-acid and lactol tautomers. 2‴-*O*-Ethyl-β-alectoronic acid (**120**) from *Alectoria sarmentosa* (*327*) is probably an artefact formed during extraction by reaction of β-alectoronic acid with ethanol.

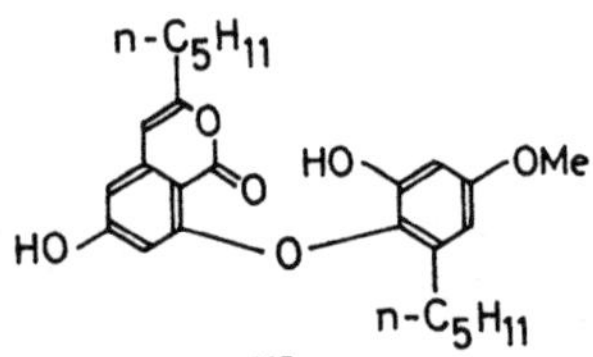

117

Micareic acid (**121**) and methoxymicareic acid (**122**) have been found in chemical races of *Micarea prasina* and have been synthesized from 2-*O*-methyl-4-hydroxy-6-*n*-heptylbenzoate (*236*) according to Scheme 89.

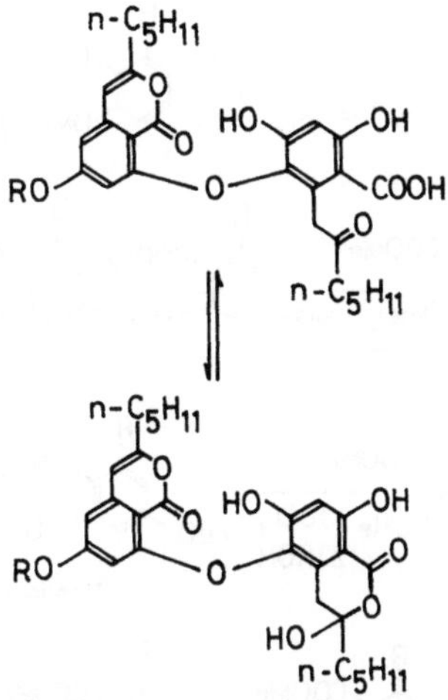

118: ß-Alectoronic acid, R = H

119: ß-Collatolic acid, R = Me

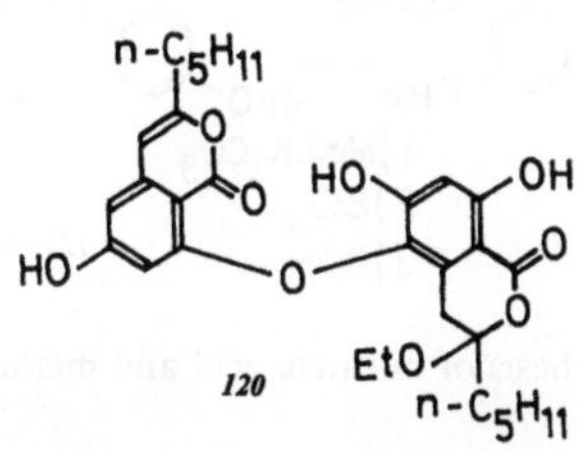

120

Epiphorellic acid 1 (**123**) and 2 (**124**) are diphenyl ethers from *Cornicularia epiphorella* (*296*). Elix and Jenie (*225*) described the synthesis of epiphorellic acid 1 (see Scheme 90) and Comber *et al.* (*77*) the synthesis of epiphorellic acid 2 (see Scheme 91).

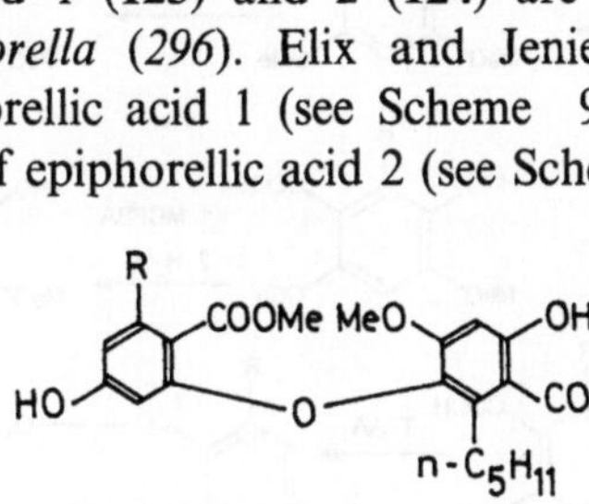

123 : Epiphorellic acid 1, R = n-C_5H_{11}

124 : Epiphorellic acid 2, R = -$(CH_2)_2$-CO-Et

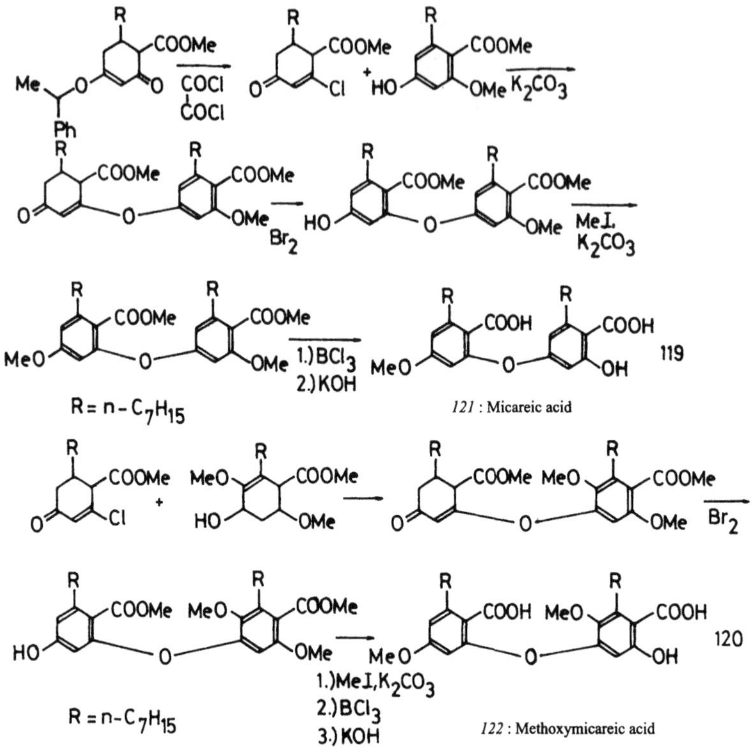

119

120

Scheme 89. Synthesis of micareic acid and methoxymicareic acid

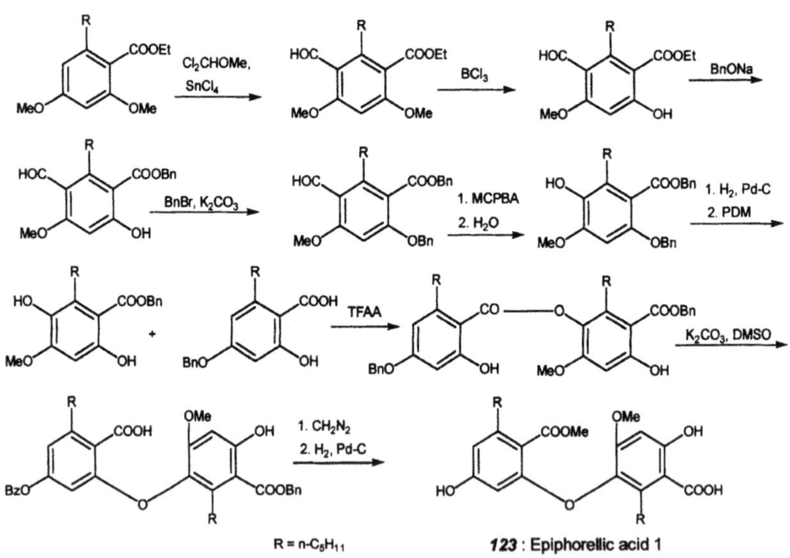

Scheme 90. Synthesis of epiphorellic acid 1

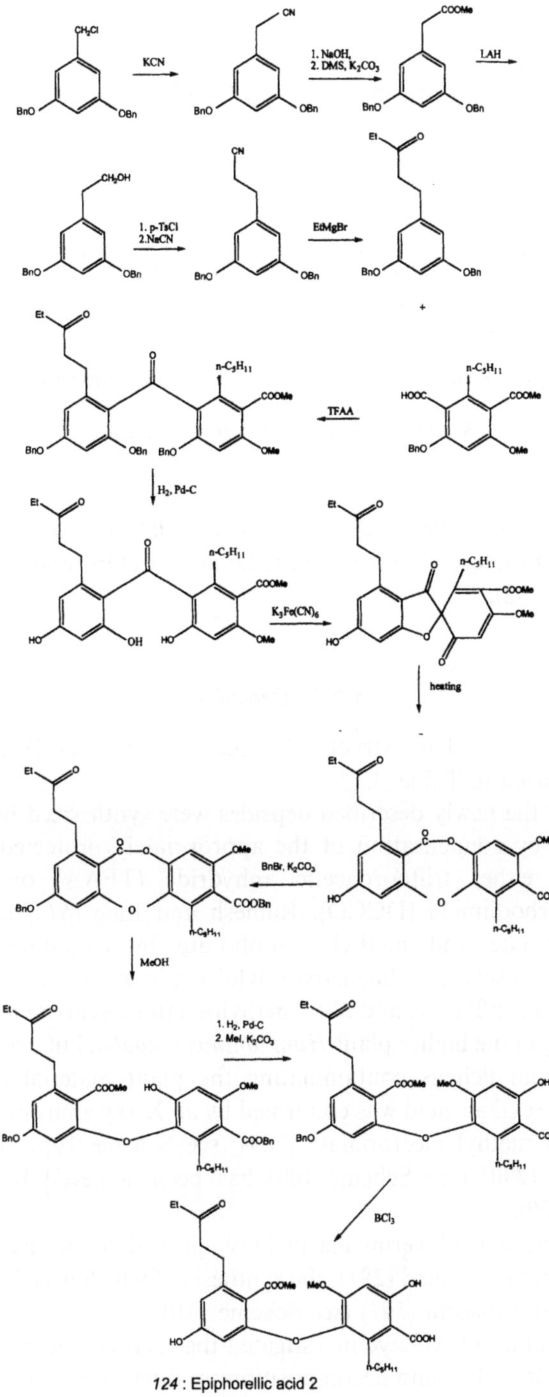

124 : Epiphorellic acid 2

Scheme 91. Synthesis of epiphorellic acid 2

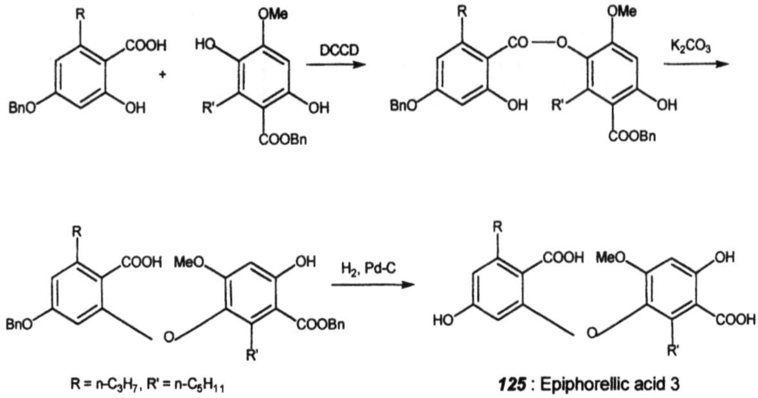

Scheme 92. Synthesis of epiphorellic acid 3

Elix and McCaffery (*246*) reported isolation from *Coelopogon abraxes* of another isomer, epiphorellic acid 3 (**125**) and its synthesis (see Scheme 92).

4.5.6. Depsides

Depsides isolated and structurally elucidated between 1983 and 1999 are summarized in Table 4.

Most of the newly described depsides were synthesized by Elix and co-workers by condensation of the appropriately protected S and A parts with either trifluoroacetic anhydride (TFAA) or *N,N′*-dicyclohexylcarbodiimide (DCCD). Ramesh and Baig (*612*) synthesized methyl evernate and methyl gyrophorate by oxidation–reduction condensation using 2,2′-dibenzoxazolyldisulfide as an oxidizing agent. The depsides confluentic and 2′-*O*-methylperlatolic acids were described as occurring in the higher plant *Himatanthus sucuuba*, but originate very probably from lichens contaminating the plant material (*280*). The structure of sekikaic acid was confirmed by an X-ray analysis (*451*). The synthesis of methyl alectorialate (**289**) (see Scheme 108) and methyl barbatolate (**290**) (see Scheme 109) has been achieved by Elix and Jayanthi (*220*).

A depside named vermicularin (*349*) proved to be identical with decarboxythamnolic acid (**291**), the synthesis of which was described by Pulgarin and Tabacchi (*591*) (see Scheme 110).

Huneck *et al.* (*408*) have investigated the thermal decomposition of lichen depsides. The main decomposition products were decarboxylated

Table 4. Depsides from lichens

Compound	Formula No.	M.p. (°C)	Lichen	Synth. Scheme	Ref.
5-*O*-Acetyl-4-*O*-metylhiascic acid	**126**	165–167	*Koerberiella wimmeriana*	93	(*277*)
Arthoniaic acid	**127**	167		94	(*266*)
Brialmontin 1	**128**	102–104	*Lecania brialmontii*		(*711*)
				From 2,4-dimethoxy-3,5,6-trimethylbenzoic acid and 3-methoxy-2,5,6-trimethylphenol with TFAA	(*192*)
Brialmontin 2	**129**	106–109	*Lecania brialmontii*		(*711*)
				From 2-hydroxy-4-methoxy-3,5,6-trimethylbenzoic acid and 3-methoxy-2,5,6-trimethylphenol with TFAA	(*192*)
3-Chlorodivaricatic acid	**130**	158–159	*Thelomma mammosum*		
			Ramalina seyreana	95	(*251*)
5-Chlorodivaricatic acid	**131**	142–143	*Dimelaena* cf. *radiata*		(*245*)
3-Chloroimbricaric acid	**132**	128–130		From 3-chloro-2-hydroxy-4-methoxy-6-*n*-pentylbenzoic acid and benzyl 2,4-dihydroxy-6-*n*-propylbenzoate with TFAA and subsequent debenzylation	(*208*)
5-Chloro-2′-*O*-methylanziaic acid	**133**	142–143	*Lecanora lividocinerea*	From 4,6-dibenzyloxy-3-chloro-2-*n*-pentyl-benzoic acid and benzyl 4-hydroxy-2-methoxy-6-*n*-pentylbenzoate with TFAA and subsequent debenzylation	(*194*)
1′-Chloronephroarctin	**134**	181	*Pseudocyphellaria pickeringii*	96	(*275*)

Table 4 (continued)

Compound	Formula No.	M.p. (°C)	Lichen	Synth. Scheme	Ref.
3-Chloroperlatolic acid	**135**	128	*Dimelaena californica*	From 3-chloro-2-hydroxy-4-methoxy-6-*n*-pentylbenzoic acid and benzyl 2,4-dihydroxy-6-*n*-pentylbenzoate with TFAA and subsequent debenzylation	(*208*)
3-Chlorostenosporic acid	**136**	138	*Dimlaena californica*	From 3-chloro-2-hydroxy-4-methoxy-6-propylbenzoic acid and benzyl 2,4-dihydroxy-6-*n*-pentylbenzoate with TFAA and subsequent debenzylation	(*208*)
Conechinocarpic acid	**137**	>290	*Relicina samoensis*		(*264*)
Crustinic acid	**138**	178–180, dec.	*Umbilicaria crustulosa*		(*401*, *582*)
				By condensation of 2,4-dibenzyloxy-6-methylbenzoic acid and benzyl 4-benzyloxy-3-(2′,4′-dihydroxy-6′-methylbenzoyloxy)-6-hydroxy-2-methylbenzoate with TFAA and subsequent debenzylation	(*223*)
Decarboxyanziaic acid	**139**	102–104	*Neofuscelia depsidella*	From 4-benzyloxy-2-hydroxy-6-*n*-pentylbenzoic acid and olivetol with DCCD and subsequent debenzylation	(*267*)
Decarboxydivaricatic acid	**140**	Oil	*Neofuscelia depsidella*	From 2-hydroxy-4-methoxy-6-*n*-propylbenzoic acid and divarol with DCCD	(*267*)
Decarboxy-2′-*O*-methyl-divaricatic acid	**141**	Oil	*Neofuscelia depsidella*	From 2-hydroxy-4-methoxy-6-*n*-propylbenzoic acid and 3-methoxy-5-*n*-propylphenol with DCCD	(*267*)
Decarboxy-2′-*O*-methyl-norimbricaric acid	**142**	96–97	*Neofuscelia depsidella*	From 4-benzyloxy-2-hydroxy-6-*n*-pentylbenzoic acid and 3-methoxy-5-*n*-propylphenol with DCCD and subsequent debenzylation	(*267*)

References, pp. 239–276

Decarboxynorimbricaric acid	**143**	99–100	*Neofuscelia depsidella*	From 4-benzyloxy-2-hydroxy-6-*n*-pentyl-benzoic acid and 1,3-dihydroxy-5-*n*-propylbenzene with DCCD and subsequent debenzylation	(*267*)
Decarboxynorstenosporic acid	**144**	84–85	*Neofuscelia depsidella*	From 2,4-dibenzyloxy-6-*n*-propylbenzoic acid and 1,3-dihydroxy-5-*n*-pentylbenzene with DCCD and subsequent debenzylation	(*267*)
Decarboxyperlatolic acid	**145**	73–74	*Neofuscelia depsidella*	From 2-hydroxy-4-methoxy-6-*n*-pentyl-benzoic acid and 1,3-dihydroxy-5-*n*-pentyl-benzene with DCCD	(*267*)
Decarboxystenosporic acid	**146**	Oil	*Neofuscelia depsidella*	From 2-hydroxy-4-methoxy-6-*n*-propyl-benzoic acid and 1,3-dihydroxy-5-*n*-pentyl-benzene with DCCD	(*267*)
4-*O*-Demethylbaeomycesic acid	**147**	171–172	*Dibaeis arcuatus*		(*336*)
4-*O*-Demethyldiffractaic acid	**148**		*Xanthoparmelia duplicata*	From 4-benzyloxy-2-methoxy-3,6-di-methylbenzoic acid and benzyl 2,4-dihydroxy-3,6-dimethylbenzoate with TFAA and subsequent debenzylation	(*199*)
4-*O*-Demethylglomellic acid	**149**	152	*Neofuscelia loxodes*	97	(*273*)
4-*O*-Demethylglomelliferic acid	**150**	156	*Neofuscelia loxodes*	97	(*273*)
4-*O*-Demethylimbricaric acid	**151**	169	*Cetrelia sanguinea*	From 2,4-dibenzyloxy-6-*n*-pentylbenzoic acid and benzyl 2,4-dihydroxy-6-*n*-propylbenzoate with TFAA and subsequent debenzylation	(*251*)
4-*O*-Demethylloxodellic acid	**152**	146	*Neofuscelia loxodes*	97	(*273*)
4-*O*-Demethylmicrophyllinic acid	**153**	162	*Parmotrema demethyl-microphyllinum*	98	(*221*)
4-*O*-Demethylperlatolic acid	**154**		*Ramalina americana*		(*96*)
4-*O*-Demethylplanaic acid	**155**	122–124	*Lecidea plana*		(*405*)
4′-*O*-Demethylsekikaic acid	**156**		*Ramalina americana*		(*96*)

Table 4 (continued)

Compound	Formula No.	M.p. (°C)	Lichen	Synth. Scheme	Ref.
4-*O*-Demethylsphaerophorin	**157**	136	*Sphaerophorus melanocarpus*	From 2,4-dibenzyloxyorsellinic acid and benzyl 2-*n*-heptyl-4,6-dihydroxy-benzoate with TFAA and subsequent debenzylation	(*262*)
4-*O*-Demethylsquamatic acid	**158**	128–130	*Pertusaria squamatica*		(*336*)
4-*O*-Demethylstenosporic acid	**159**	108	*Neofuscelia pulla*	From 4-benzyloxy-2-hydroxy-6-*n*-propyl-benzoic acid and benzyl 2,4-dihydroxy-6-*n*-pentylbenzoate with TFAA and subsequent debenzylation	(*251*)
4-*O*-Demethylsuperconfluentic acid	**160**	165	*Porpidia glaucophaea*	99	(*265*, *103*)
Depsidellin A	**161**	Oil	*Neofuscelia depsidella*	From 2-hydroxy-4-methoxy-6-*n*-propyl-benzoic acid and olivetol with DCCD	(*267*)
Depsidellin B	**162**	114–115	*Neofuscelia depsidella*	From 2-hydroxy-4-benzyloxy-6-*n*-pentyl-benzoic acid and olivetol with DCCD and subsequent debenzylation	(*267*)
Depsidellin C	**163**	Oil	*Neofuscelia depsidella*	From 2-hydroxy-4-methoxy-6-*n*-propyl-benzoic acid and 3′-hydroxy-5′-*n*-pentyl-4-benzyloxy-2-hydroxy-6-*n*-pentylbenzoate with DCCD and subsequent debenzylation	(*267*)
3,5-Dichloro-4-*O*-methylplanaic acid	**164**	119–129	*Lecanora jamesii*		(*46*)
			Lecanora lividocinerea	From 4-benzyloxy-3,5-dichloro-2-meth-oxy-6-*n*-pentylbenzoic acid and benzyl-2-*O*-methyl-olivetolcarboxylate with TFAA and subsequent debenzylation	(*194*)

3,5-Dichloro-2′-*O*-methylanziaic acid	**165**	154–155	*Lecanora sulphurella*	From 3,5-dichloroolivetolcarboxylic acid and benzyl 4-hydroxy-2-methoxy-6-*n*-pentylbenzoate with TFAA and subsequent debenzylation	(*251*)
3,5-Dichloro-2′-*O*-methylnor-hyperlatolic acid	**166**	154–155	*Lecanora lividocinerea*	From 3,5-dichloro-2,4-dihydroxy-6-*n*-pentylbenzoic acid and benzyl 2-*n*-heptyl-4-hydroxy-6-methoxybenzoate with DCCD and subsequent debenzylation	(*194*)
3,5-Dichloro-2′-*O*-methyl-norstenosporic acid	**167**	165–166	*Lecanora lividocinerea*	From 3,5-dichloro-2,4-dihydroxy-6-*n*-propylbenzoic acid and benzyl 4-hydroxy-2-methoxy-6-*n*-pentylbenzoate with DCCD and subsequent debenzylation	(*194*)
2,2′-Di-*O*-methyldivaricatic acid	**168**	129–131	*Pertusaria subplanaica*	From 2,4-dimethoxy-6-*n*-propylbenzoic acid and benzyl 4-hydroxy-2-methoxy-6-*n*-propyl-benzoate with TFAA and subsequent debenzylation	(*192*)
2,4-Di-*O*-methylhiascic acid	**169**			By condensation of 3-benzyloxy-4,6-dimethoxy-2-methylbenzoic acid and benzyl lecanorate with TFAA and subsequent debenzylation	(*223*)
2,5-Di-*O*-methylhiascic acid	**170**			By condensation of 4-benzyloxy-3,6-dimethoxy-2-methylbenzoic acid and benzyl lecanorate with TFAA and subsequent debenzylation	(*223*)
2,2′-Di-*O*-methylimbricaric acid	**171**	137–138		From 2,4-dimethoxy-6-*n*-pentylbenzoic acid and benzyl 4-hydroxy-2-methoxy-6-*n*-propylbenzoate with TFAA and subsequent debenzylation	(*192*)

Table 4 (continued)

Compound	Formula No.	M.p. (°C)	Lichen	Synth. Scheme	Ref.
2,2′-Di-*O*-methylstenosporic acid	**172**	135	*Pertusaria subplanaica*	From 2,4-dimethoxy-6-*n*-propylbenzoic acid and benzyl 4-hydroxy-2-methoxy-6-*n*-pentylbenzoate with TFAA and subsequent debenzylation	(*192*)
Dissectic acid	**173**	>285	*Heterodermia dissecta*	From 2,4,5-trihydroxy-3,6-dimethyl-benzoate and haematommic acid with DCCD	(*263*)
Echinocarpic acid	**174**	146–148	*Parmelia norcrambidiocarpa*		(*241*)
Elatinic acid	**175**	212	*Haematomma ochrophaeum*		(*95*)
			Loxospora elatina	From 2-hydroxy-4-methoxy-6-methyl-3-formylbenzoic acid and methyl 2,4-dihydroxy-3,6-dimethylbenzoate with TFAA and subsequent oxidation with $NaOCl_2$	(*199*)
Glaucophaeic acid	**176**	107	*Porpidia glaucophaea*	99	(*265*)
Glomellic acid	**177**	132		97	(*273*)
Glomelliferic acid	**178**	145		97	(*273*)
8-Hydroxybarbatic acid	**179**	192–194		By reduction of baeomycesic acid with $NaBH_4$	(*192, 96*)
8-Hydroxydiffractaic acid	**180**	156—158	*Usnea longissima*		(*565*)
3-Hydroxygyrophoric acid	**181**	154–160		By condensation of 2,3,4-tribenzyloxy-6-methylbenzoic acid and benzyl lecanorate with TFAA and subsequent debenzylation	(*223*)
5-Hydroxylecanoric acid	**182**				(*206*)

3-Hydroxy-4-*O*-methylgyrophoric acid	**183**	183–184		By condensation of 2,3-dibenzyloxy-4-methoxy-6-methylbenzoic acid and benzyl lecanorate with TFAA and subsequent debenzylation	(*223*)
3-Hydroxy-4-*O*-methylumbilicaric acid	**184**	168–169		By condensation of 3-benzyloxy-2,4-dimethoxy-6-methylbenzoic acid and benzyl lecanorate with TFAA and subsequent debenzylation	(*223*)
3-Hydroxyovoic acid	**185**	155–157		By condensation of 2,3,4-tribenzyloxy-6-methylbenzoic acid and benzyl 2-*O*-methyllecanorate with TFAA and subsequent debenzylation	(*223*)
3-Hydroxyumbilicaric acid	**186**	186–187	*Parmelinopsis bonariensis*	100a	(*234*)
Hyperconfluentic acid	**187**	148	*Pseudobaeomyces pachycarpa*	94	(*266*)
Hyperhomosekikaic acid	**188**	125–126	*Cladonia chlorophaea*	From 2-hydroxy-4-methoxy-6-*n*-pentylbenzoic acid and benzyl 2,3-dihydroxy-4-methoxy-6-*n*-pentylbenzoate with TFAA and subsequent debenzylation	(*192*)
			Ramalina americana		(*96*)
Hyperlatolic acid	**189**	113	*Ropalospora viridis*		(*101*)
				From 4-*O*-methylolivetolcarboxylic acid and benzyl 2,4-dihydroxy-6-*n*-heptylbenzoate with TFAA and subsequent debenzylation	(*262*)
Hyperplanaic acid	**190**	128	*Lecanora planaica*	From 2,4-dimethylolivetolcarboxylic acid and benzyl 2-methoxy-4-hydroxy-6-*n*-heptylcarboxylate with TFAA and subsequent debenzylation	(*193*)

Table 4 (continued)

Compound	Formula No.	M.p. (°C)	Lichen	Synth. Scheme	Ref.
Hypoalectorialic acid	**191**	190–195	*Hypotrachyna hypoalectorica*		(*264*)
Hyponephroarctin (NA-3)	**192**	165–170	*Nephroma arcticum*		(*453*)
Isohyperlatolic acid	**193**	103	*Ropalospora viridis*		(*101*)
				From 2-hydroxy-4-methoxy-6-*n*-heptylbenzoic acid and benzyl 2,4-dihydroxy-6-*n*-pentylbenzoate with TFAA and subsequent debenzylation	(*262*)
Isohyperplanaic acid	**194**	122–123	*Lecanora planaica*	From 2,4-dimethoxy-6-*n*-heptylbenzoic acid and benzyl 2-*O*-methyl-4-hydroxy-6-*n*-pentylbenzoate with TFAA and subsequent debenzylation	(*193*)
Isolecanoric acid	**195**	>300, dec.	*Parmotrema tinctorum*		(*627*)
Isomerochlorophaeic acid	**196**	146	*Ramalina* spec. from Fiji	From 2,4-dimethoxy-6-*n*-pentylbenzoic acid and 4-benzyloxy-2,3-dihydroxy-6-*n*-propylbenzoate withTFAA and subsequent debenzylation	(*192*)
Isonorobtusatic acid	**197**	168–169	*Ramalina americana*	From 2,4-dibenzyloxy-6-methylbenzoic acid and benzyl 2,4-dihydroxy-3,6-dimethylbenzoate with TFAA and subsequent debenzylation	(*192*)
Isopatagonic acid	**198**	146–147	*Bunodophoron patagonicum*	101	(*260*)

Isopseudocyphellarin A	**199**	201	*Pseudocyphellaria pickeringii*	From 2,4-dihydroxy-3-formyl-5,6-dimethylbenzoic acid and methyl 2,4-dihydroxy-3,5,6-trimethylbenzoate with TFAA	(*275*)
Isosphaeric acid	**200**	142	*Dimelaena oreina*	From 2-hydroxy-4-methoxy-6-*n*-heptylbenzoic acid and benzyl orsellinate with TFAA and subsequent debenzylation	(*262*)
Lasallic acid	**201**	181–183	*Lasallia asiae-orientalis*		(*556, 558*)
				By condensation of 2,4-dibenzyloxy-6-methylbenzoic acid and benzyl 4-benzyloxy-3-(2′,4′-dihydroxy-6′-methylbenzoyloxy)-2-hydroxy-6-methylbenzoate with TFAA and subsequent debenzylation	(*223*)
Loxodellic acid	**202**	137		97	(*273*)
3-Methoxygyrophoric acid	**203**	175–177		By condensation of 2,4-dibenzyloxy-3-methoxy-6-methylbenzoic acid and benzyl lecanorate with TFAA and subsequent debenzylation	(*223*)
3-Methoxy-4-*O*-methylgyrophoric acid	**204**	192–193		By condensation of 2-benzyloxy-3,4-dimethoxy-6-methylbenzoic acid and benzyl lecanorate with TFAA and subsequent debenzylation	(*223*)
3-Methoxyumbilicaric acid	**205**	170–171	*Parmelinopsis bonariensis*	100b	(*234*)
Methyl alectorialate	**289**	204–205, dec.		108	(*220*)
2-*O*-Methylanziaic acid	**206**	116–117	*Ramalina americana*		(*96*)
2′-*O*-Methylatranorin	**207**	146–147	*Oropogon loxensis*	From 3-formyl-2,4-dihydroxy-6-methylbenzoic acid and methyl 4-hydroxy-2-methoxy-3,6-dimethylbenzoate with TFAA	(*199*)
Methyl barbatate	**208**	173	*Haematomma ochrophaeum*		(*95*)

Table 4 (continued)

Compound	Formula No.	M.p. (°C)	Lichen	Synth. Scheme	Ref.
Methyl barbatolate	**290**	185–186, dec.	*Alectoria nigricans*	109	(*220*)
Methyl 5-chloro-4-*O*-demethyl-barbatate	**209**	162–164	*Erioderma* spec.	From 2,4-dihydroxy-3,6-dimethyl-5-chlorobenzoic acid and methyl β-orsellinate with TFAA	(*244*)
Methyl 5-chloronorobtusate	**210**	164	*Erioderma* spec.	From 2,4-dibenzyloxy-3,6-dimethyl-5-chlorobenzoic acid and methyl orsellinate with TFAA and subsequent debenzylation	(*199*)
4′-*O*-Methylcryptochlorophaeic acid	**211**	152	*Ramalina asahinae*	From 4-benzyloxy-2-methoxy-6-*n*-pentyl-benzoic acid and 2,3-dihydroxy-4-methoxy-6-*n*-pentylbenzoate with TFAA and subsequent debenzylation	(*261*)
Methyl 2,2′-di-*O*-methyldivarica-tate	**212**	63–63	*Pertusaria oraraensis*		(*269*)
Methyl 2,2′-di-*O*-methylerioder-mate	**213**	168	*Erioderma* spec.	From 2-methoxy-3,6-dimethyl-4-benzyloxy-5-chlorobenzoic acid and methyl 2-methoxy-3,5,6-trimethyl-4-hydroxybenzoate with TFAA and subsequent debenzylation	(*244*)
Methyl 2,2′-di-*O*-methylsteno-sporate	**214**	56–57	*Pertusaria subplanaica*		(*269*)
2-*O*-Methyldivaricatic acid	**215**	118	*Ramalina seyreana*	From 2,4-dimethoxy-6-*n*-propylbenzoic acid and benzyl 2,4-dihydroxy-6-*n*-propyl-benzoate with TFAA and subsequent debenzylation	(*251*)

2′-*O*-Methyldivaricatic acid	**216**	142–143	*Pertusaria velloziae*	From benzyl 4-hydroxy-2-methoxy-6-*n*-propylbenzoate and 2-benzyloxy-4-methoxy-6-*n*-propylbenzoic acid with TFAA and subsequent debenzylation	(*191*)
Methyl eriodermate	**217**	165	*Erioderma* spec.	From 2,4-dibenzyloxy-5-chloro-3,6-dimethylbenzoic acid and methyl 2,4-dihydroxy-3,5,6-trimethylbenzoate with TFAA and subsequent debenzylation	(*244*)
2-*O*-Methylglomelliferic acid	**218**	104–105	*Ramalina americana*	94	(*96*, *266*)
2″-*O*-Methylgyrophoric acid	**219**	176–178	*Diploschistes gyrophoricus*, *Rinodina alba*	102	(*190*)
2-*O*-Methylhiascic acid	**220**	137–139	*Parmelinopsis neodamaziana*	From 4,5-dibenzyloxy-2-methoxy-6-methylbenzoic acid and benzyl lecanorate with TFAA and subsequent debenzylation	(*224*)
2′-*O*-Methylhiascic acid	**221**	207–209	*Catillaria corymbosa*		(*712*)
4-*O*-Methylhiascic acid	**222**		*Parmelinopsis schindleri*		(*277*)
Methyl 8-hydroxybarbatate	**223**	148–152	*Erioderma* spec.	By reduction of methyl baeomycesate with $NaBH_4$	(*192*)
Methyl 8-hydroxy-4-*O*-demethylbarbatate	**224**	156	*Oropogon loxodensis*	By reduction of atranorin with $NaBH_4$	(*192*)
2-*O*-Methylhyperlatolic acid	**225**	82–83	*Pertusaria follmanniana*	From 2,4-dimethoxy-6-*n*-pentylbenzoic acid and benzyl 2,4-dihydroxy-6-*n*-heptylbenzoate with TFAA and subsequent debenzylation	(*191*)
2′-*O*-Methylhyperlatolic acid	**226**		*Lecanora jamesii*		(*46*)
4-*O*-Methylhyperolivetoric acid	**227**		*Pseudobaeomyces pachycarpa*		(92)
				94	(*266*)
2′-*O*-Methylhyperphyllinic acid A	**228**	148	*Pseudobaeomyces pachycarpa*	94	(*266*)

Table 4 (continued)

Compound	Formula No.	M.p. (°C)	Lichen	Synth. Scheme	Ref.
2′-*O*-Methylhyperphyllinic acid B	**229**	140	*Pseudobaeomyces pachycarpa*	94	(*266*)
Methyl hyperplanaiate	**230**	53	*Pertusaria manamensis*		(*269*)
2′-*O*-Methylimbricaric acid	**231**	126	*Lecidea cyanosarca*	From 2-hydroxy-4-methoxy-6-*n*-pentylbenzoic acid and benzyl 4-hydroxy-2-methoxy-6-*n*-propylbenzoate with TFAA and subsequent debenzylation	(*192*)
4-*O*-Methylisocryptochlorophaeic acid	**232**	152–153	*Pertusaria paradoxica*	From 2,4-dimethoxy-6-*n*-pentylbenzoic acid and benzyl 4,6-dibenzyloxy-3-hydroxy-2-*n*-pentylbenzoate with TFAA and subsequent debenzylation	(*192*)
2-*O*-Methylisohyperlatolic acid	**233**	84	*Pertusaria follmaniana*	From 2-*n*-heptyl-4,6-dimethoxybenzoic acid and benzyl 2,4-dihydroxy-6-*n*-pentylbenzoate with TFAA and subsequent debenzylation	(*191*)
2′-*O*-Methylisohyperlatolic acid	**234**	127–128.5	*Biatoraarediosa*	From 2-*n*-heptyl-4-methoxy-6-hydroxybenzoic acid and benzyl 2-*O*-methylolivetolcarboxylate with TFAA and subsequent debenzylation	(*204*)
Methyl isohyperplanaiate	**235**	60–61	*Pertusaria manamensis*		(*269*)
2′-*O*-Methylisopseudocyphellarin A	**236**	171	*Pseudocyphellaria pickeringii*	From 5-formyl-2,4-dihydroxy-3,6-dimethylbenzoic acid and 4-hydroxy-2-methoxy-3,5,6-trimethylbenzoate with TFAA	(*275*)
Methyl 2-*O*-methyldivaricatate	**237**	80–82	*Pertusaria oraraenis*		(*269*)

Methyl 2-*O*-methyleriodermate	**238**	167	*Erioderma* spec.	From 4-benzyloxy-3-chloro-6-methoxy-2,5-dimethylbenzoic acid and methyl 2,4-dihydroxy-3,5,6-trimethylbenzoate with TFAA and subsequent debenzylation	(*244*)
Methyl 2′-*O*-methyleriodermate	**239**	172	*Erioderma* spec.	From 2,4-dibenzyloxy-5-chloro-3,6-dimethylbenzoate and methyl 4-hydroxy-2-methoxy-3,5,6-trimethylbenzoate with TFAA and subsequent debenzylation	(*244*)
Methyl 4-*O*-methyleriodermate	**240**	124	*Erioderma* spec.	From 2-benzyloxy-5-chloro-4-methoxy-3,6-dimethylbenzoic acid and methyl 2,4-dihydroxy-3,5,6-trimethylbenzoate with TFAA and subsequent debenzylation	(*244*)
Methyl 2′-*O*-methylmicrophyllinate	**241**		*Porpidia diversa*		(*334*)
Methyl 2-*O*-methylperlatolate	**242**	Oil	*Pertusaria xenisomata*		(*269*)
2′-*O*-Methylmicrophyllinic acid	**243**	152	*Porpidia leptocarpa*	94	(*266*)
2′-*O*-Methylnorbarbatic acid	**244**	162	*Pseudocyphellaria norvergica*	From 2,4-dibenzyloxy-3,6-dimethylbenzoate and benzyl 4-hydroxy-2-methoxy-3,6-dimethylbenzoate with TFAA and subsequent debenzylation	(*200*)
4′-*O*-Methylnorcryptochlorophaeic acid	**245**	160	*Cladonia merochlorophaea*	From 2,4-dibenzyloxy-6-*n*-pentylbenzoic acid and benzyl 2,3-dihydroxy-4-methoxy-6-*n*-pentylbenzoate with TFAA and subsequent debenzylation	(*261*)
2-*O*-Methylnordivaricatic acid	**246**		*Ramalina americana*		(*96*)
4′-*O*-Methylnorhomosekikaic acid	**247**	176	*Ramalina luciae*	From 2,4-dibenzyloxy-6-*n*-propylbenzoic acid and benzyl 2,3-dihydroxy-4-methoxy-6-*n*-pentylbenzoate with TFAA and subsequent debenzylation	(*261*)

Table 4 (continued)

Compound	Formula No.	M.p. (°C)	Lichen	Synth. Scheme	Ref.
2-*O*-Methylnorimbricaric acid	**248**	124–125	not yet found in lichens	From 2-methoxy-4-benzyloxy-6-*n*-pentylbenzoic acid and 2,4-dihydroxy-6-*n*-propylbenzoate with TFAA and subsequent debenzylation	(*197*)
2′-*O*-Methylnorobtusatic acid	**249**	191–193	*Pseudocyphellaria norvegica*	From 4-benzyloxy-2-hydroxy-3,6-dimethylbenzoic acid and benzyl 4-hydroxy-2-methoxy-6-methylbenzoate with TFAA and subsquent debenzylation	(*200*)
4′-*O*-Methylnorsekikaic acid	**250**	159–161, 182	*Ramalina farinacea*, *R. luciae*	From 2,4-dibenzyloxy-6-*n*-propylbenzoic acid and benzyl 2,3-dihydroxy-4-methoxy-6-*n*-propylbenzoate with TFAA and subsequent debenzylation	(*261*)
2-*O*-Methylnorstenosporic acid	**251**	111–113	*Ramalina americana*		(*96*)
2-*O*-Methylnorsuperlatolic acid	**252**	85–87	not yet found in lichens	From 2-methoxy-4-benzyloxy-6-*n*-heptylbenzoic acid and 2,4-dihydroxy-6-*n*-heptylbenzoate with TFAA and subsequent debenzylation	(*197*)
2′-*O*-Methylnorsuperphyllinic acid	**253**	165	*Stirtonia ramosa*		(*103*)
				94	(*266*)
4-*O*-Methyl-2″-oxo-cryptochlorophaeic acid	**254**	75	*Ramalina subfraxinea*		(*97*)
				94	(*266*)
2′-*O*-Methylphenarctin	**255**	166	*Pseudocyphellaria pickeringii*	From 3,5-diformyl-2,4-dihydroxy-6-methylbenzoic acid and methyl 4-hydroxy-2-methoxy-3,5,6-trimethylbenzoate with TFAA	(*275*)

Methyl planaiate	**256**	58–60	*Pertusaria manamenensis*		(*269*)
2′-*O*-Methylpseudo-cyphellarin A	**257**	190	*Pseudocyphellaria pickeringii*	From 3-formyl-2,4-dihydroxy-5,6-dimethyl-benzoic acid and methyl 2,4-dihydroxy-3,5,6-trimethylbenzoate with TFAA	(*275*)
Methyl sekikate	**258**	124	*Protousnea dusenii*		(*698*)
2-*O*-Methylsquamatic acid	**259**	216–220	*Protousnea magellanica*		(*313*)
				By oxidation of 2-*O*-methylbaeomycesic acid with $NaOCl_2$ sulfamic acid	(*199*)
2-*O*-Methylstenosporic acid	**260**	108, 112	*Ramalina seyreana*, *Cetrelia sanguiana*	From 2,4-dimethoxy-6-*n*-propylbenzoic acid and benzyl 2,4-dihydroxy-6-*n*-pentylbenzoate with TFAA and subsequent debenzylation	(*251*)
2′-*O*-Methylstenosporic acid	**261**	127	*Physcidia cylindrophora*	From 2-hydroxy-4-methoxy-6-*n*-propyl-benzoic acid and benzyl 4-hydroxy-2-methoxy-6-*n*-pentylbenzoate with TFAA and subsequent debenzylation	(*192*)
2-*O*-Methylsulphurellin	**262**		*Lecanora jamesii*		(*46*)
2-*O*-Methylsuperlatolic acid	**263**	71–72	*Pertusaria follmanniana*	From 2-*n*-heptyl-4,6-dimethoxybenzoic acid and benzyl 2,4-dihydroxy-6-*n*-heptylbenzoate with TFAA and subsequent debenzylation	(*191*)
2′-*O*-Methylsuperlatolic acid	**264**	114–115	*Biatora sorediosa*	From 2-*n*-heptyl-6-hydroxy-4-methoxyben-zoic acid and benzyl 2-*n*-heptyl-4-hydroxy-6-methoxybenzoate with TFAA and subsequent debenzylation	(*204*)
4-*O*-Methylsuperolivetoric acid	**265**	112	*Pseudobaeomyces pachycarpa*	94	(*266*)
2′-*O*-Methylsuperphyllinic acid	**266**	152	*Porpidia glaucophaea*	99	(*265*)
Microphyllinic acid	**267**	116	*Cetrelia japonica*	98	(*221*)
Miriquidic acid	**268**	140–141	*Miriquidica leucophaea*	103	(*219*)
Norisoobtusatic acid	**269**		*Ramalina americana*		(*96*)
Normiriquidic acid	**270**	137	*Miriquidica leucophaea*	103	(*219*)

Table 4 (continued)

Compound	Formula No.	M.p. (°C)	Lichen	Synth. Scheme	Ref.
Norobtusatic acid	**271**		*Hypotrachyna chicitae*		(*206*)
Oxostenosporic acid	**272**	119–1230	*Neofuscelia verruculifera*	From divaricatinic acid and benzyl 2,4-dihydroxy-6-(2′-oxo-*n*-pentyl)-benzoate with DCCD and subsequent debenzylation	(*273*)
Patagonic acid	**273**	104–105	*Bunodophoron patagonicum*	From 2-hydroxy-4-methoxy-6-methylbenzoic acid and benzyl 2,4-dihydroxy-6-(2′-oxo-*n*-nonyl)-benzoate with DCCD and subsequent debenzylation	(*260*)
Prasinic acid (Superlatolic acid)	**274**	93–94	*Micarea prasina*, *Dimealaena oreina*		(*92*, *101*)
Pseudocyphellarin A	**275**	174–175	*Pseudocyphellaria vaccina*		(*383*)
				104	(*239*)
				105	(*590*)
Pseudocyphellarin B	**276**	168–169	*Pseudocyphellaria vaccina*		(*383*)
				104	(*239*)
				105	(*590*)
Ramalinaic acid	**277**	220–221	*Ramalina americana*		(*96*)
para-Scrobiculin	**278**		*Lobaria scrobiculata*	106	(*211*)
Sphaerophorin	**279**	141		From 2-benzyloxy-4-methoxy-6-methylbenzoic acid and 2-*n*-heptyl-4,6-dihydroxybenzoate with TFAA and subsequent debenzylation	(*262*)
Subconfluentic acid	**280**	163–164	*Lecidella* cf. *cyanosarca*		(*92*)
				107	(*271*)
Subdivaricatic acid	**281**	153–154	*Ramalina americana*		(*96*)

				From 2-benzyloxy-4-methoxy-6-methylbenzoic acid and benzyl 2,4-dihydroxy-6-*n*-propylbenzoate	
Submerochlorophaeic acid	**282**	156–157	*Cladonia merochlorophaea*		(*93*)
				From 4-benzyloxy-2-methoxy-6-*n*-propylbenzoic acid and 4-benzyloxy-2,3-dihydroxy-6-*n*-propylbenzoate with TFAA and subsequent debenzylation	(*192*)
Subpaludosic acid	**283**	181–182	*Ramalina paludosa*		(*93*)
				From 4-benzyloxy-2-methoxy-6-*n*-propylbenzoic acid and 4-benzyloxy-2,3-dihydroxy-6-*n*-propylbenzoate with TFAA and subsequent debenzylation	(*192*)
Subsekikaic acid	**284**	159–160	*Ramalina americana*		(*96*)
				From 2-benzyloxy-4-methoxy-6-methylbenzoic acid and benzyl 2,3-dihydroxy-4-methoxy-6-*n*-propylbenzoate with TFAA and subsequent debenzylation	(*192*)
Subsphaeric acid	**285**	120	*Dimelaena thysanota*	From 2-benzyloxy-4-methoxy-6-methylbenzoic acid and benzyl 2,4-dihydroxy-6-*n*-pentylbenzoate with TFAA and subsequent debenzylation	(*208*)
Superconfluentic acid	**286**	139–140	*Porpidia glaucophaea*		(*92*)
				99	(*265*)
Superplanaic acid	**287**	108	*Lecanora planaica*	From 2-*n*-heptyl-4,6-dimethoxybenzoic acid and benzyl 2-*n*-heptyl-4-hydroxy-2-methoxybenzoate with TFAA and subsequent debenzylation	(*193*)
Wrightiin	**288**	216.5	*Erioderma wrightii*		(*518*)

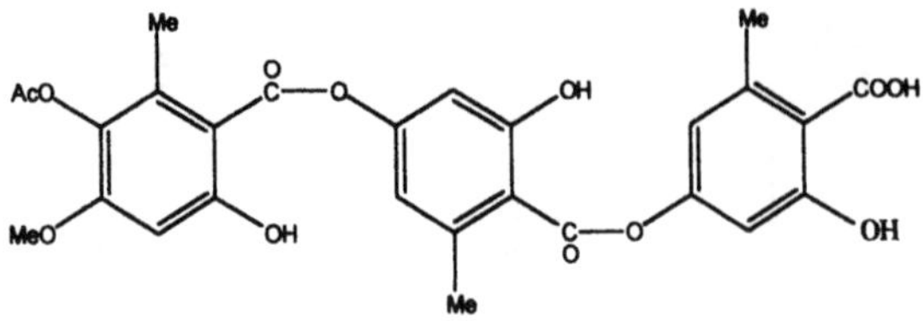

126

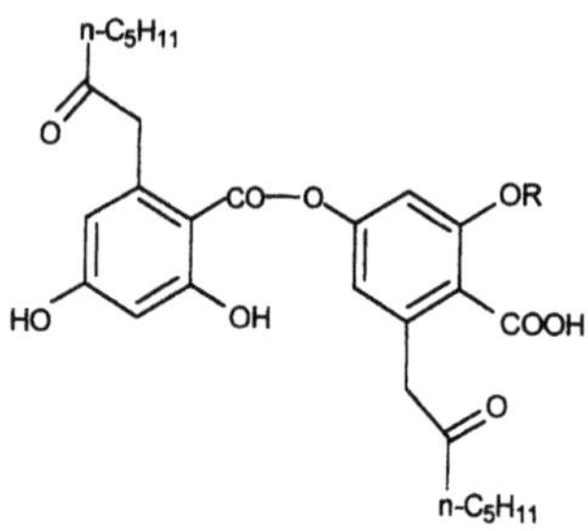

127: R = Me
153: R = H

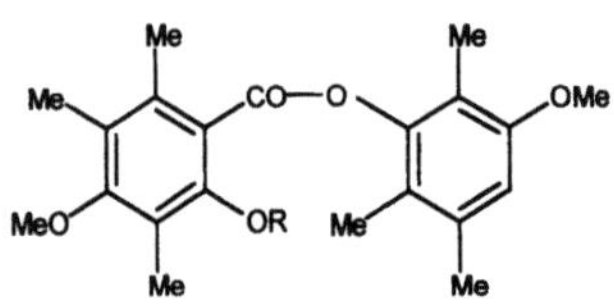

128 : R = Me
129 : R = H

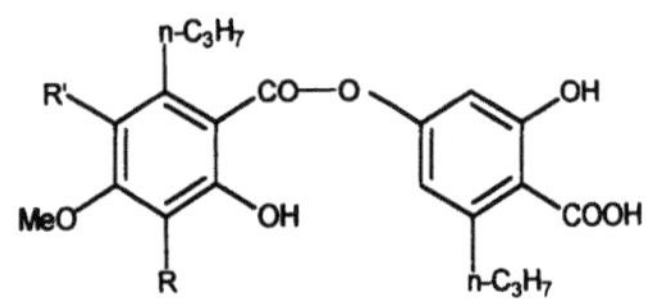

130 : R = Cl, R' = H
131 : R = H, R' = Cl

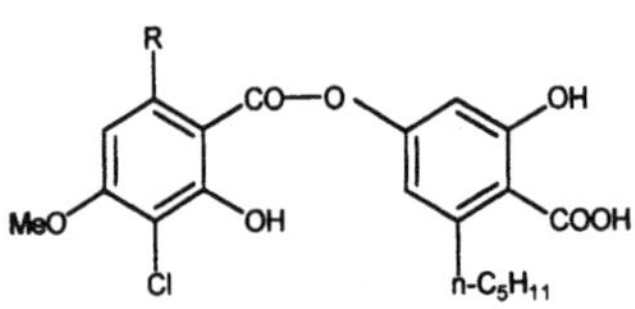

135 : R = n-C_5H_{11}
136 : R = n-C_3H_7

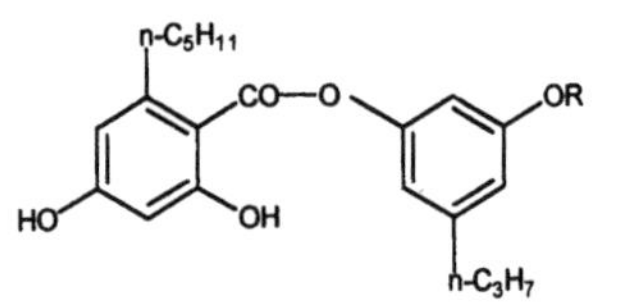

142 : R = Me
143 : R = H

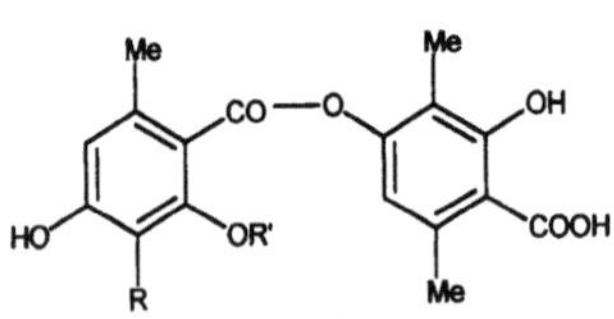

147 : R = COH, R' = H
148 : R = R' = Me

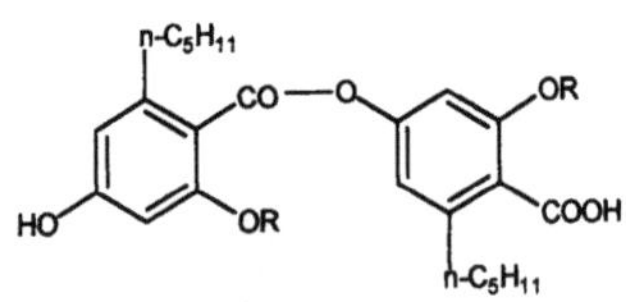

154 : R = H
155 : R = Me

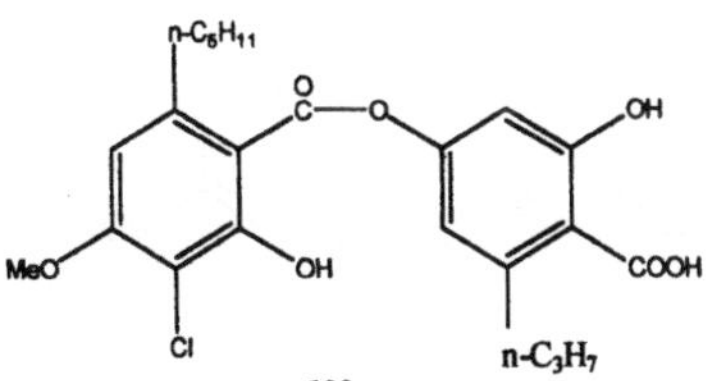

132

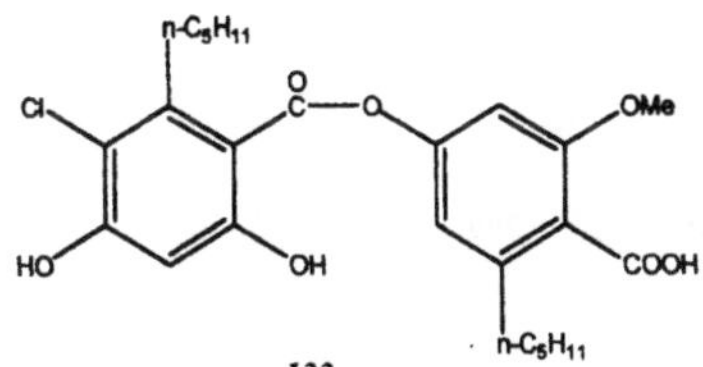

133

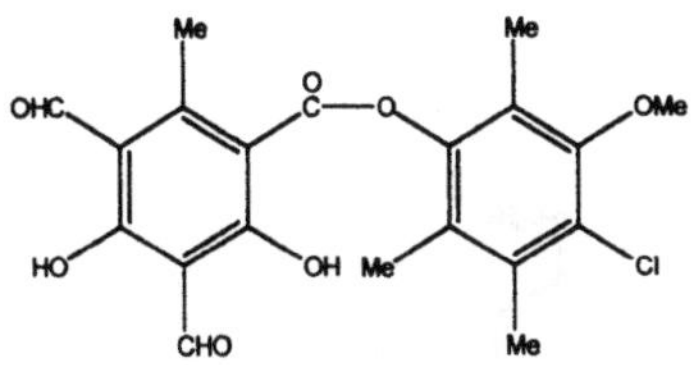

134

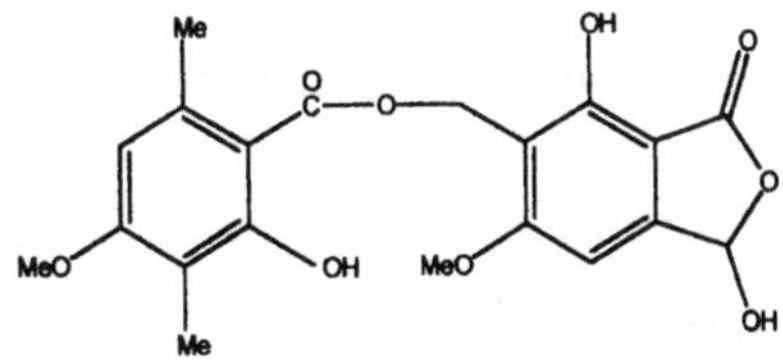

137

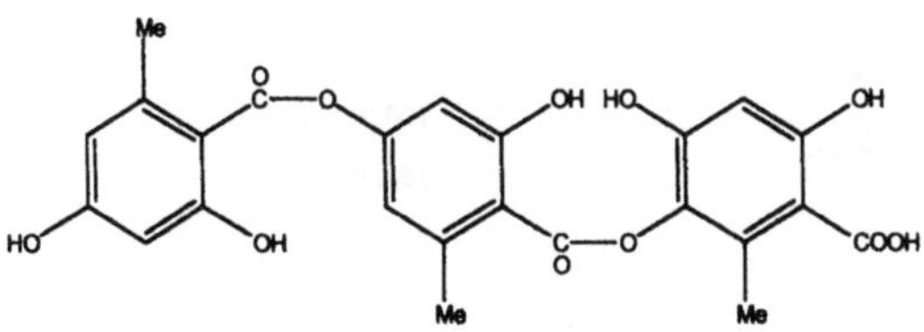

138

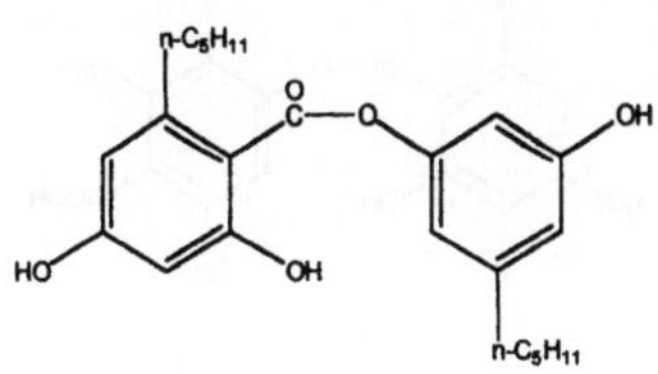

139

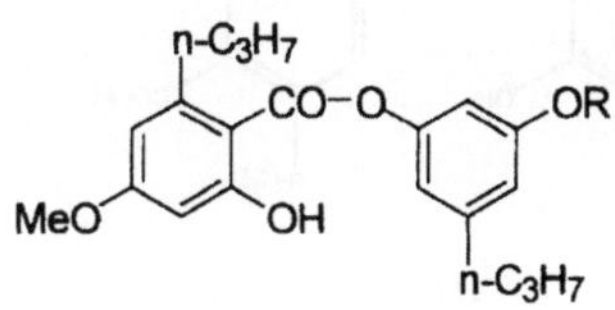

140: R = H

141: R = Me

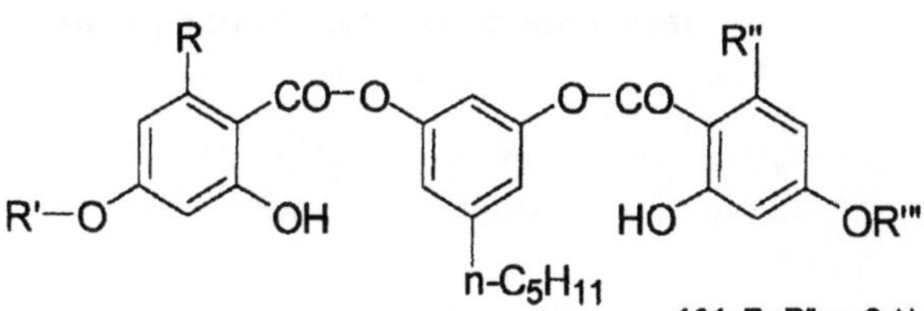

161: R=R"=n-C_3H_7,R'=R'"=Me

162: R=R"=n-C_5H_{11},R'=R'"=H

163: R=n-C_3H_7,R'=Me,R"=n-C_5H_{11},R'"=H

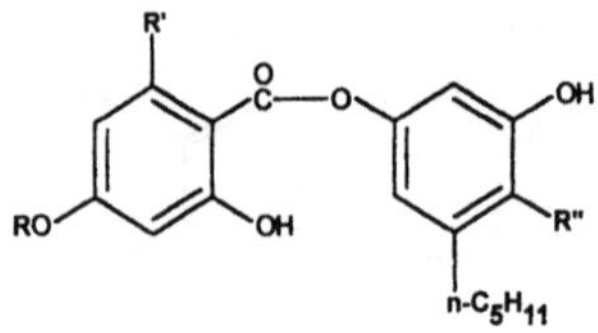

144: R=R"=H,R'=n-C_3H_7
145: R=Me,R'=n-C_5H_{11},R"=H
146: R=Me,R'=n-C_3H_7,R"=H
159: R=H,R'=n-C_3H_7,R"=COOH

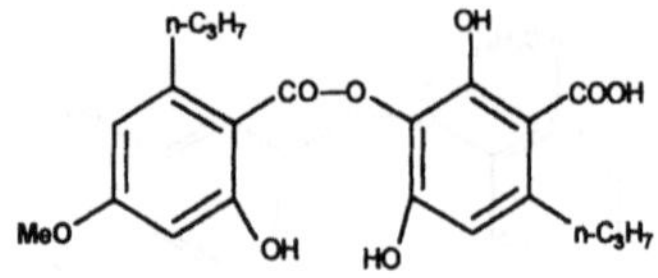

156

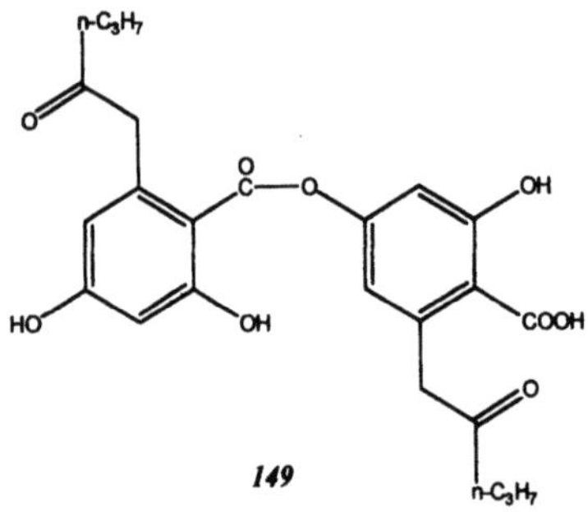

149

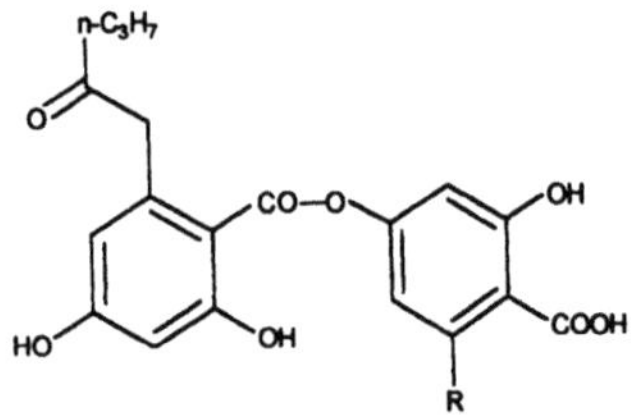

150: R=n-C_5H_{11}
152: R=n-C_3H_7

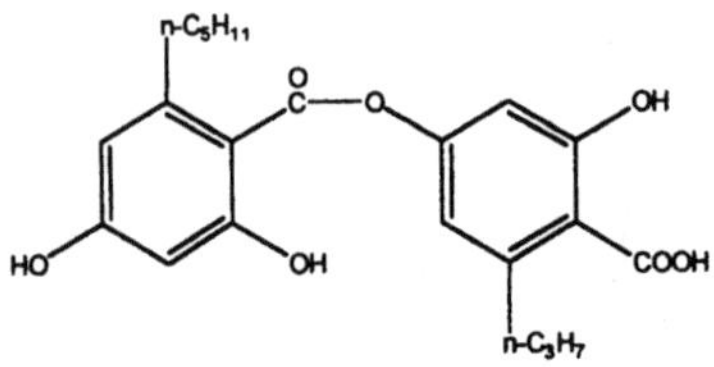

151

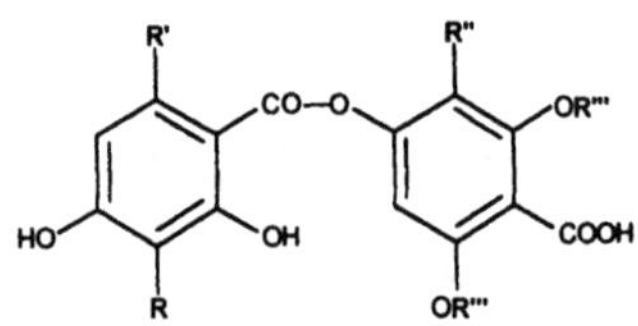

157: R=R"=R""=H,R'=Me,R'"=n-C_7H_{15}
158: R=COOH,R'=R"=R'"=Me,R""=H
160: R=R"=H,R'=CH_2-CO-n-C_7H_{15},R'"=n-C_7H_{15},R""=Me

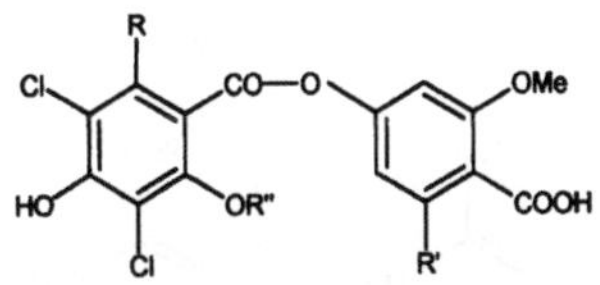

164: R = R' = n-C_5H_{11}, R" = Me
165: R = R' = n-C_5H_{11}, R" = H
166: R = n-C_5H_{11}, R' = n-C_7H_{15}, R" = H
167: R = n-C_3H_7, R' = n-C_5H_{11}, R" = H

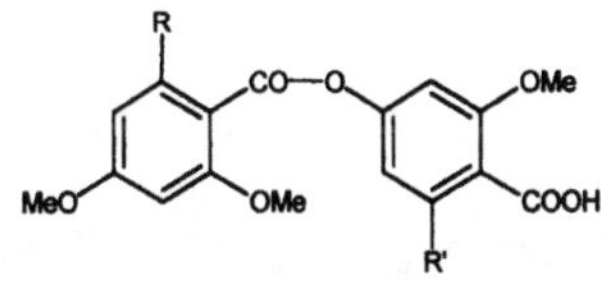

168: R = R' = n-C_3H_7
171: R = n-C_5H_{11}, R' = n-C_3H_7
172: R = n-C_3H_7, R' = n-C_5H_{11}

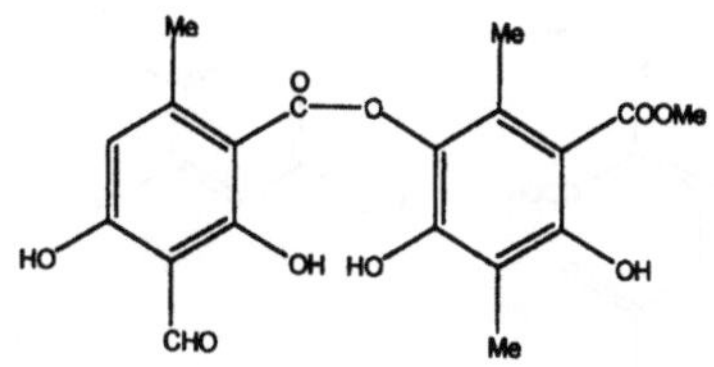

173

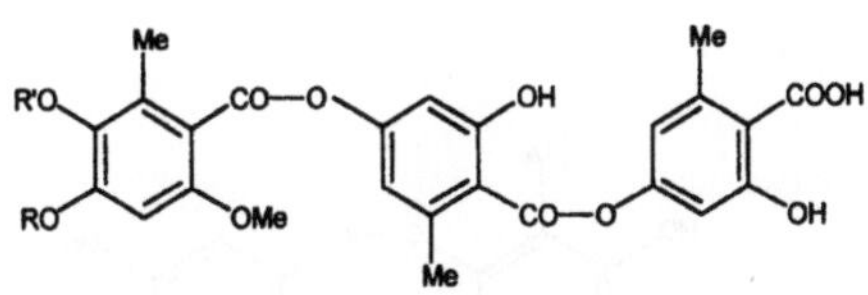

169: R = Me, R' = H
170: R = H, R' = Me

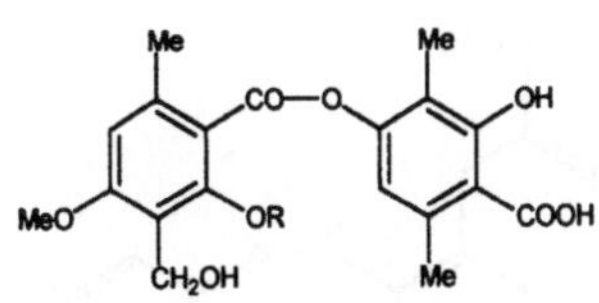

179: R = H
180: R = Me

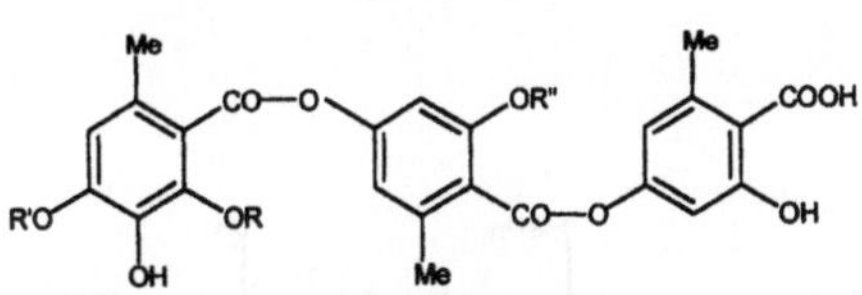

181: R = R' = R" = H
183: R = R" = H , R' = Me
184: R = R' = Me, R" = H
185: R = R' = H, R" = Me
186: R = Me, R' = R" = H

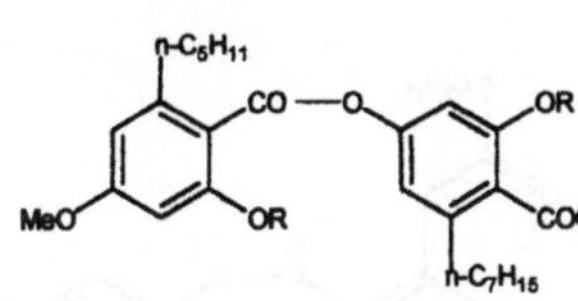

189: R = H
190: R = Me

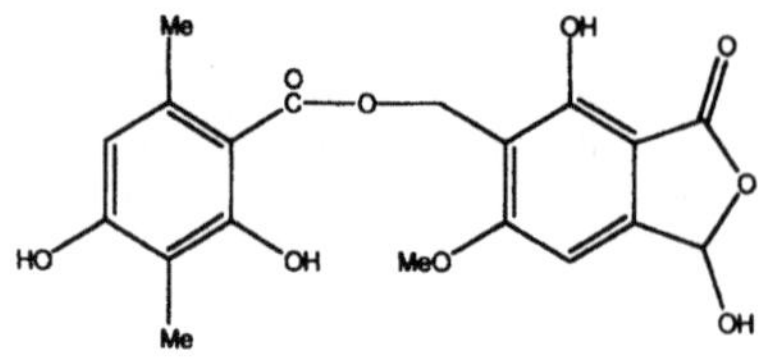

174

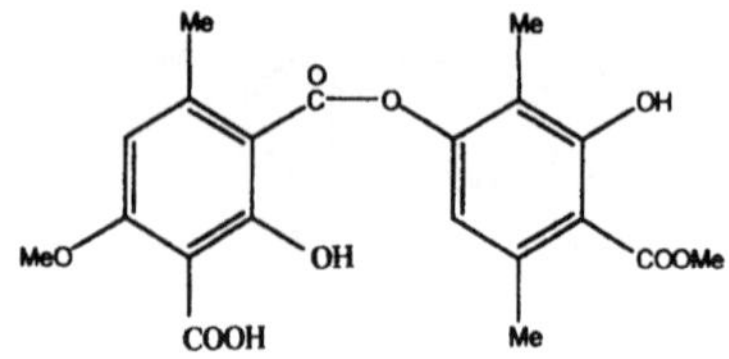

175

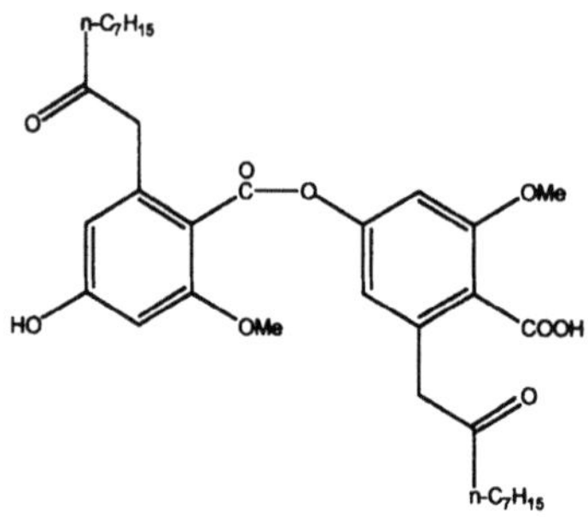

176

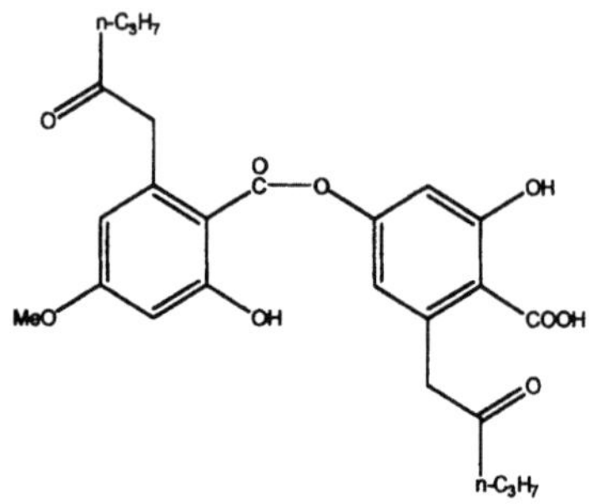

177

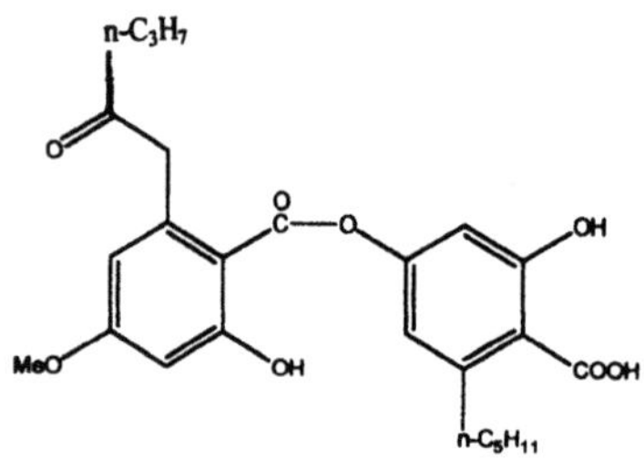

178

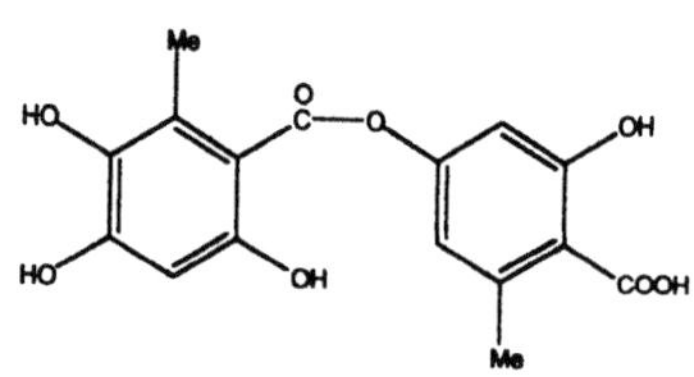

182

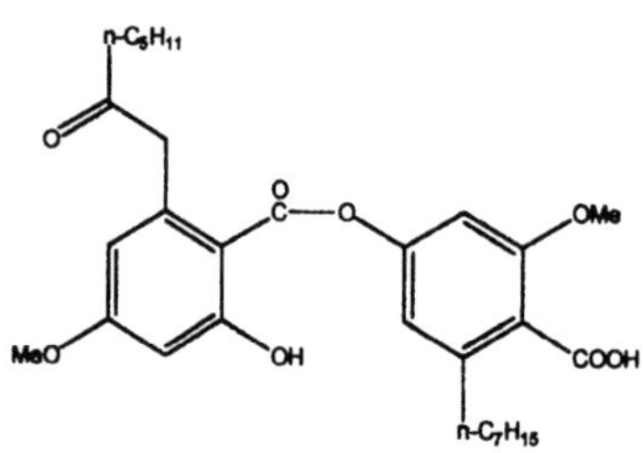

187

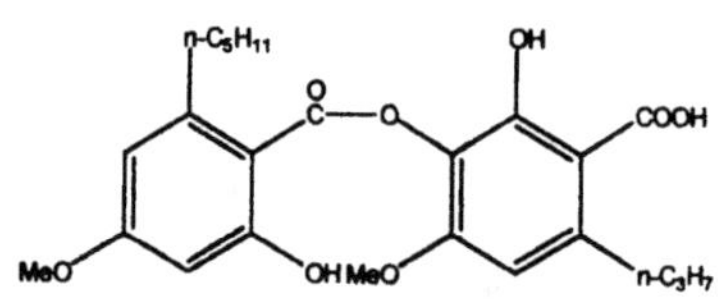

188

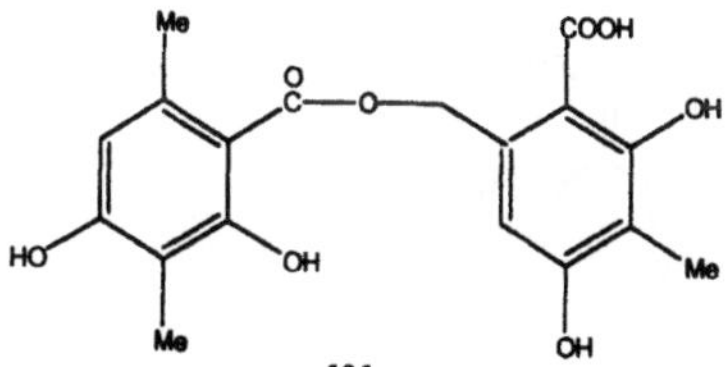

191

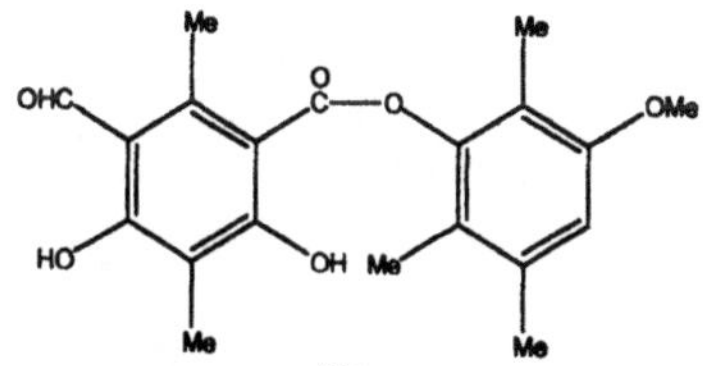

192

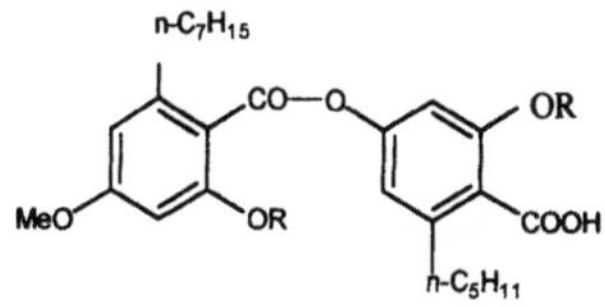

193 : R = H
194 : R = Me

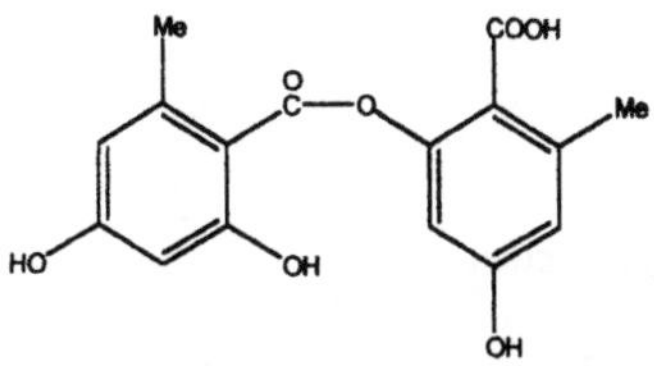

195

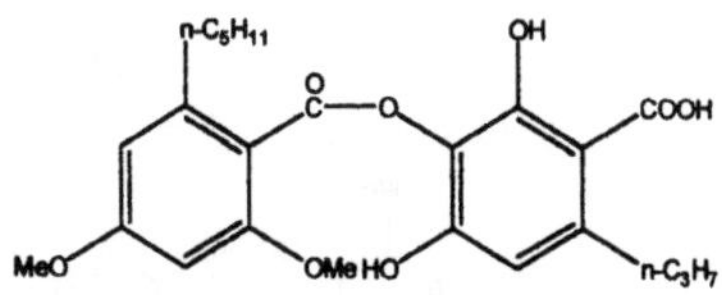

196

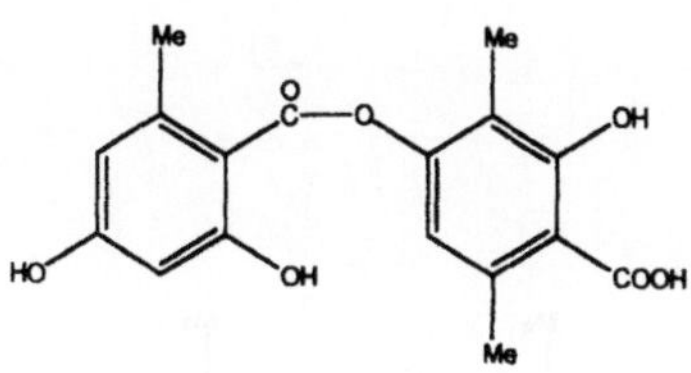

197

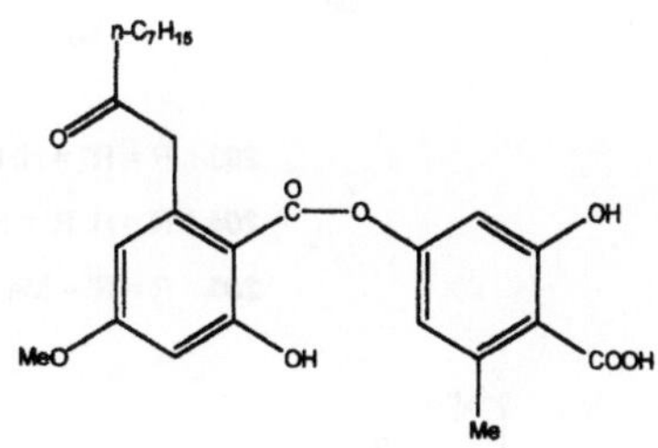

198

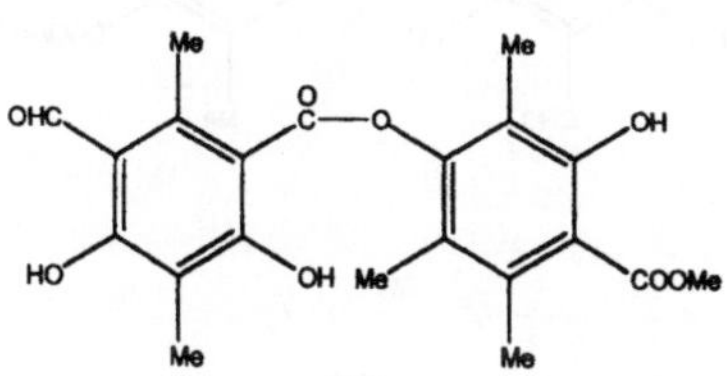

199

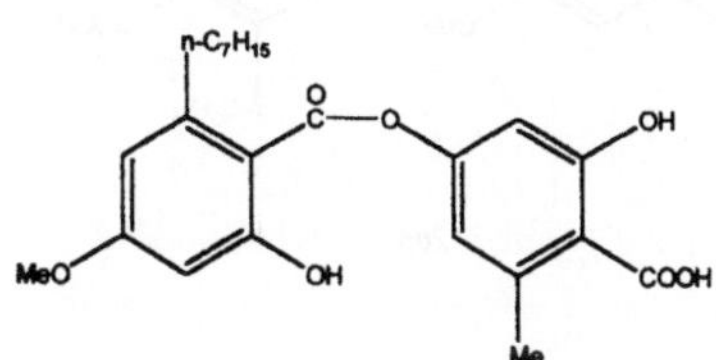

200

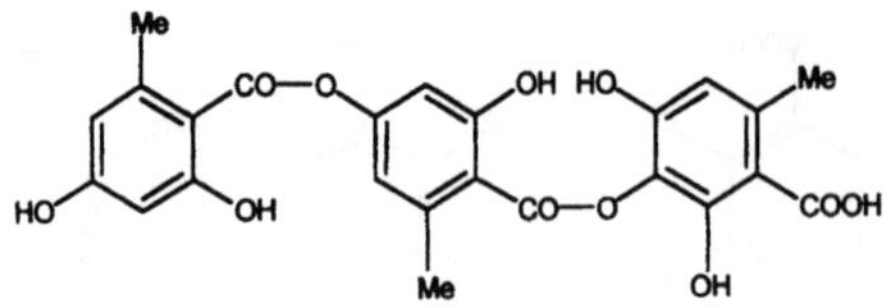

201

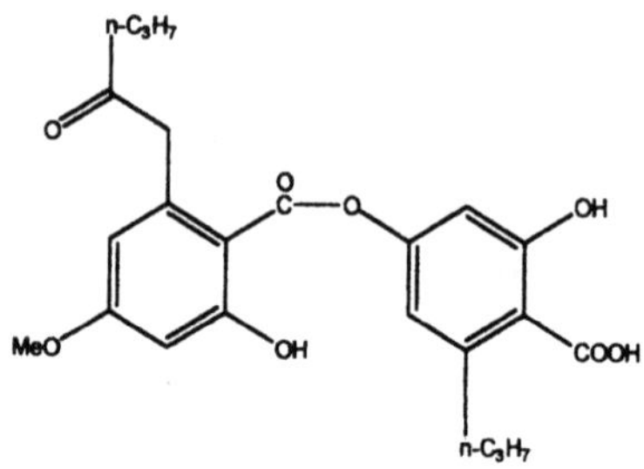

202

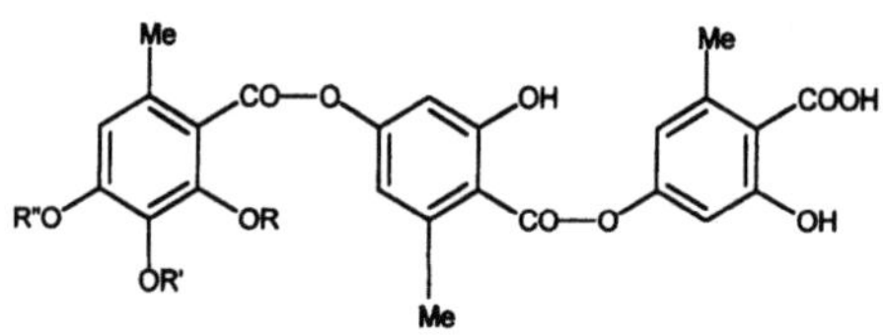

203 : R = R" = H, R' = Me

204 : R = H, R' = R" = Me

205 : R = R' = Me, R" = H

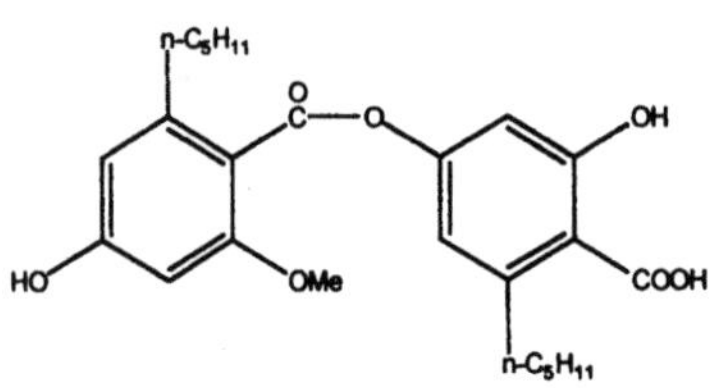

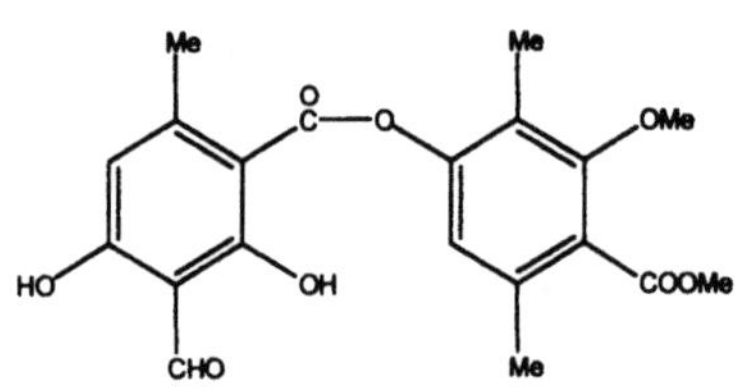

206

207

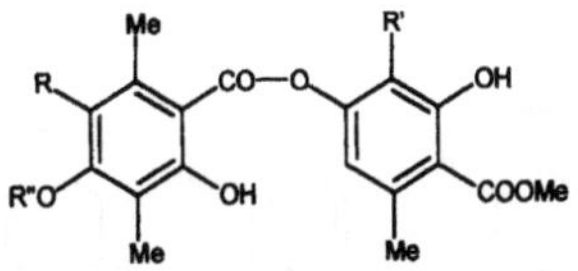

208: R = H, R' = R" = Me

209: R = Cl, R' = Me, R" = H

210: R = Cl, R' = R" = H

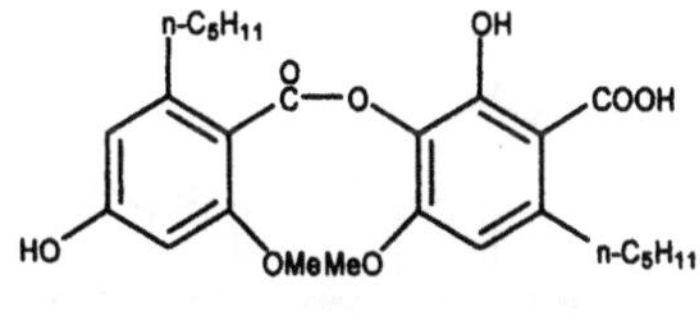

211

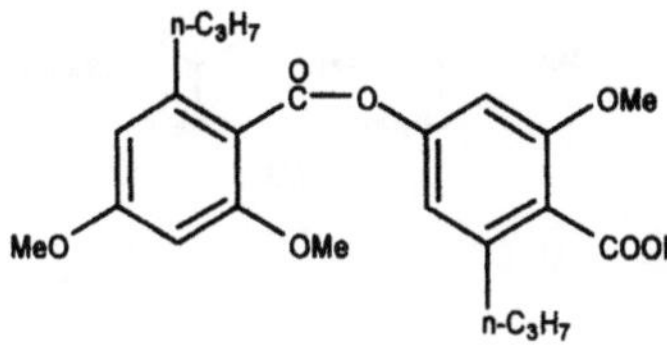

212

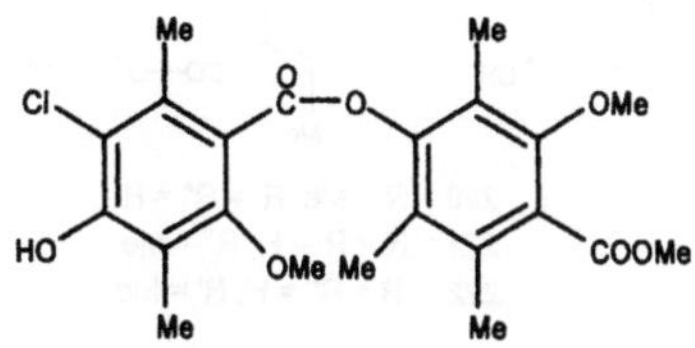

213

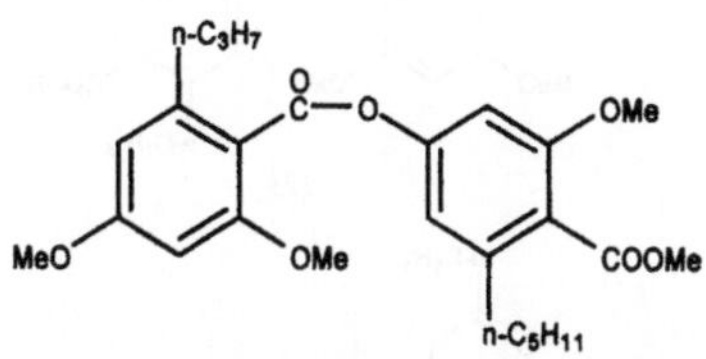

214

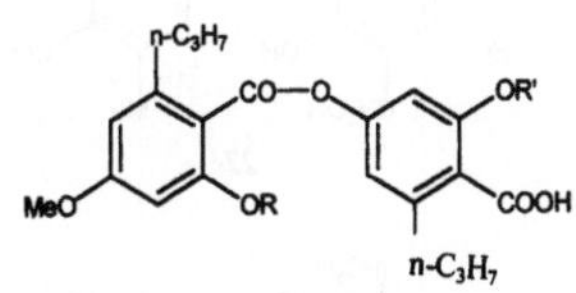

215: R = Me, R' = H

216: R = H, R' = Me

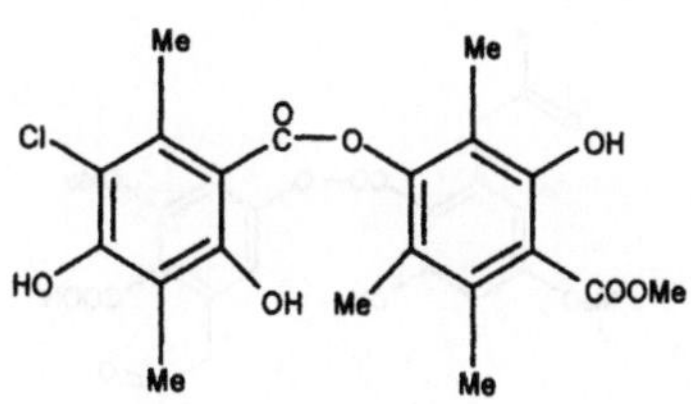

217

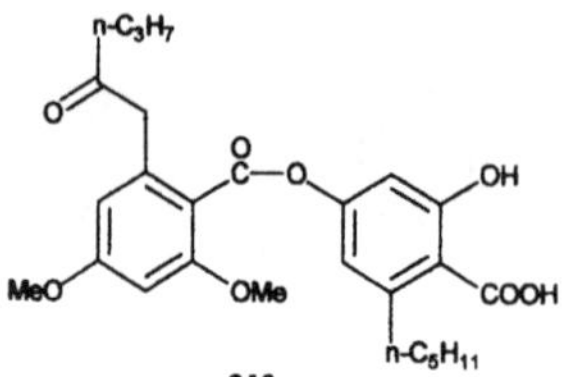

218

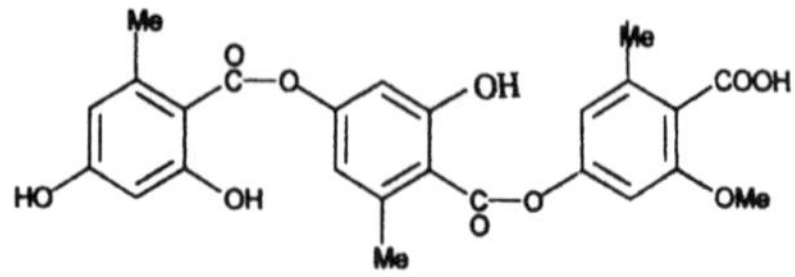

219

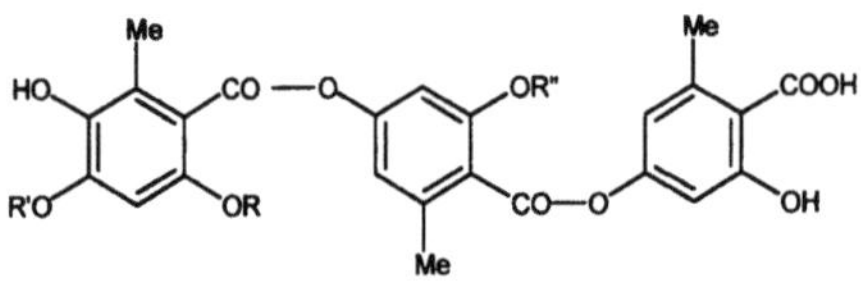

220: R = Me, R' = R'' = H
221: R = R' = H, R'' = Me
222: R = R'' = H, R' = Me

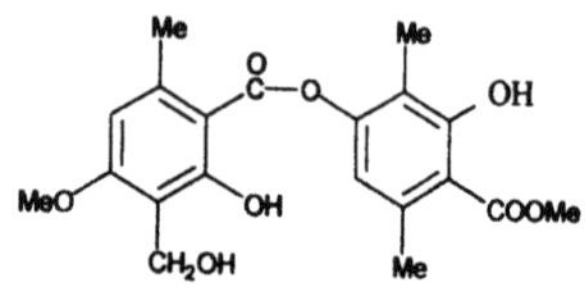

223

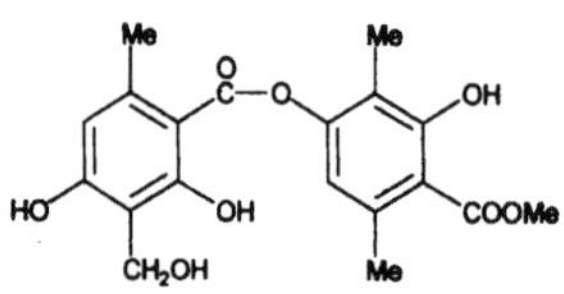

224

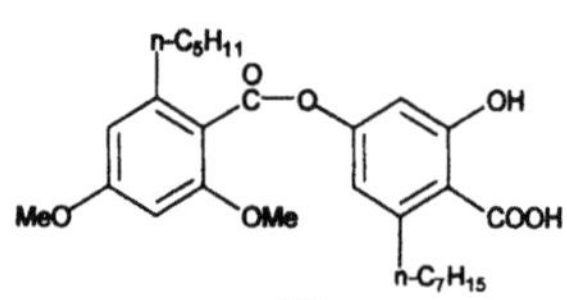

225

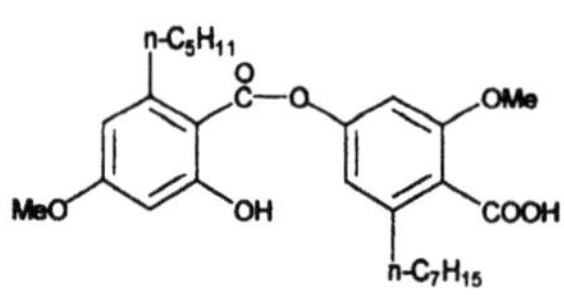

226

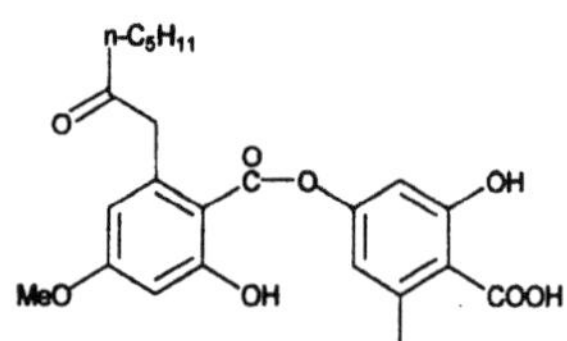

227

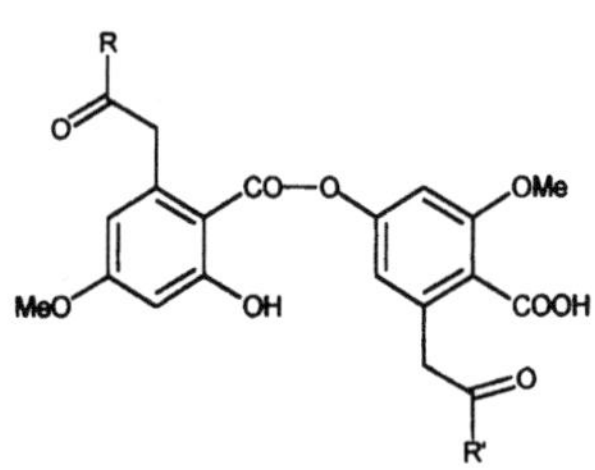

228: R = n-C_7H_{15}, R' = n-C_5H_{11}
229: R = n-C_5H_{11}, R' = n-C_7H_{15}

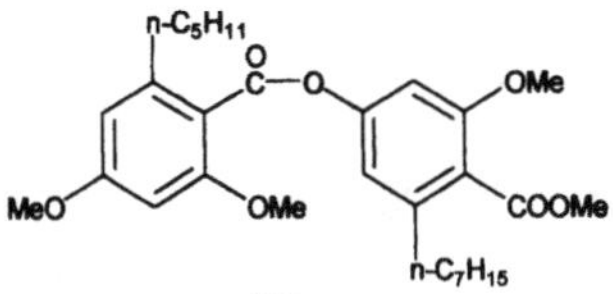

230

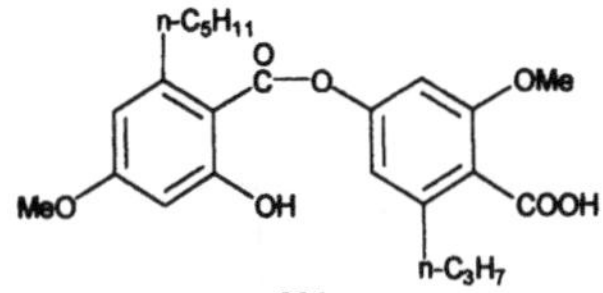

231

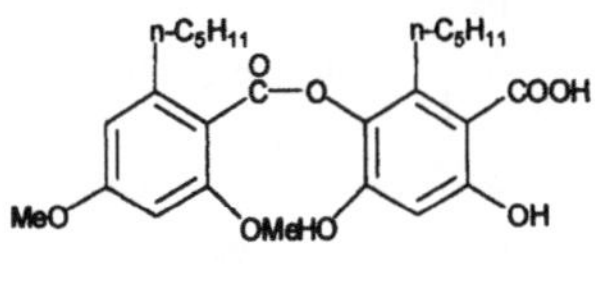

232

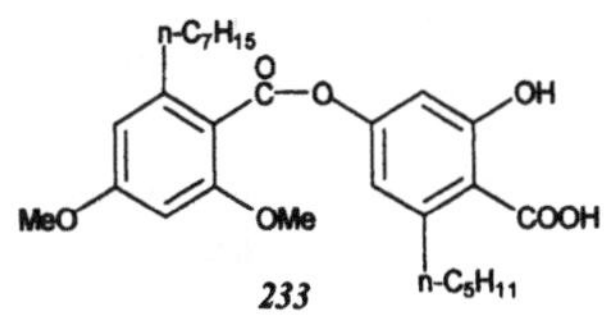

233

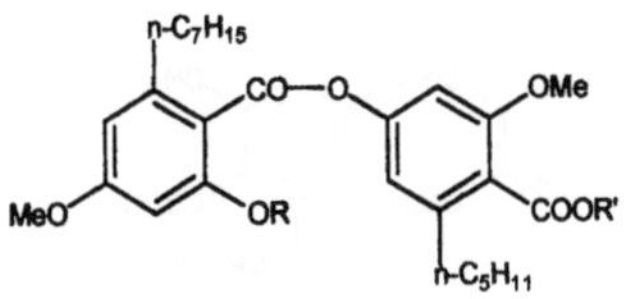

234: R = R' = H
235: R = R' = Me

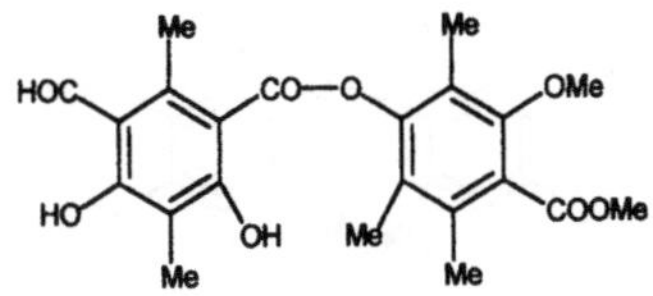

236

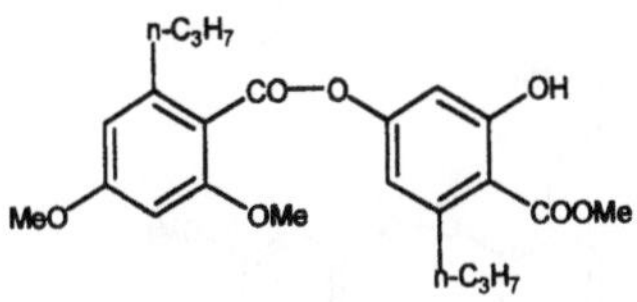

237

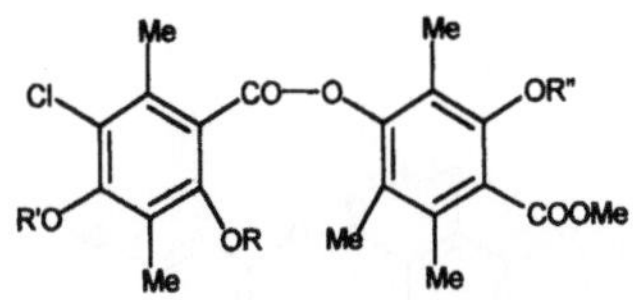

238: R = Me, R' = R" = H
239: R = R' = H, R" = Me
240: R = R" = H, R' = Me

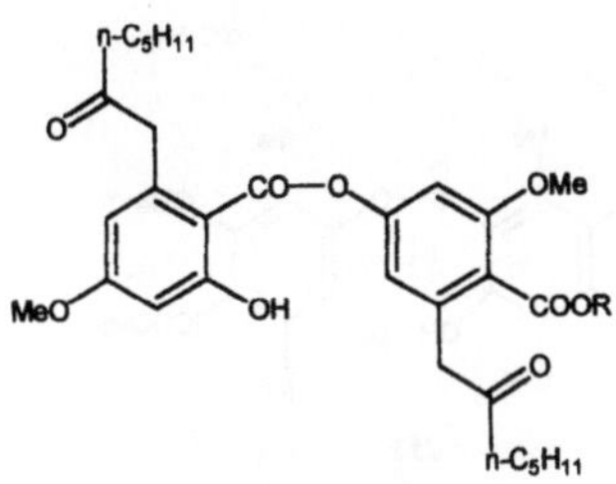

241: R = Me
243: R = H

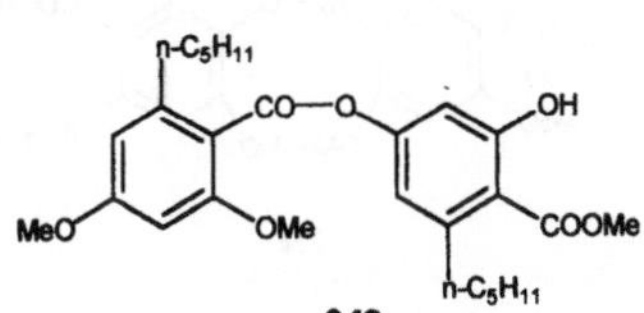

242

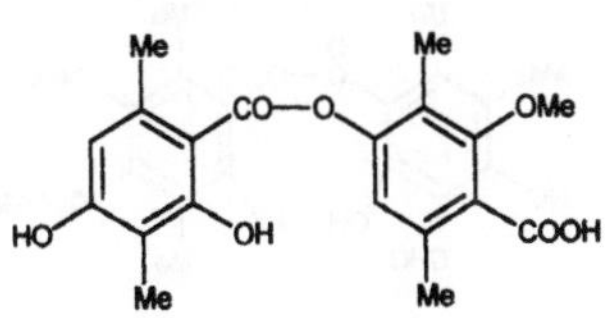

244

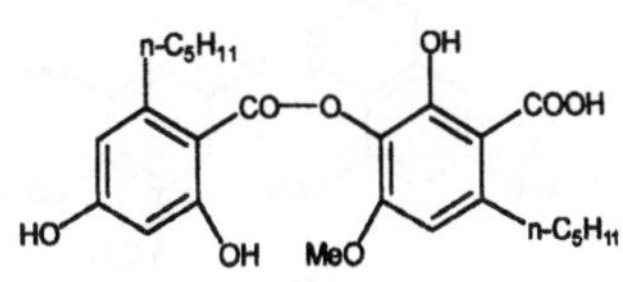

245

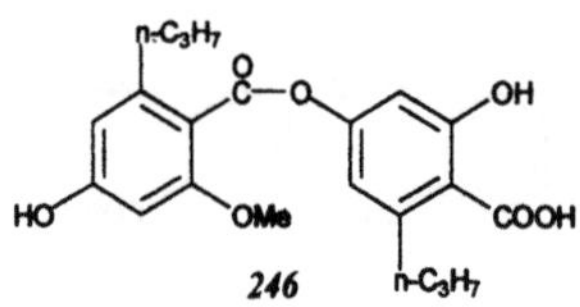

246

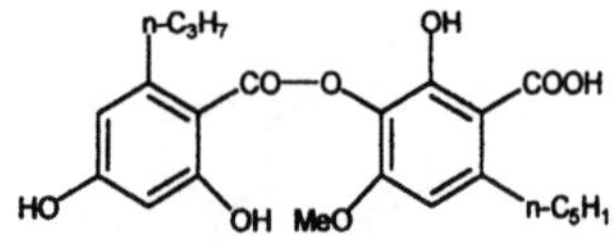

247

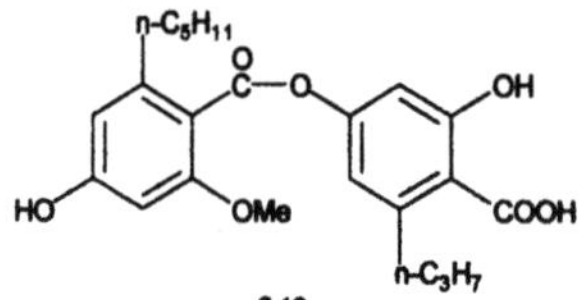

248

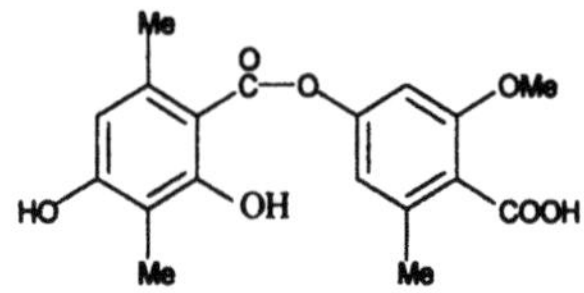

249

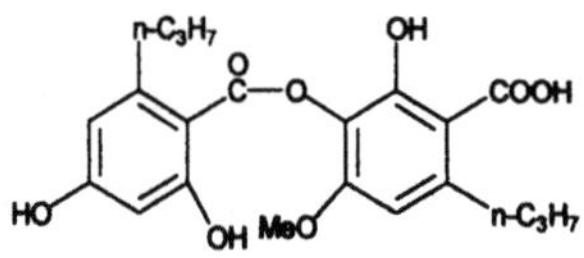

250

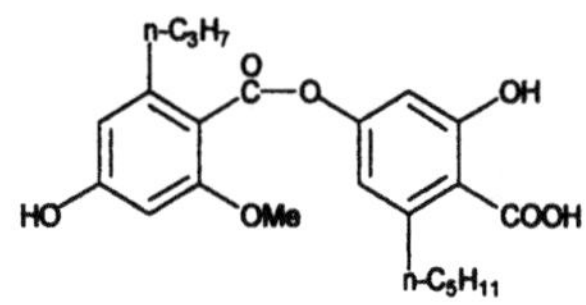

251

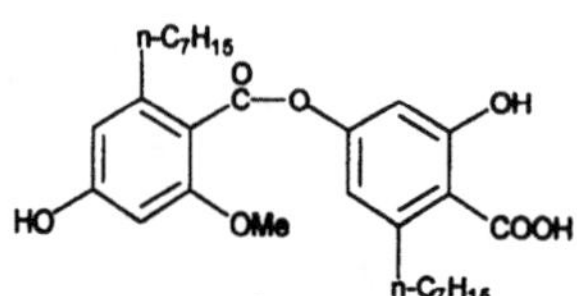

252

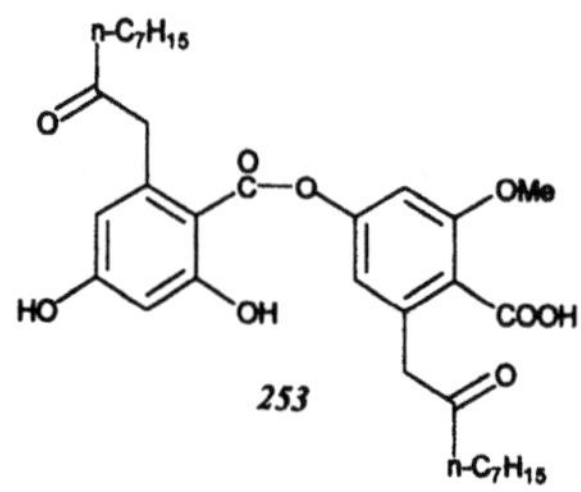

253

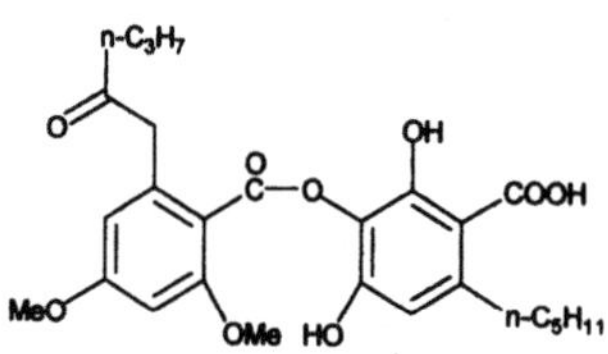

254

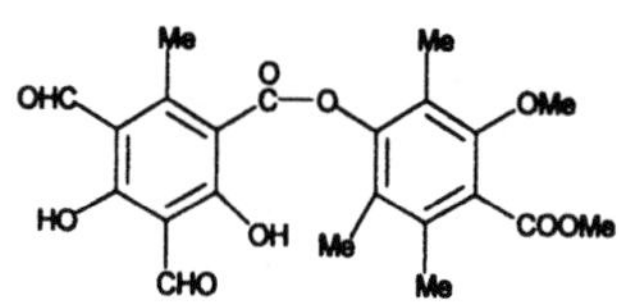

255

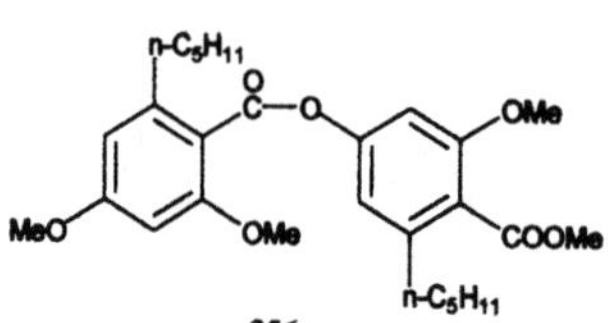

256

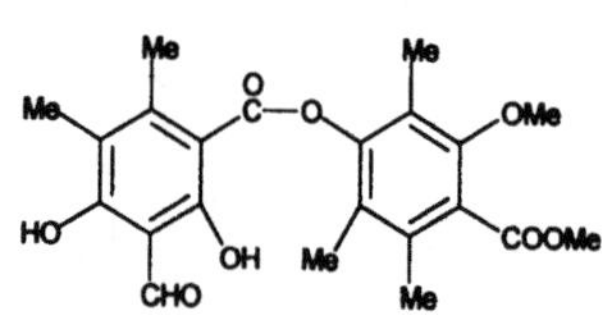

257

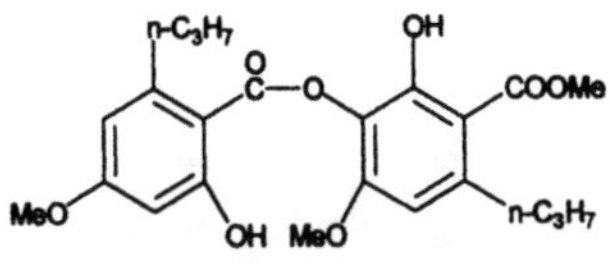

258

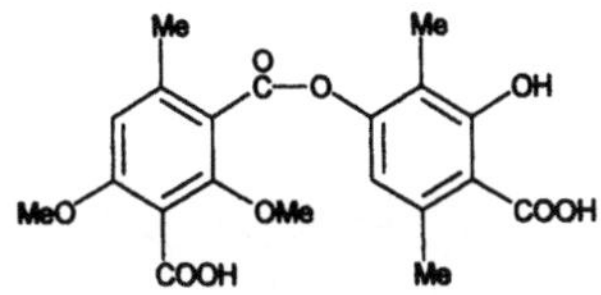

259

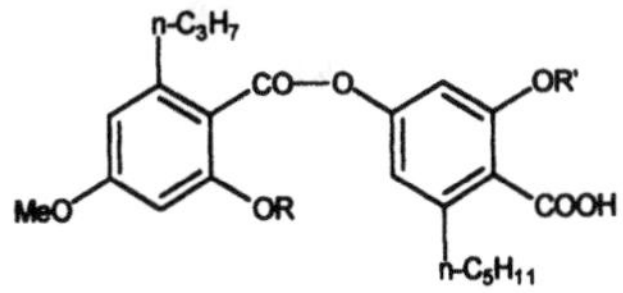

260 : R = Me, R' = H
261 : R = H, R' = Me

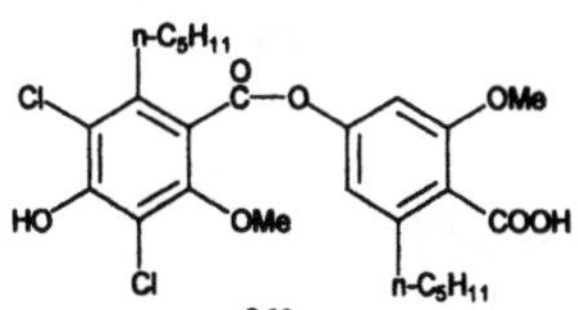

262

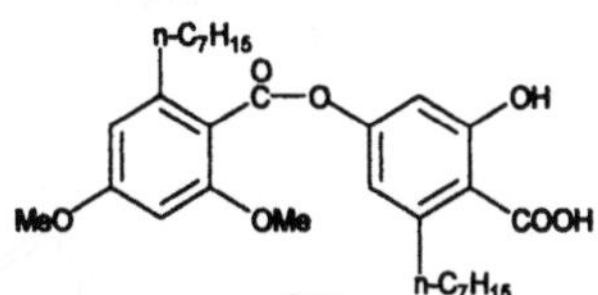

263

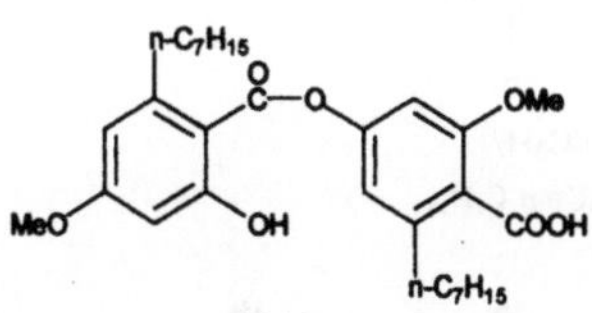

264

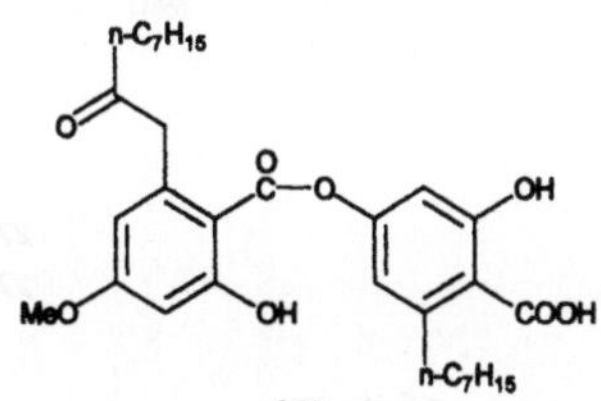

265

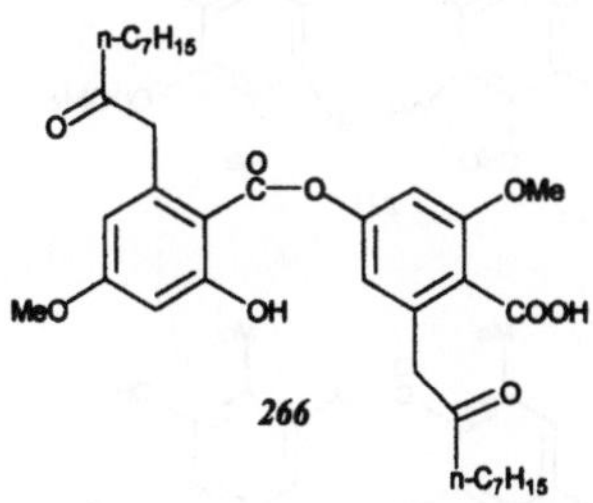

266

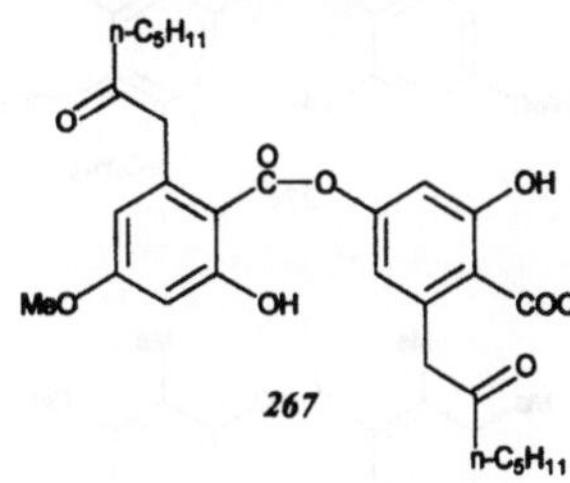

267

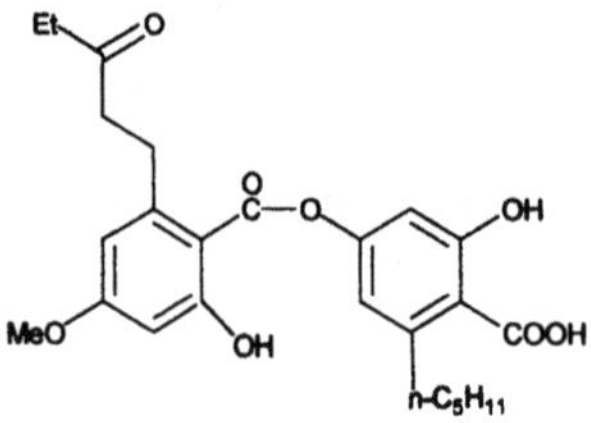

268

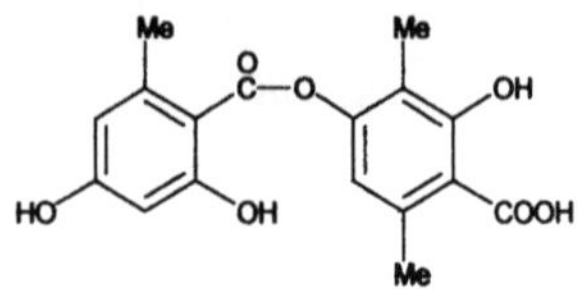

269

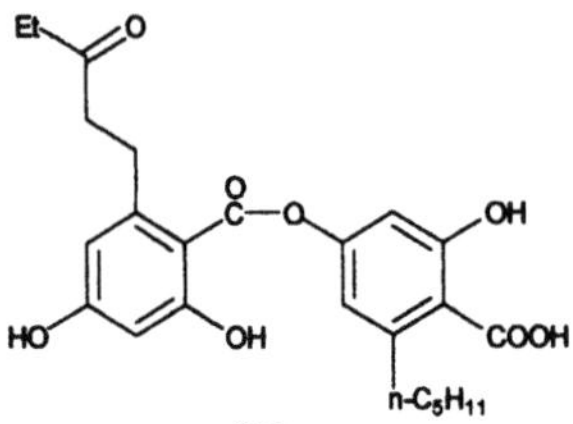

270

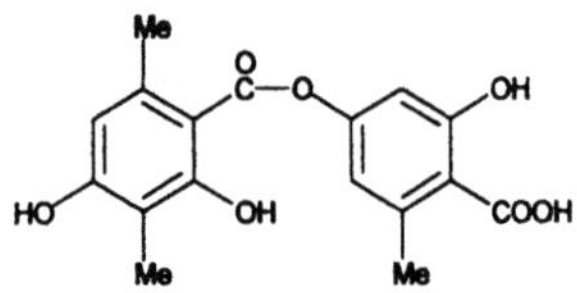

271

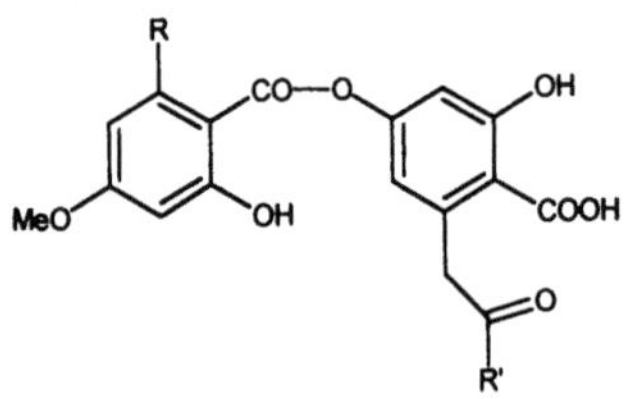

272: R = R' = $n\text{-}C_3H_7$

273: R = Me, R' = $n\text{-}C_7H_{15}$

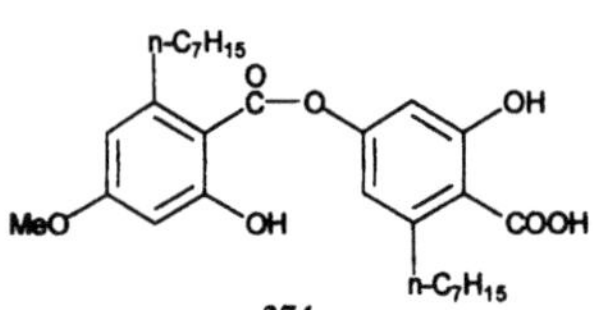

274

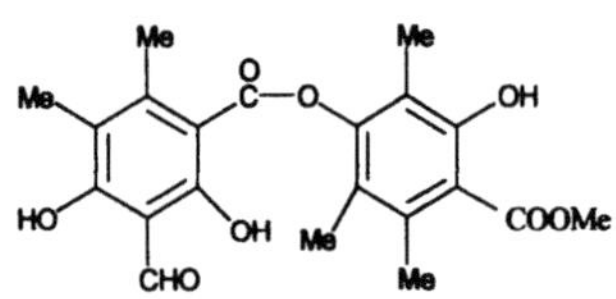

275

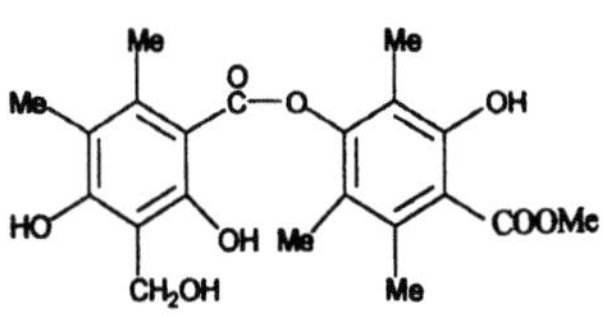

276

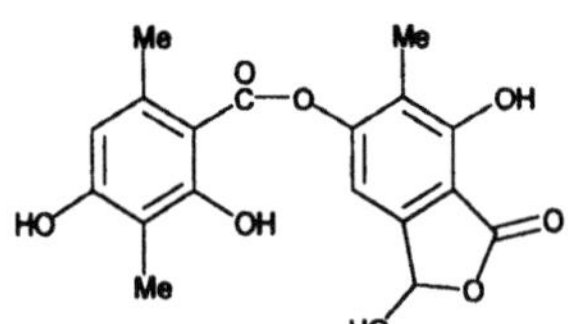

277

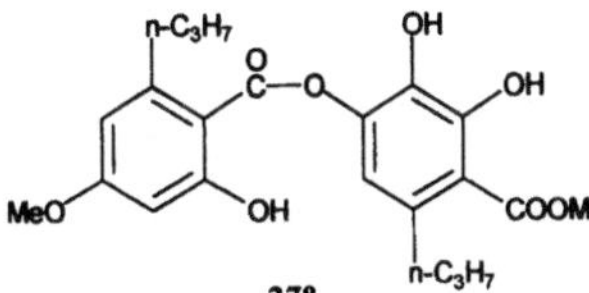

278

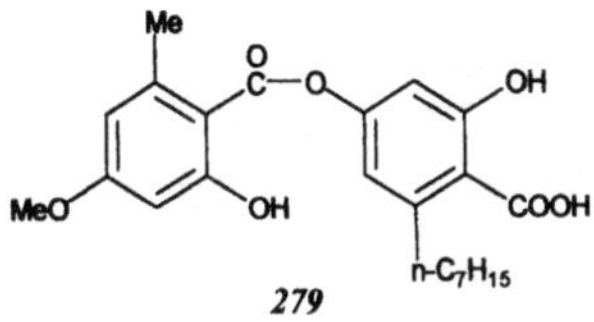

279

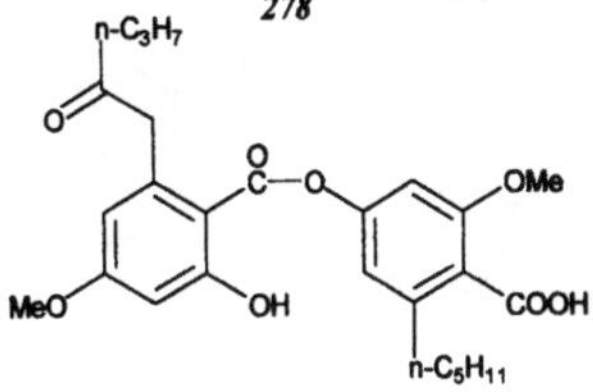

280

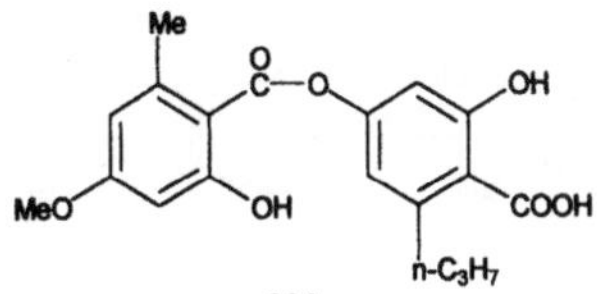

281

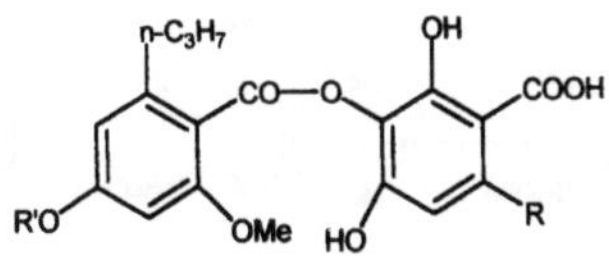

282 : R = n-C_5H_{11}, R' = Me
283 : R = n-C_3H_7, R' = H

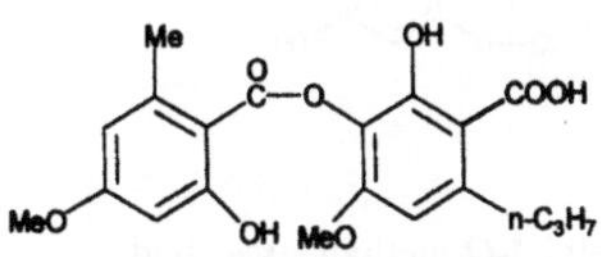

284

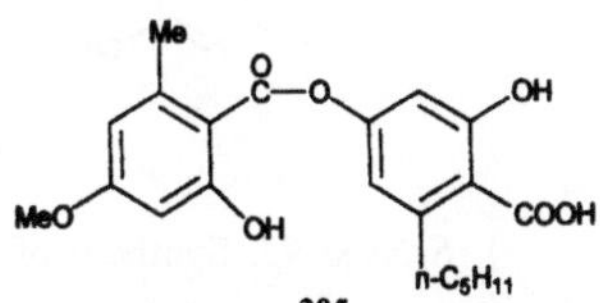

285

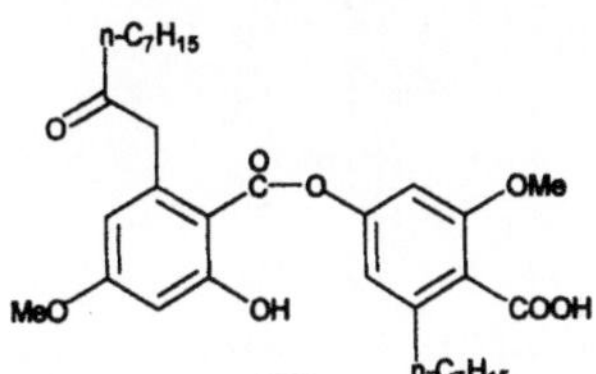

286

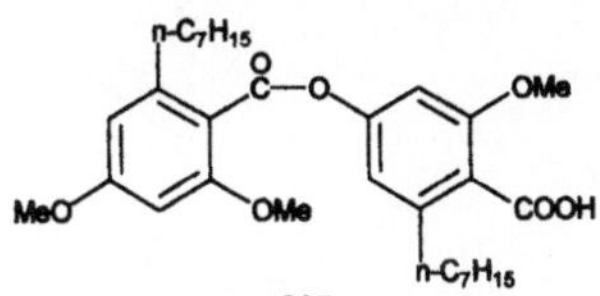

287

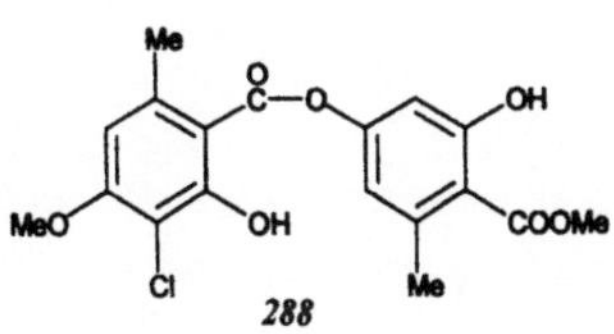

288

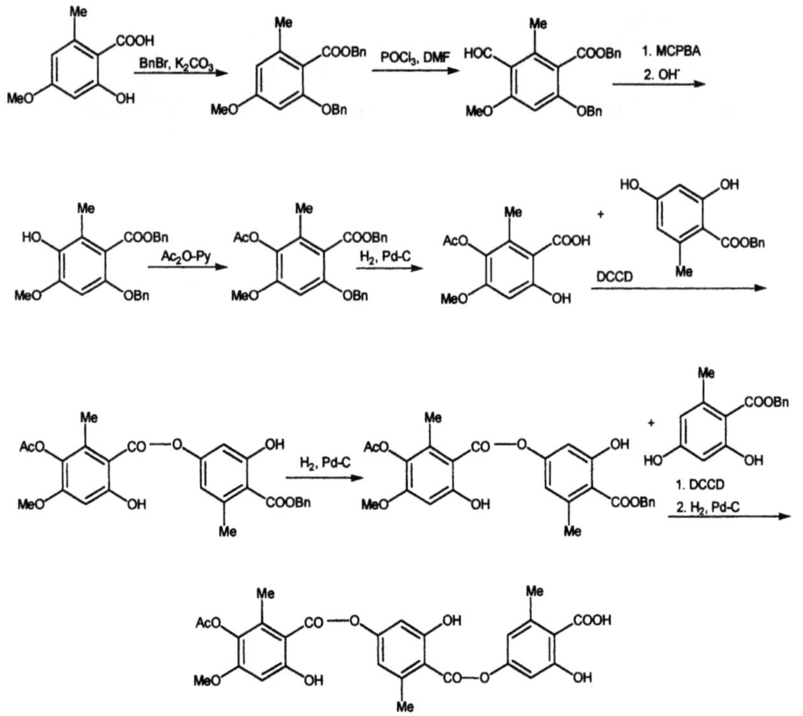

126

Scheme 93. Synthesis of 5-*O*-acetyl-4-*O*-methylhiascic acid

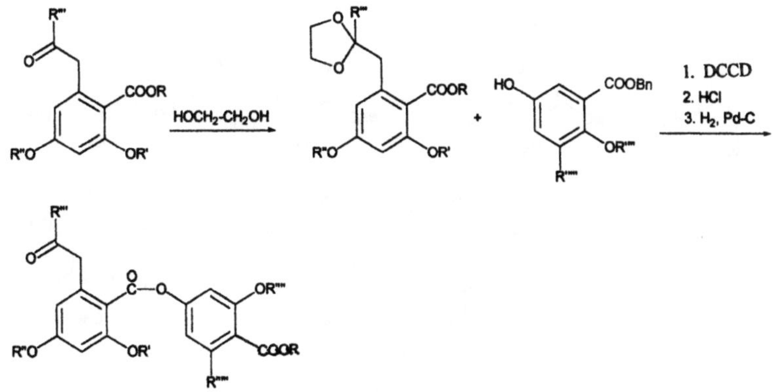

	R	R'	R''	R'''	R''''	R'''''
127: Arthoniaic acid	H	H	H	n-C_5H_{11}	Me	-CH_2-CO-n-C_5H_{11}
187: Hyperconfluentic acid	H	H	Me	„	„	n-C_7H_{15}
218: 2-O-Methylglomelliferic acid	H	Me	Me	n-C_3H_7	H	n-C_5H_{11}
228: 2′-O-Methylhyperphyllinic acid A	H	H	Me	n-C_7H_{15}	Me	-CH_2-CO-n-C_5H_{11}
229: 2′-O-Methylhyperphyllinic acid B	H	H	Me	n-C_5H_{11}	Me	-CH_2-CO-n-C_7H_{15}
243: 2′-O-Methylmicrophyllinic acid	H	H	Me	„	„	-CH_2-CO-n-C_5H_{11}
253: 2′-O-Methylnorsuperphyllinic acid	H	H	H	n-C_7H_{15}	„	-CH_2-CO-n-C_7H_{15}
265: 4-O-Methylsuperolivetoric acid	H	H	Me	„	H	n-C_7H_{15}
267: Microphyllinic acid	H	H	Me	n-C_5H_{11}	H	-CH_2-CO-n-C_5H_{11}

Scheme 94. Synthesis of arthoniaic acid and related compounds

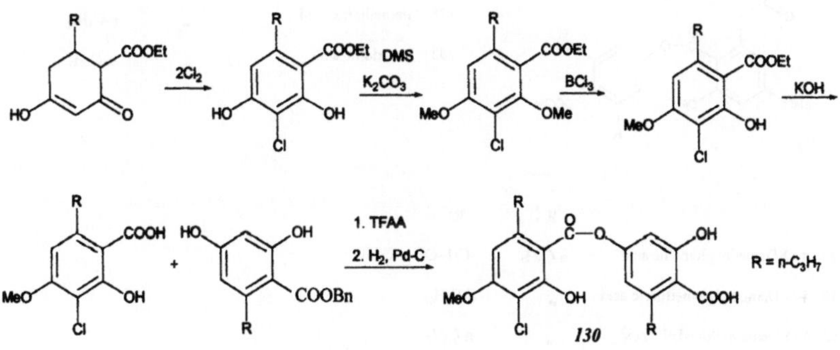

Scheme 95. Synthesis of 3-chlorodivaricatic acid

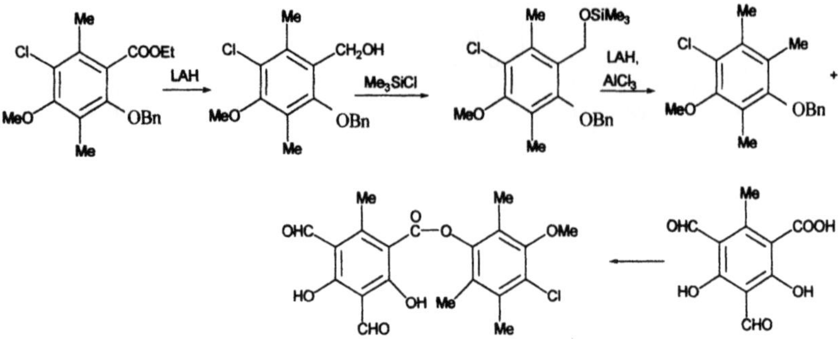

134

Scheme 96. Synthesis of 1′-chloronephroarctin

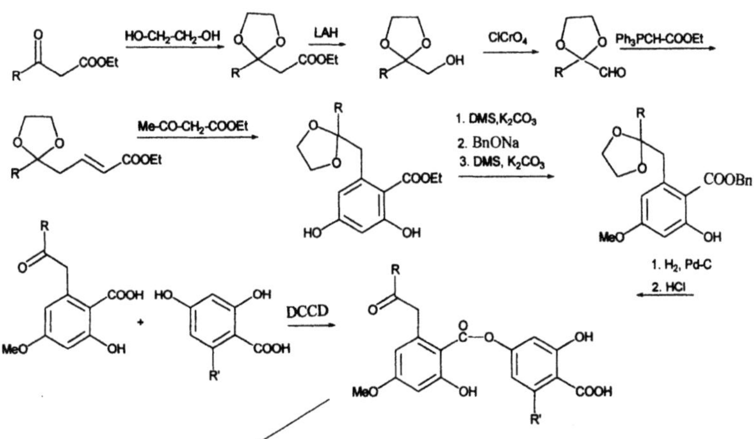

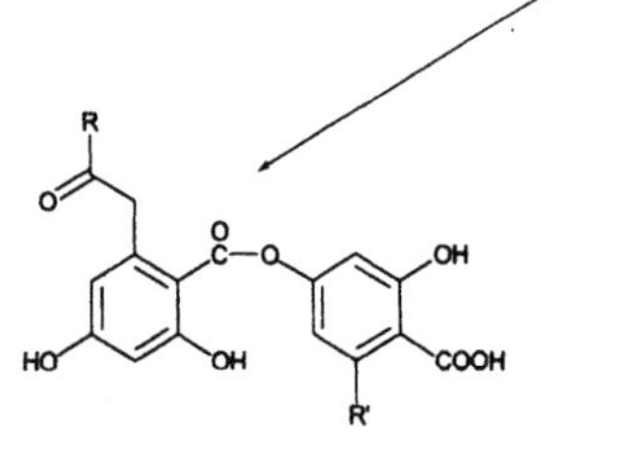

	R	R'
177 : Glomellic acid	$n\text{-}C_3H_7$	$\text{-}CH_2\text{-}CO\text{-}n\text{-}C_3H_7$
178 : Glomelliferic acid	"	$n\text{-}C_5H_{11}$
202 : Loxodellic acid	"	$n\text{-}C_3H_7$

	R	R'
149 : 4-O-Demethylglomellic acid	$n\text{-}C_3H_7$	$\text{-}CH_2\text{-}CO\text{-}n\text{-}C_3H_7$
150 : 4-O-Demethylglomelliferic acid	"	$n\text{-}C_5H_{11}$
152 : 4-O-Demethylloxodellic acid	"	$n\text{-}C_3H_7$

Scheme 97. Synthesis of glomellic acid and related compounds

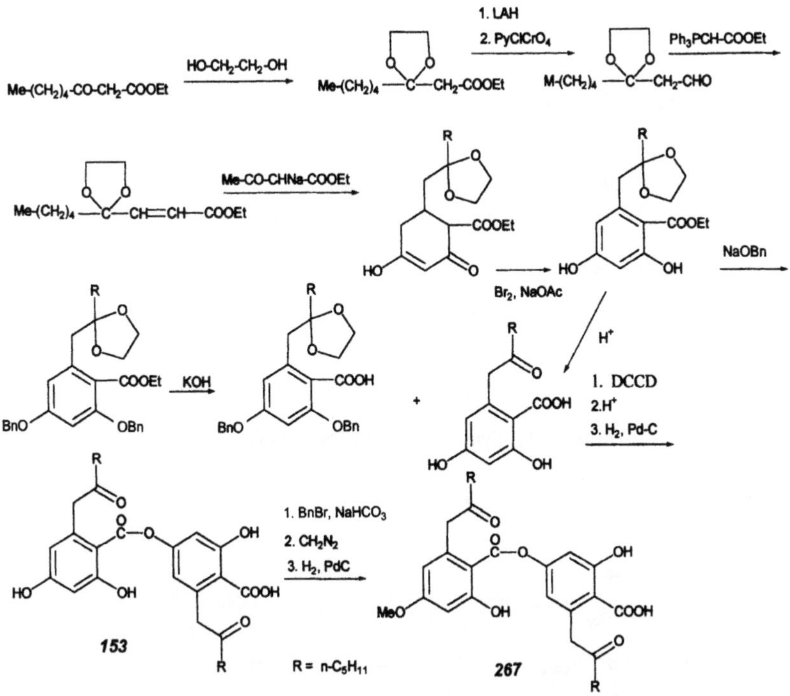

Scheme 98. Synthesis of 4-*O*-demethylmicrophyllinic acid (*153*)

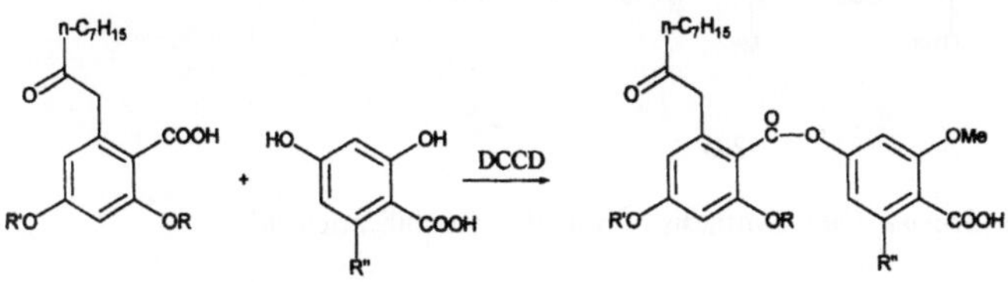

	R	R'	R''
160 : 4-O-Demethylsuperconfluentic acid	H	H	n-C_7H_{15}
176 : Glaucophaeic acid	Me	H	-CH_2-CO-n-C_7H_{15}
266 : 2'-O-Methylsuperphyllinic acid	H	Me	"
286 : Superconfluentic acid	H	Me	n-C_7H_{15}

Scheme 99. Synthesis of 4-*O*-demethylsuperconfluentic acid and related compounds

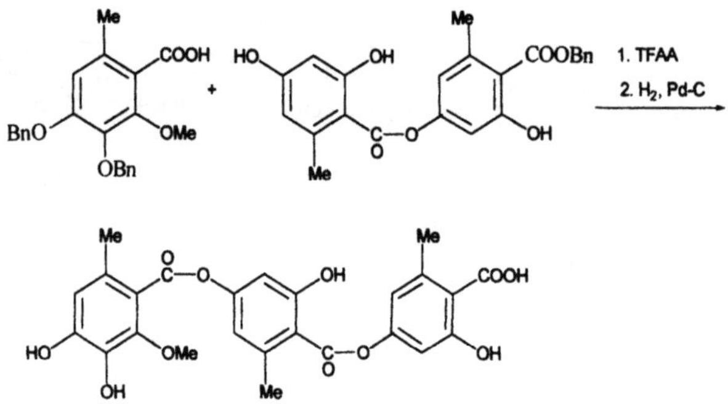

186

Scheme 100a. Synthesis of 3-hydroxyumbilicaric acid

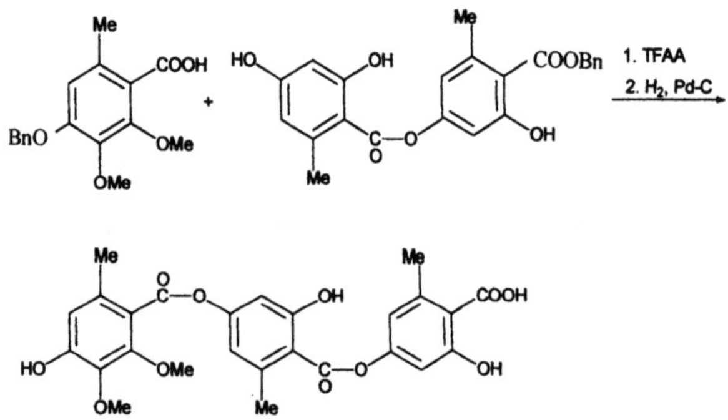

205

Scheme 100b. Synthesis of 3-methoxyumbilicaric acid

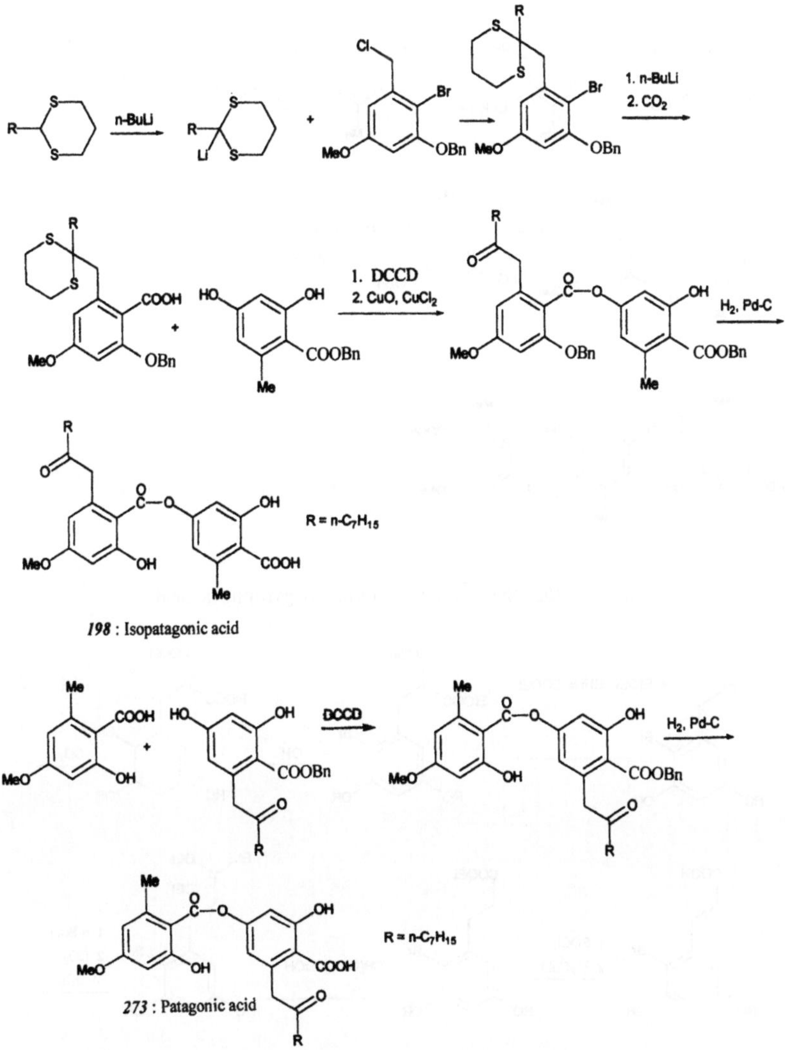

Scheme 101. Synthesis of isopatagonic acid and patagonic acid

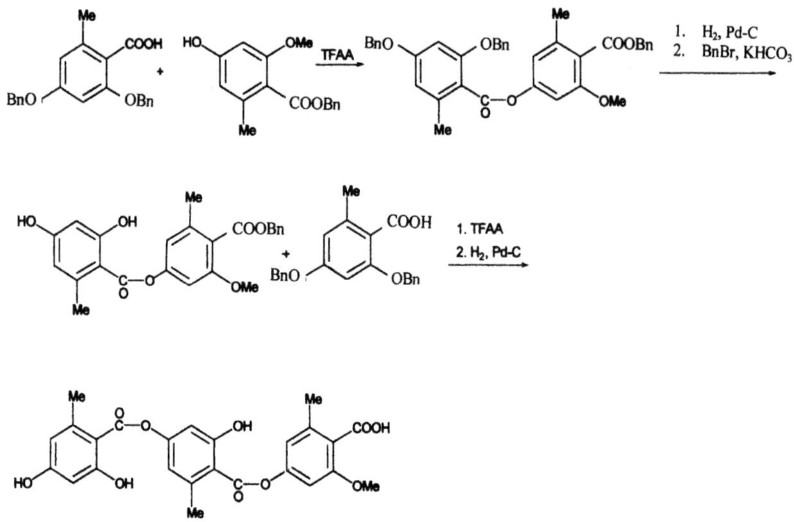

219

Scheme 102. Synthesis of 2″-*O*-methylgyrophoric acid

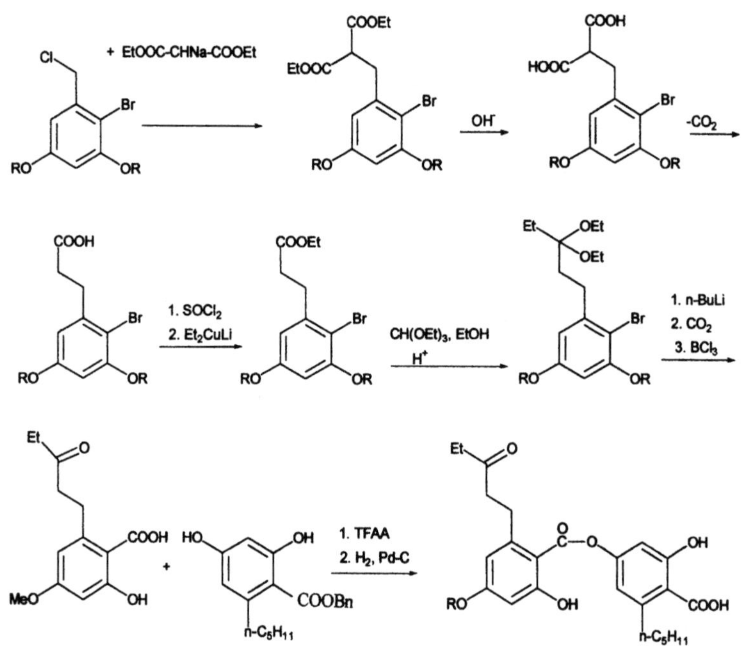

268 : Miriquidic acid, R = Me

270 : Normiriquidic acid, R = H

Scheme 103. Synthesis of miriquidic acid and normiriquidic acid

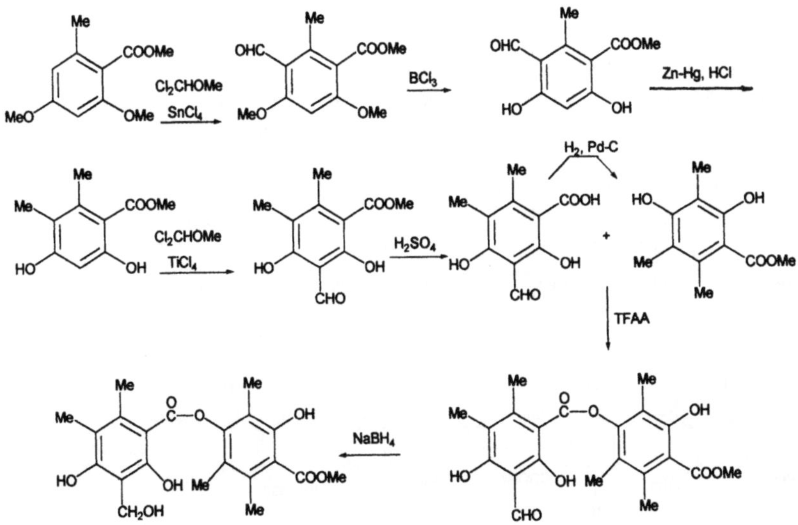

276 *275*

Scheme 104. Synthesis of pseudocyphellarin A and B after Elix and Lajide

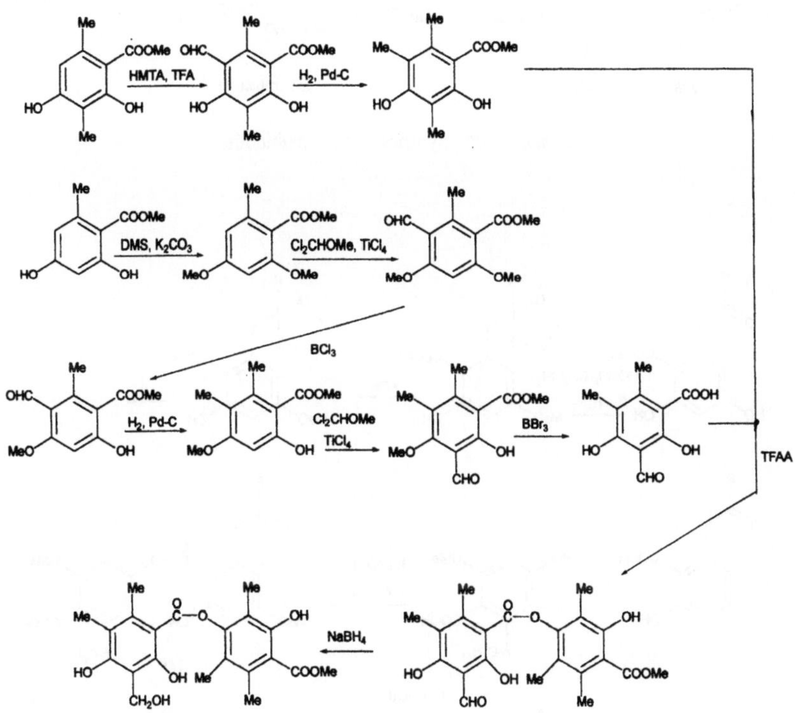

276 *275*

Scheme 105. Synthesis of pseudocyphellarin A and B after Pulgarin *et al.*

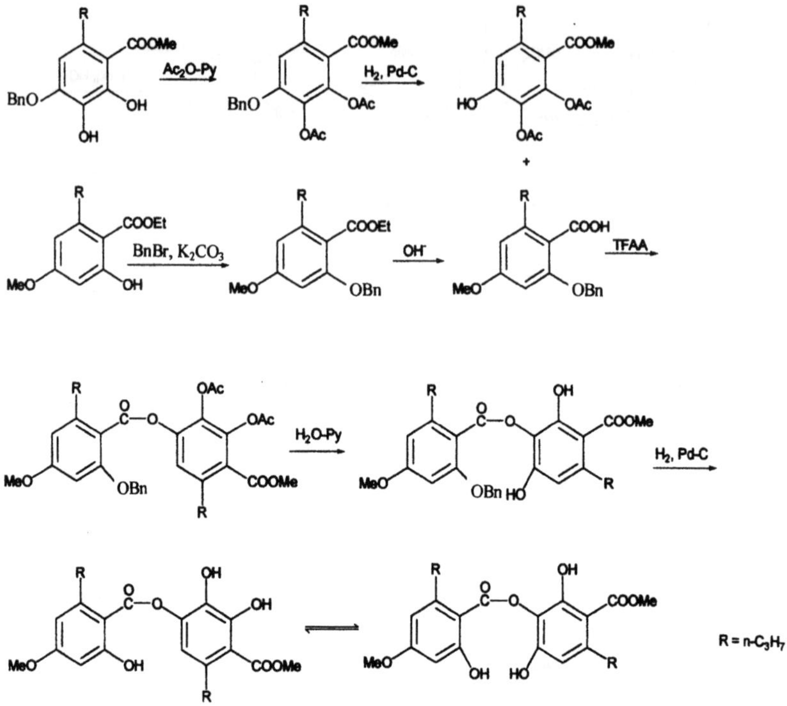

278 : *p*-Scrobiculin *m*-Scrobiculin

Scheme 106. Synthesis of *p*-scrobiculin

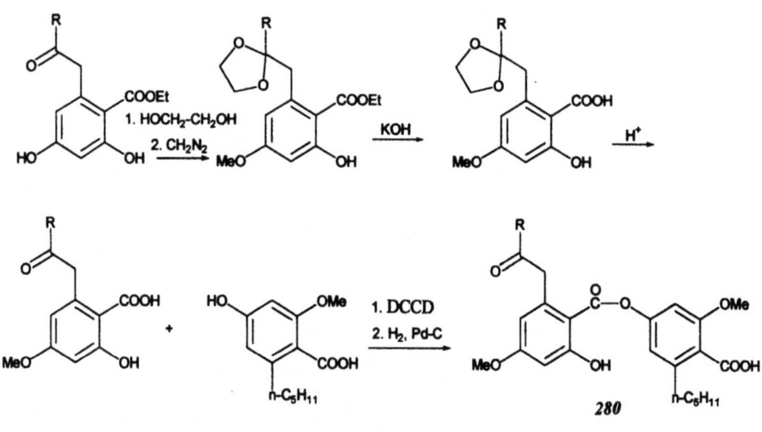

Scheme 107. Synthesis of subconfluentic acid

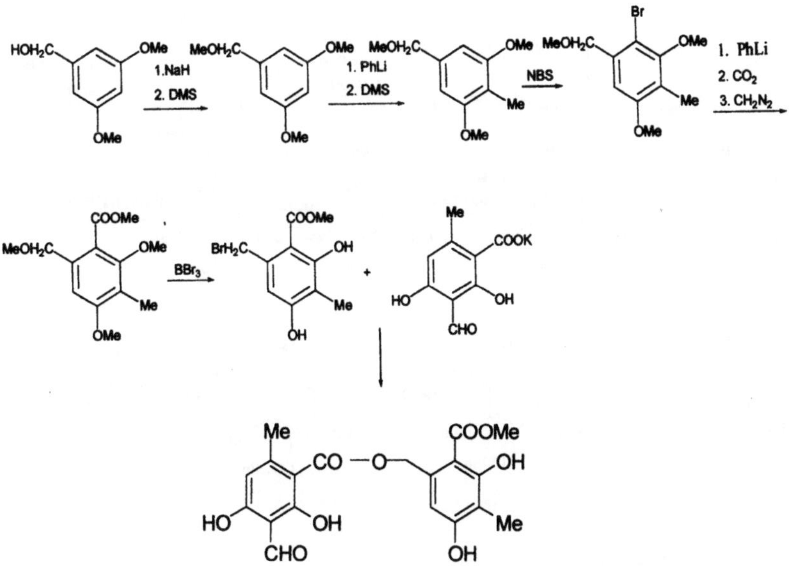

289: Methyl alectorialate

Scheme 108. Synthesis of methyl alectorialate

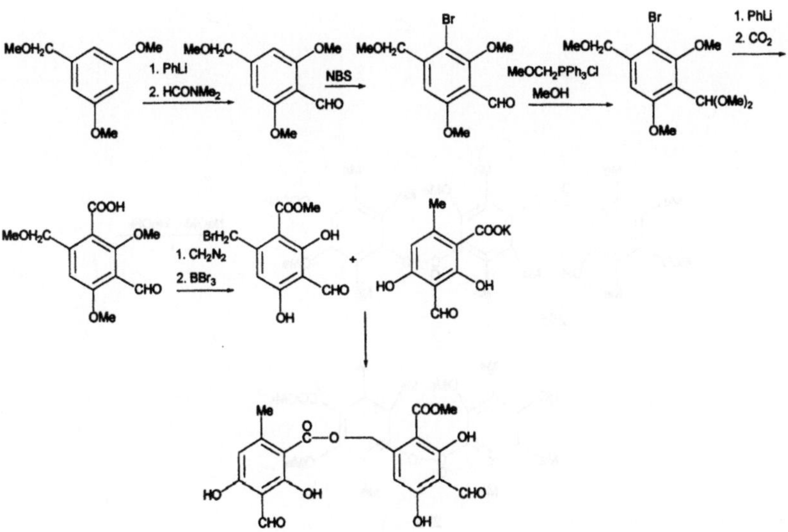

290: Methyl barbatolate

Scheme 109. Synthesis of methyl barbatolate

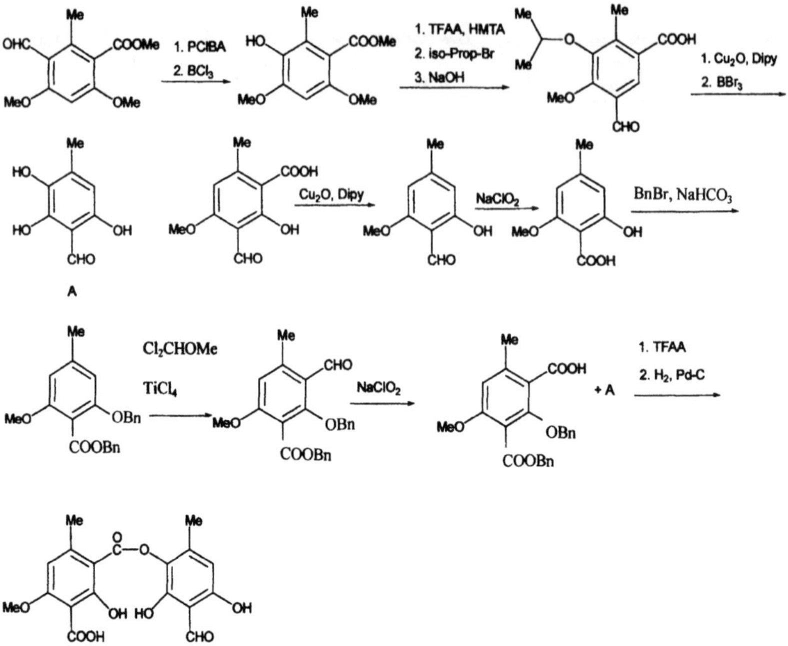

291 : Decarboxythamnolic acid

Scheme 110. Synthesis of decarboxythamnolic acid

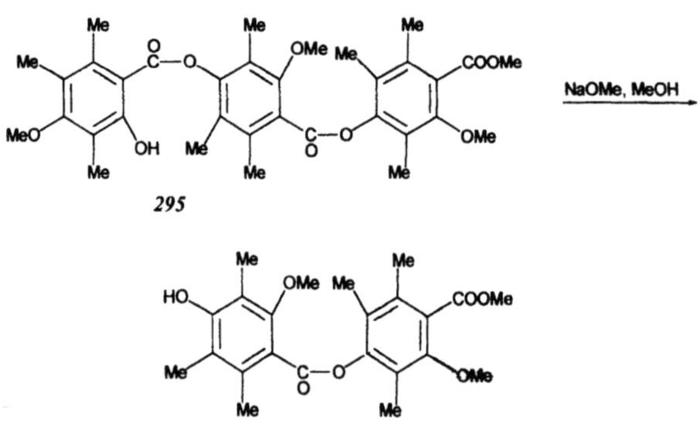

295

296

Scheme 111. Methanolysis of methyl thielavinate B

compounds, phenolic units, rearranged depsides and xanthones. Trimethylammonium salts of depside carboxylic acids decomposed at considerably lower temperature than the corresponding acids. Thielavin A (**292**), B (**293**) and C (**294**) are tridepsides from the fungus *Thielavia terricola* and are inhibitors of prostaglandin synthetase (*459*, *460*). Methyl thielavinate B (**295**) gave the depside **296** on methanolysis (see Scheme 111).

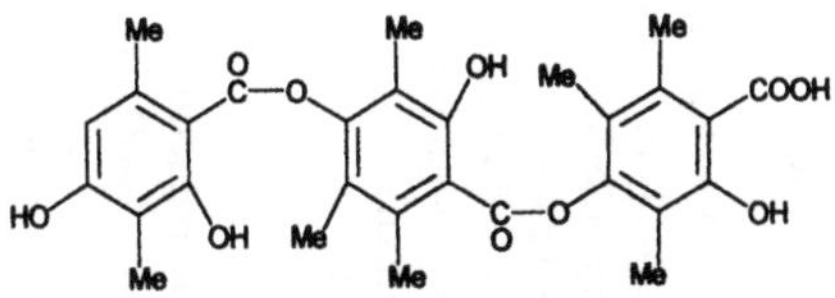

292 : Thielavin A, mp 235-236°C

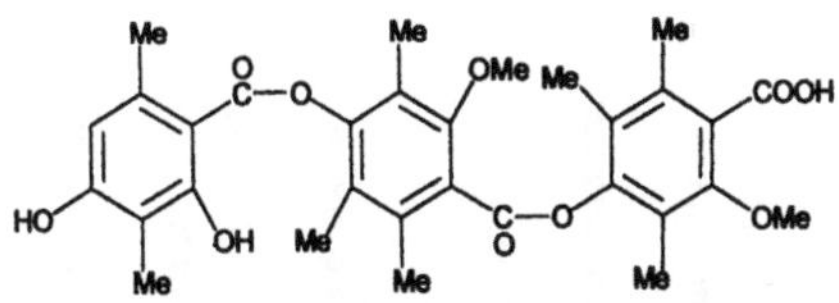

293 : Thielavin B, mp 250°C

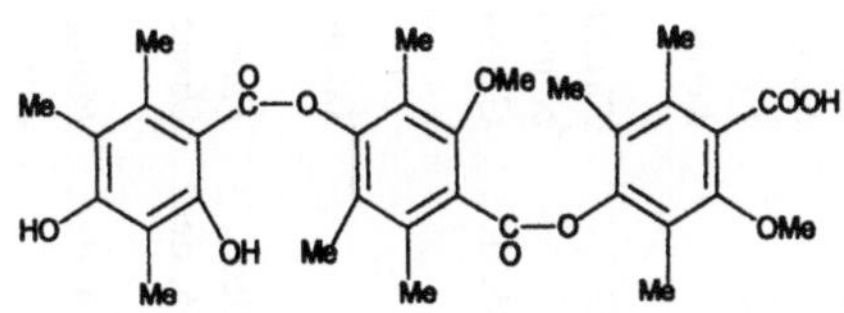

294 : Thielavin C, mp 107-109°C

4.5.7. Depsidones

The newly described depsidones, their occurrence and synthesis are listed in Table 5. The synthesis of depsidones with a lactole ring like those in norstictic, stictic or salazinic acids have caused considerable difficulties. Hareed (*353*) tried to prepare a depsidone by a) oxidative coupling, b) by a Smiles rearrangement and c) by an Ullmann reaction of a diphenyl ether, but all three attempts failed (see Scheme 123). Some time later Rama (*604*) successfully synthesized the lactone depsidone **359** and the lactole depsidone **360** *via* oxidative coupling of a dibenzophenone (see Scheme 124). The structure of the precursor **358**

Table 5. Depsidones from lichens

Compound	Formula No.	M.p. (°C)	Lichen	Synth. Scheme	Ref.
α-Acetylconstictic acid	**297**	>230, dec.	*Menegazzia platytrema*, *Pseudocyphellaria faveolata*		(*214*)
α-Acetylhypoconstictic acid	**298**	220–240, dec.	*Menegazzia dispora*, *Xanthoparmelia metastrigosa*	By acetylation of hypoconstictic acid with AcOH in DMF	(*213*)
Allorhizin	**299**	230–232	*Psoroma allorhizum*	112	(*240*)
Chlorolecideoidin	**300**	234	*Lecanora sulphurescens*	By chlorination of lecideoidin with SO_2Cl_2 in dioxane	(*515*)
Chlorophyllopsorin	**301**	228–229	*Phyllopsora corallina* var. *ochroxantha*	By chlorination of phyllopsorin with Cl_2	(*258*)
Confumarprotocetraric acid	**302**	240–245, dec.	*Cladonia phyllophora*	By reduction of fumarprotocetrartic acid with $NaBH_4$ in 1,2-dimethoxyethane	(*276*)
Conhypoprotocetraric acid	**303**	>330	*Lecanora myriocarpoides*	113	(*243*)
Conphysodalic acid	**304**	>230, dec.	*Flavoparmelia springtonensis*		(*214*)
Conprotocetraric acid	**305**	230–235, dec.	*Usnea trichodeoides*	By reduction of protocetraric acid with $NaBH_4$ in 1,2-dimethyoxyethane	(*276*)
Consuccinoprotocetraric acid	**306**	237–240, dec.	*Flavoparmelia succinoprotocetrarica*	By reduction of succinoprotocetraric acid with $NaBH_4$ in 1,2-dimethoxyethane	(*276*)
Convirensic acid	**307**	235–240, dec.	*Sulcaria virens*	By reduction of virensic acid with $NaBH_4$ in 1,2-dimethoxyethane	(*276*)
Cyclographin	**308**	245–246	*Catarraphia dictyoplaca*	114	(*250*)
Decarboxyalectoronic acid	**309**	144–148	*Hypotrachyna gracilescens*		(*189*)
3-Dechlorogangaleoidin	**310**	226–228	*Tylothallia pahiensis*		(*259*)
Dechloropannarin	**311**	182–184	*Lecanora dispersa*		(*497*)

				115	*(240)*
3-*O*-Demethylscensidin	**312**	226–227.5	*Diploicia canescens*	By regioselective demethylation of scensidin with NaI in DMF	*(230)*
Didechlorolecideoidin	**313**	245		By hydrogenation of lecideoidin with H_2-PdC	*(259)*
4,2′-Di-*O*-methylcon-norstictic acid	**314**		*Lobaria pulmonaria*		*(9)*
2′-*O*-Methylphysodic acid	**315**	160	*Pseudeverina furfuracea*		*(339)*
Dinorscensidin	**316**		not yet found in lichens		*(206)*
Divaronic acid	**317**	110, 160–162	*Cladonia chlorophaea*		*(93)*
				116	*(228)*
Eriodermin	**318**	234–236	*Erioderma physcioides*		*(79)*
				117	*(589)*
8′-*O*-Ethylstictic acid (Vesuvianic acid)	**319**	245	*Stereocaulon vesuvianum*		
			(probably an artefact)		*(438, 412)*
Fulgidin	**320**	204–207	*Fulgensia fulgida*	118	*(32)*
Furfuric acid	**321**	250–254	*Pseudevernia furfuracea*		*(340)*
				Synthesis of methyl tri-*O*-methyl desoxyfurfurate 119	*(341)*
				Synthesis of furfuric acid 120	*(209)*
Gangaleoidin		214–215	*Lecanora gangaleoides*		*(633)*
Glomelliferonic acid	**322**	184–185	*Neofuscelia subincerta*	The ^{13}C NMR signals of glomelliferonic acid could be assigned by the relaxation time T_1 of the corresponding C atoms	*(227)*

Table 5 (continued)

Compound	Formula No.	M.p. (°C)	Lichen	Synth. Scheme	Ref.
Glomellonic acid	**323**	149–151	*Neofuscelia subincerta*		(*227*)
Hypophysciosporin	**324**	238–240	*Erioderma phaeorhizum*	By chlorination of methyl hypoprotocetrarate with SO_2Cl_2	(*348*)
Isidiophorin	**325**	268–270	*Lobaria isidiophora*		(*36*)
Isofulgidin	**326**	260–261	*Fulgensia canariensis*, *Hafellia fulgida*		(*32*)
Isonorpannarin	**327**	195–198		By chlorination of nordechloropannarin with SO_2Cl_2	(*226*)
Isonotatic acid	**328**	166–167	*Ramalina americana*		(*96*)
			Xanthoparmelia notata	121	(*238*)
Isovicanicin	**329**	247–249	*Psoroma athrophyllum*		(*240*)
Leoidin	**330**	232–233	*Lecanora gangaleoides*		(*206*)
Loxodellonic acid	**331**	159–161	*Neofuscelia subincerta*		(*227*)
2′-*O*-Methylconorstictic acid	**332**		*Lobaria pulmonaria*		(*9*)
4-*O*-Methyl-5-dechlorovicanicin	**333**		not yet found in lichens		(*206*)
Methyl 2′-*O*-demethylpsoromate	**334**		*Lecanora intumescens*		(*270*)
Methyl 5,1′-dichloronorpsoromate	**335**	230–232	*Phyllopsora corallina* var. *ochroxantha*		(*258*)
Methyl 5,1′-dichloropsoromate	**336**	204.5–207	*Phyllopsora corallina* var. *ochroxantha*	By chlorination of methyl psoromate with chlorine in AcOH	(*258*)
4-*O*-Methyldiploicin	**337**	220	*Diploicia canescens*	By methylation of diploicin	(*230*)
4-*O*-Methylhypophysciosporin	**338**	200–203	*Erioderma phaeorhizum*	By chlorination of methyl 4-*O*-methylprotocetrarate with SO_2Cl_2	(*348*)

3-*O*-Methyllividic acid	**339**	156.5–159	*Hypotrachyna livida*	By methylation of benzyl lividate and subsequent debenzylation	(*256*)
4-*O*-Methylpannarin	**340**		not yet found in lichens		(*206*)
Methyl pseudoalectoronate	**341**		*Parmotrema poolii*		(*268*)
Methyl pseudonorstictate	**342**		*Pertusaria falklandica*		(*270*)
Methyl pseudosalazinate	**343**	265, dec.			(*270*)
Methyl psoromate	**344**		*Lecanora intumescens*		(*270*)
O-Methylvicanicin	**345**		*Erioderma* spec.		(*230*)
Norargopsin	**346**	152–154	*Erioderma chilense*	By chlorination of norpannarin with SO_2Cl_2	(*226*)
Nordechloropannarin	**347**		*Lecanora dispersa*		(*497*)
Norisonotatic acid	**348**		*Ramalina americana*		(*96*)
Norpannarin	**349**	198–202	*Erioderma chilense*	By decarboxylation of 5-chlorovirensic acid with CuO-2,2′-bipyridyl in *bis*-(2-methoxyethyl)ether	(*226*)
Oxolobaric acid	**350**	178	*Anzia hypoleucoides*		(*272*)
Phyllopsorin	**351**	233.5	*Phyllopsora corallina* var. *ochroxantha*		(*258*)
Siphulellic acid	**352**	>240, dec.	*Siphulella coralloidea*	Structure confirmed by conversion into methyl *O*-methylpsoromate	(*216*)
Stenosporonic acid	**353**	139–140	*Cladonia grayi*	116	(*228*)
Sublobaric acid	**354**	183–184	*Anzia hypoleucoides*		(*272*)
Subnorstictic acid	**355**	>315	*Diploschistes ocellatus*		(*189*)
Subnotatic acid	**356**	223–225	*Parmelia notata*	122	(*238*)
Substictic acid	**357**	>270, dec.	*Aspicilia mashiginensis*		(*214*)
Vesuvianic acid: see 8′-*O*-Ethylstictic acid					

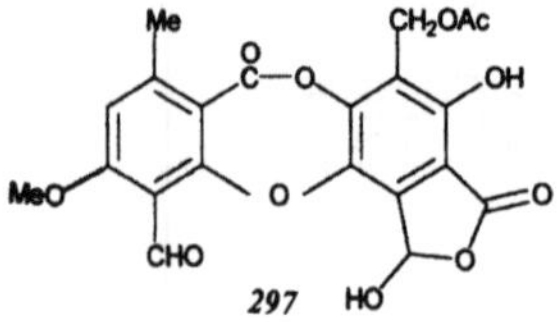

297

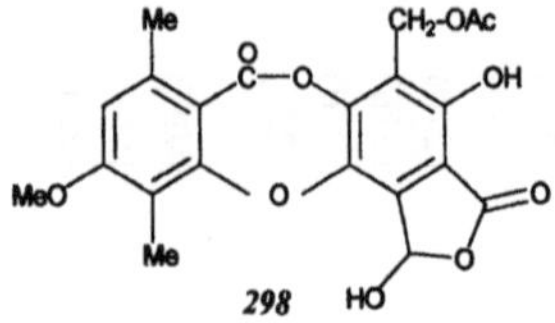

298

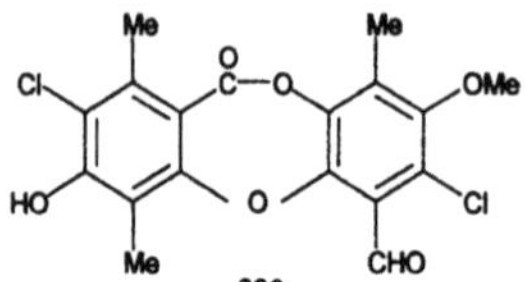

299

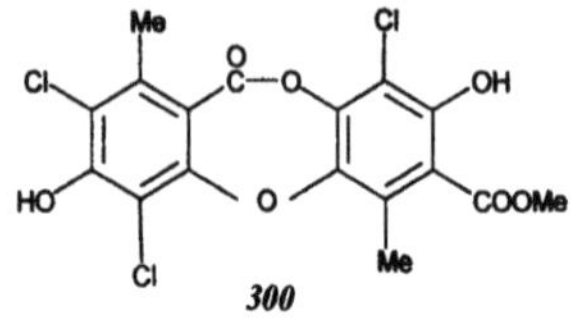

300

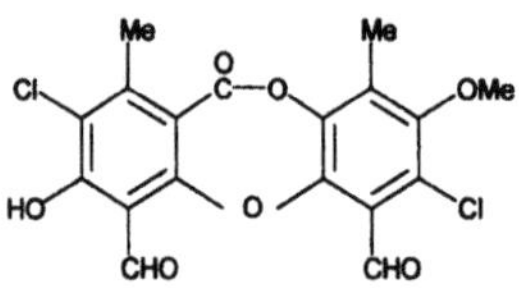

301

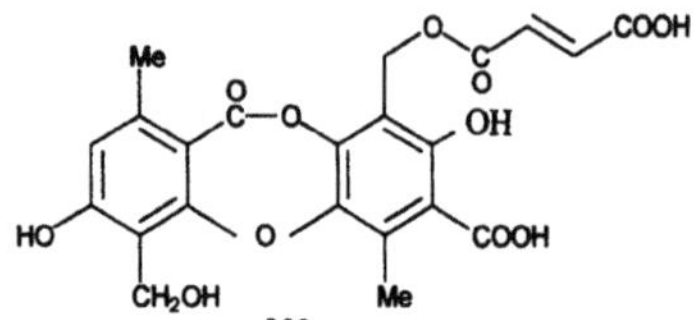

302

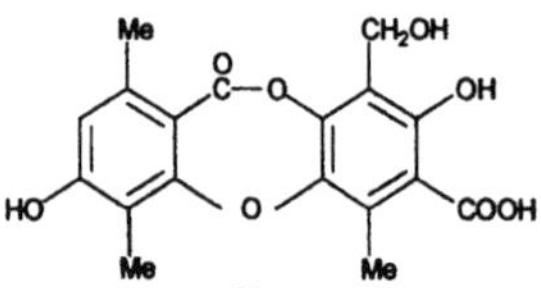

303

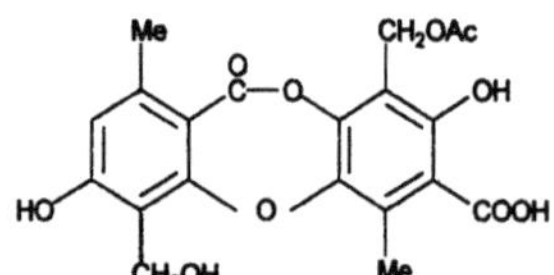

304

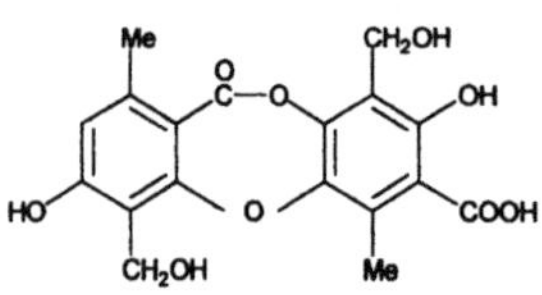

305

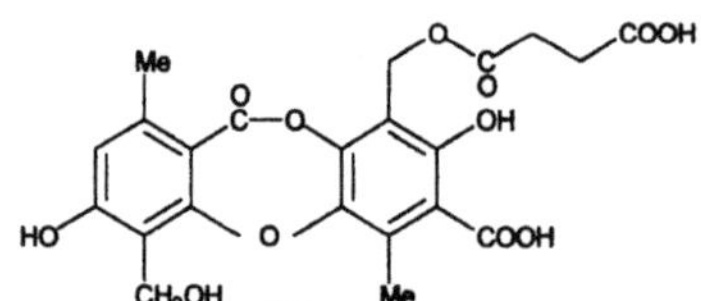

306

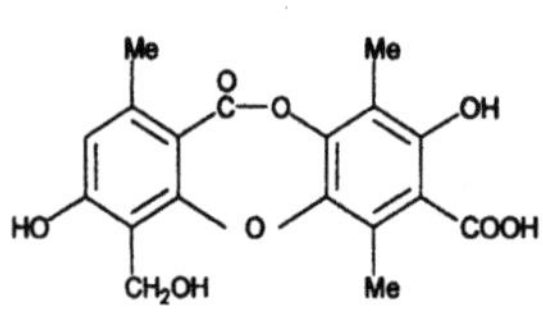

307

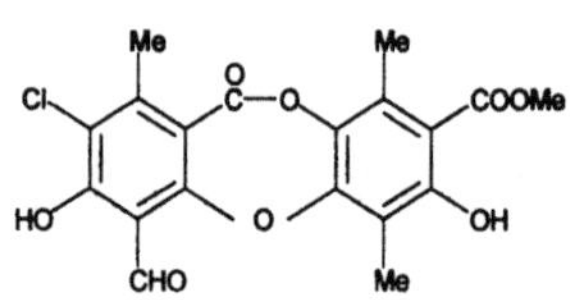

308

was determined by an X-ray analysis. Pannarin (**361**) undergoes rearrangement on treatment with sulphuric acid to isopannarin (**362**) (*176*) (see Scheme 125). The reactions of stictic, norstictic and salazinic acids with MeOH, EtOH and *tert*-BuOH have been investigated by Huneck

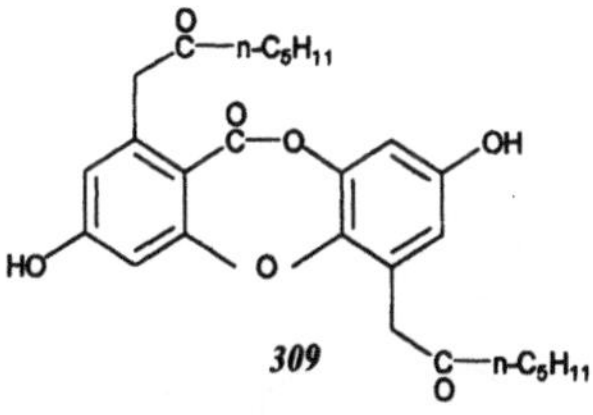

309

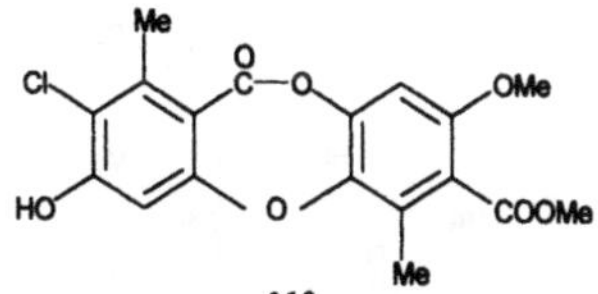

310

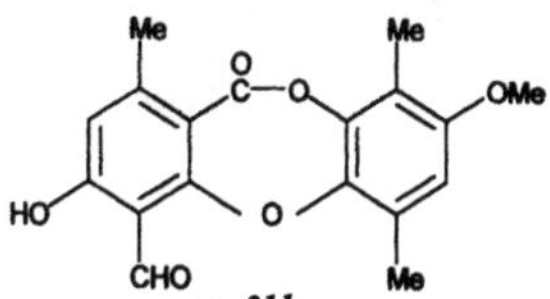

311

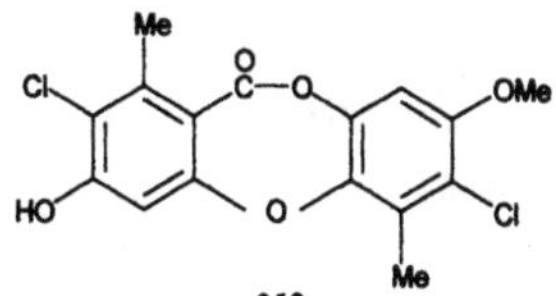

312

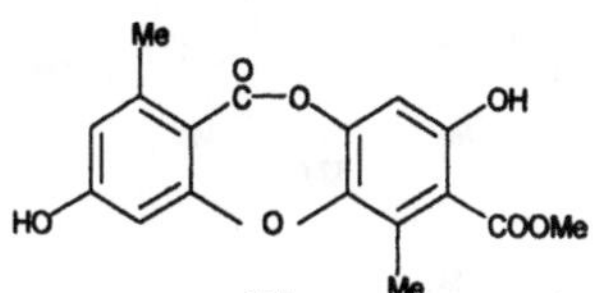

313

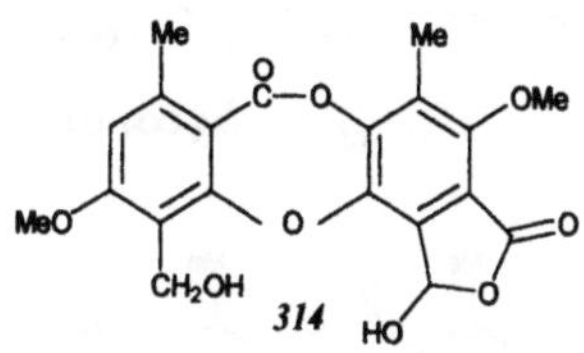

314

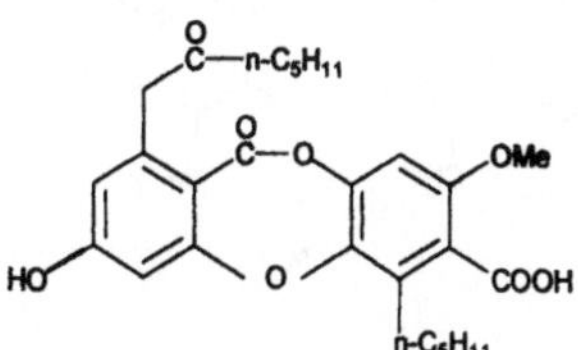

315

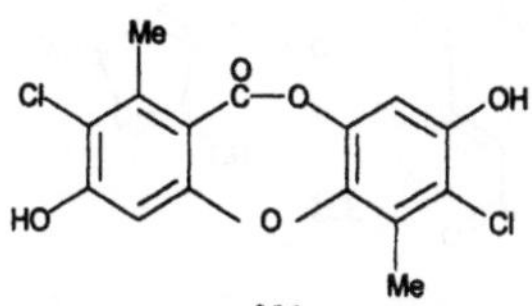

316

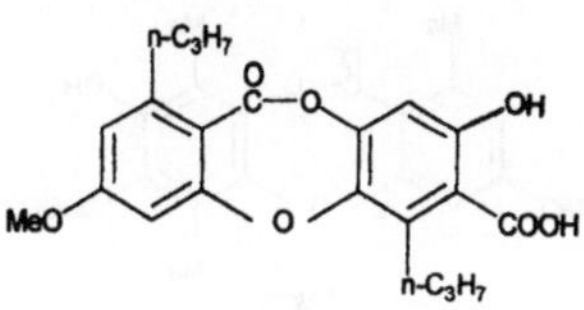

317

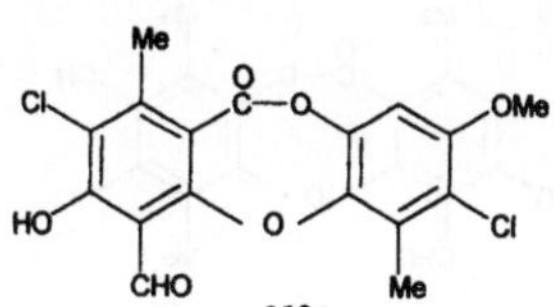

318

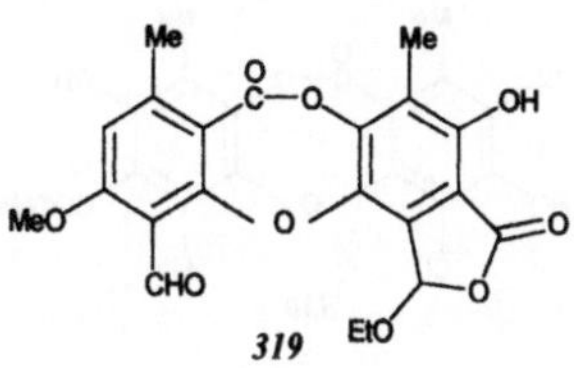

319

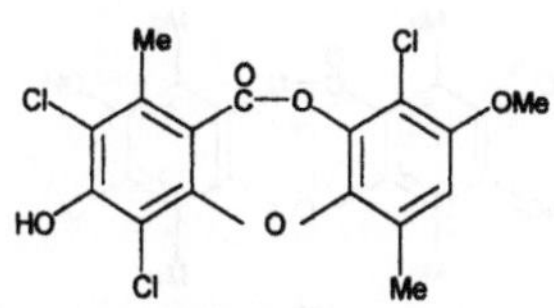

320

and Tabacchi (*412*). Methylstictic acid from *Lobaria oregana* is identical with 8′-*O*-methylstictic acid (**363**) while Handong's cetrariastrumin is identical with 8′,9′-di-*O*-ethylsalazinic acid (**364**).

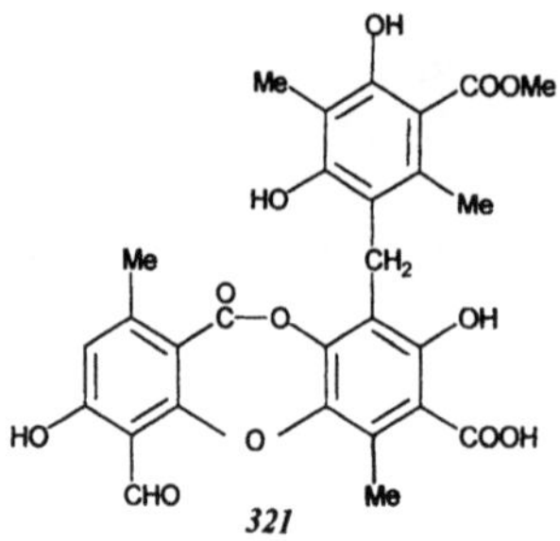

321

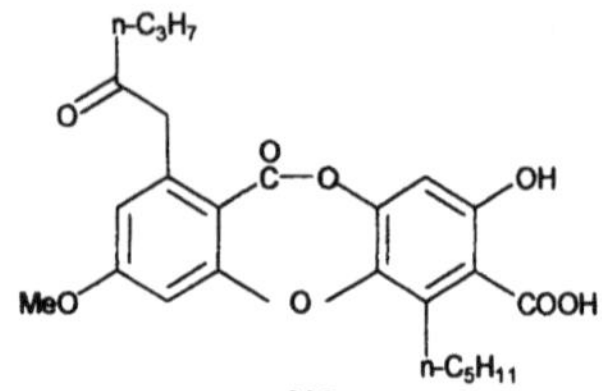

322

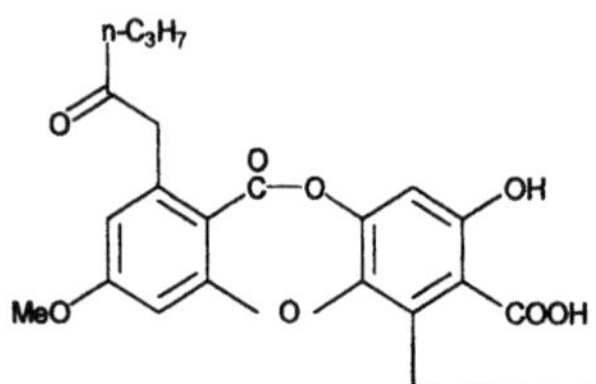

323

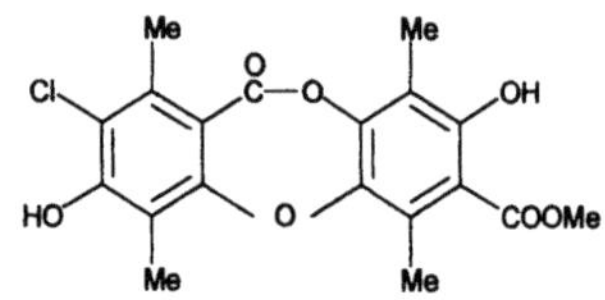

324

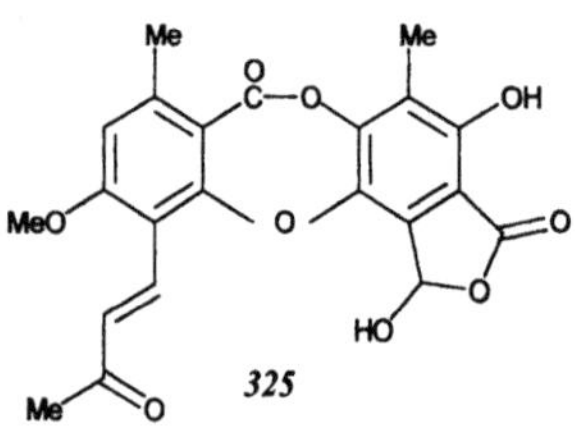

325

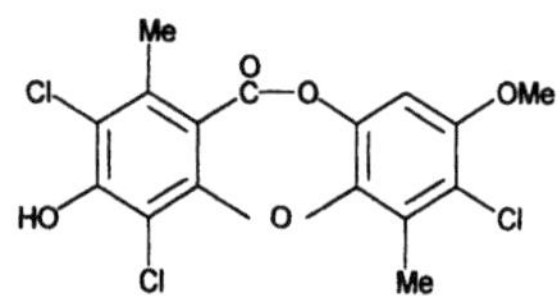

326

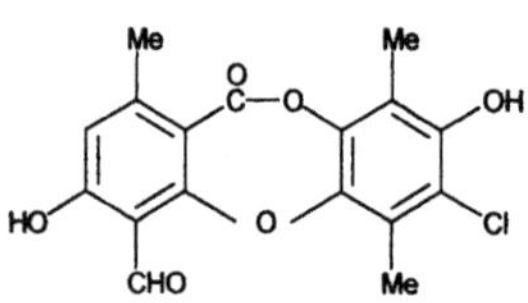

327

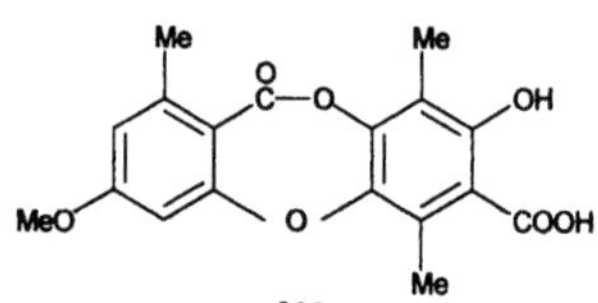

328

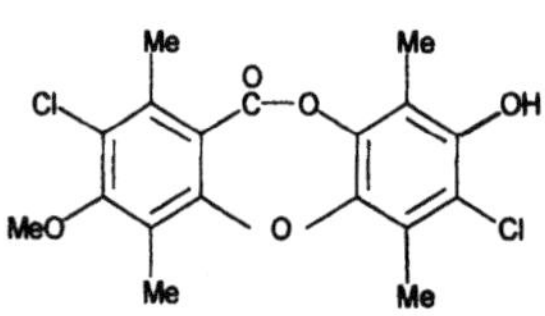

329

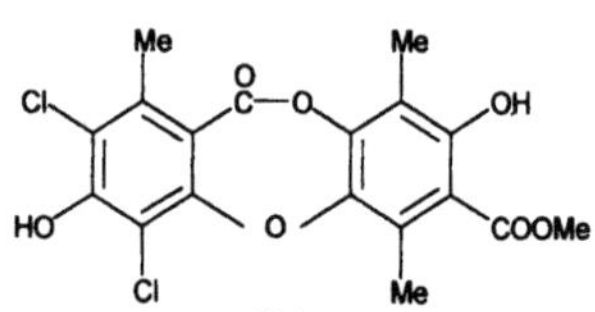

330

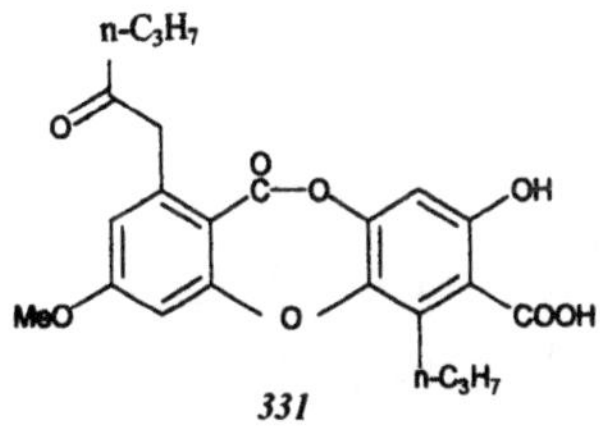

331

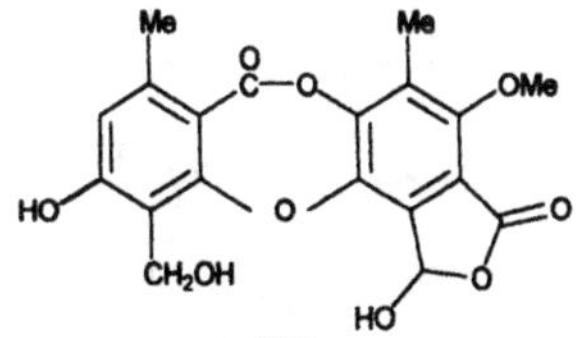

332

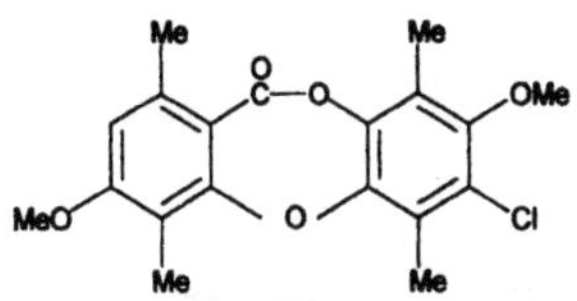

333

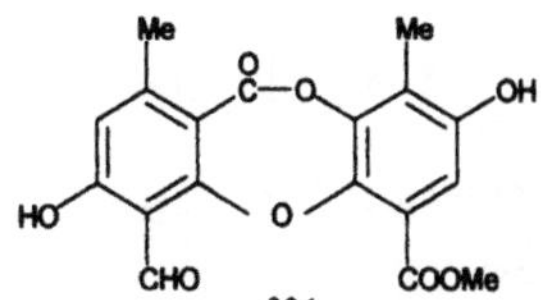

334

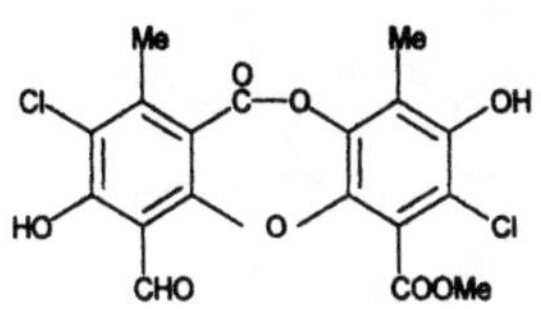

335

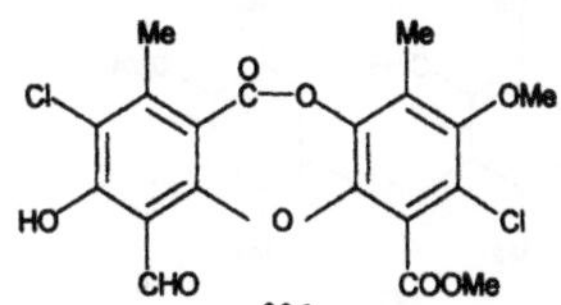

336

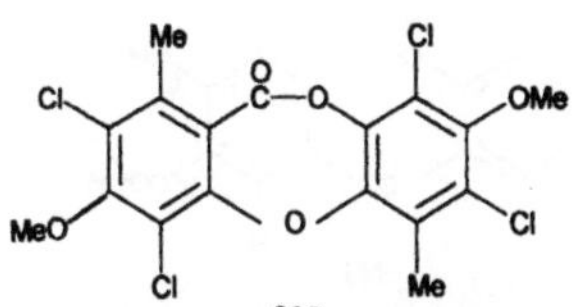

337

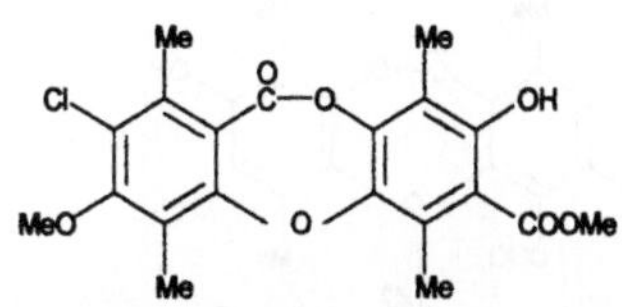

338

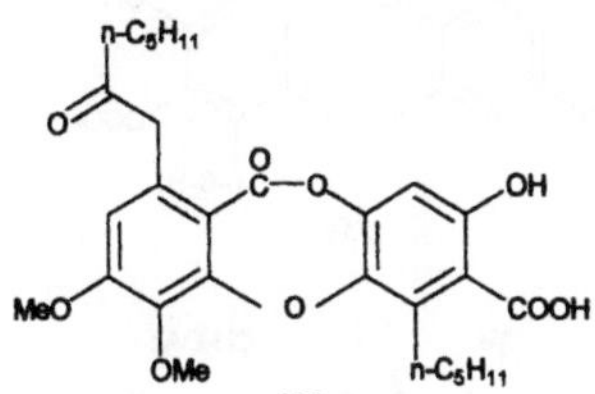

339

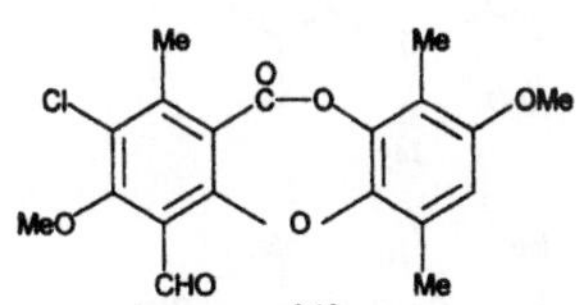

340

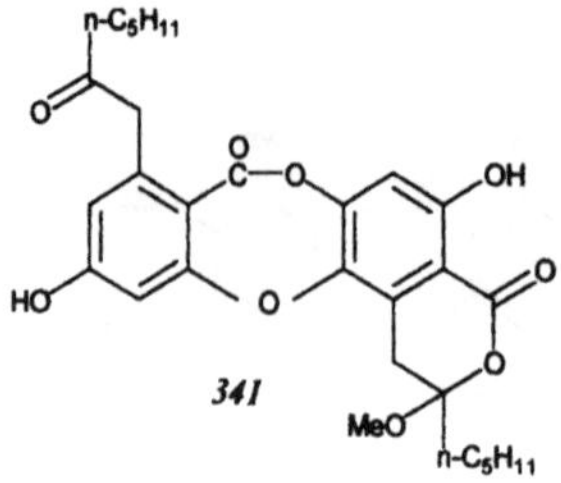

341

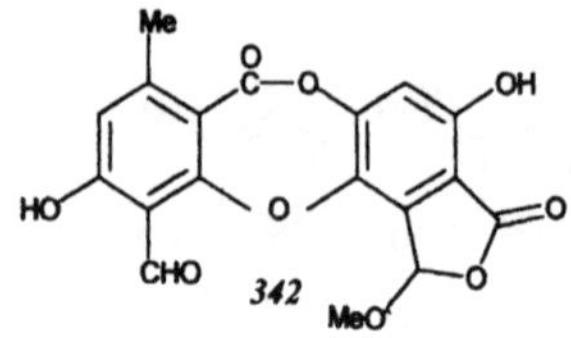

342

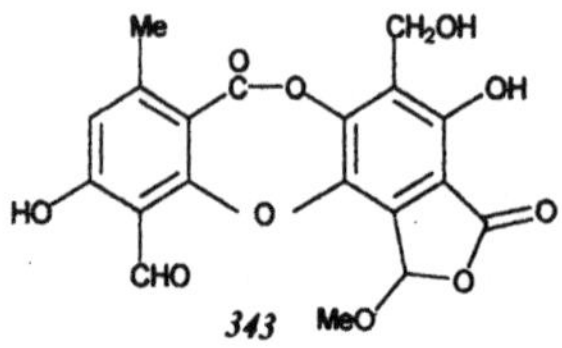

343

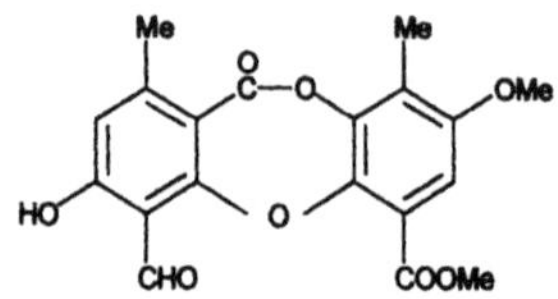

344

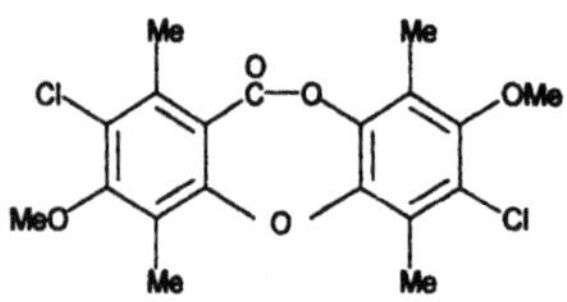

345

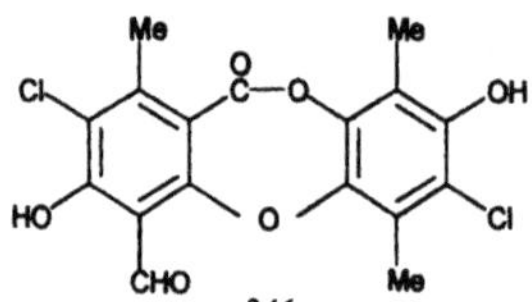

346

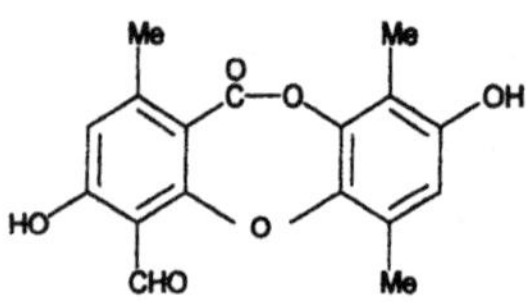

347

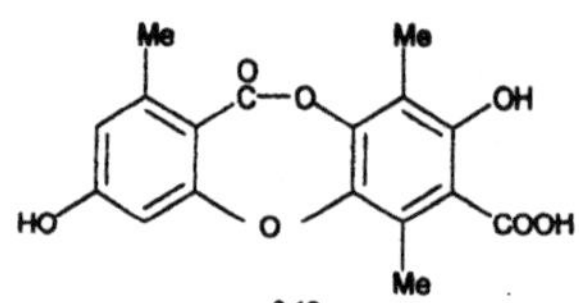

348

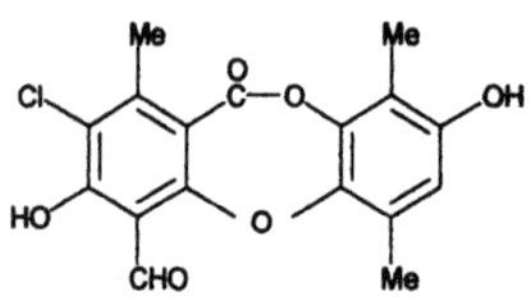

349

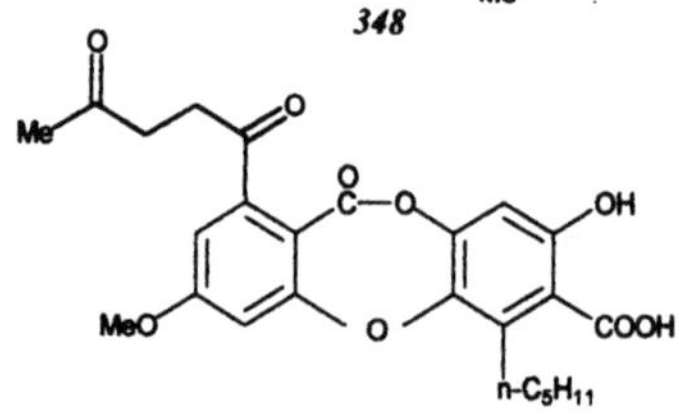

350

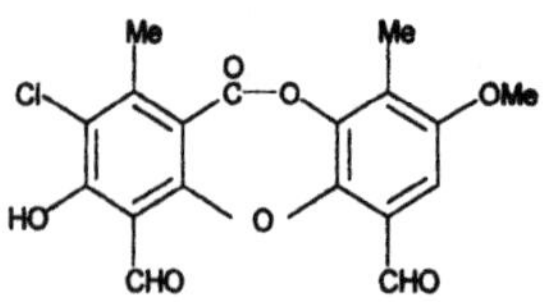

351

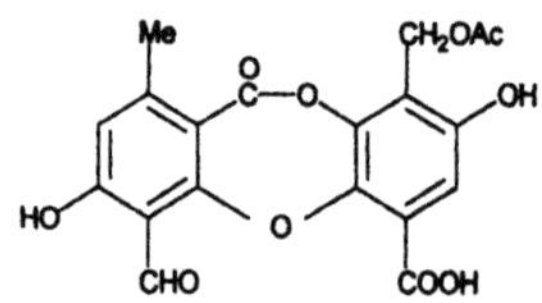

352

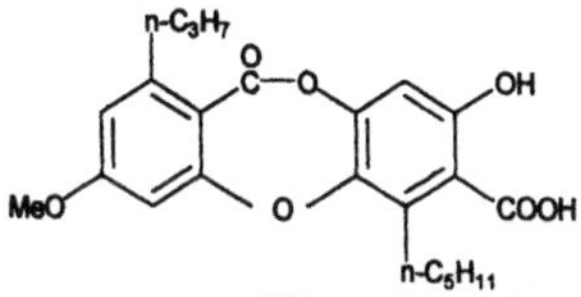

353

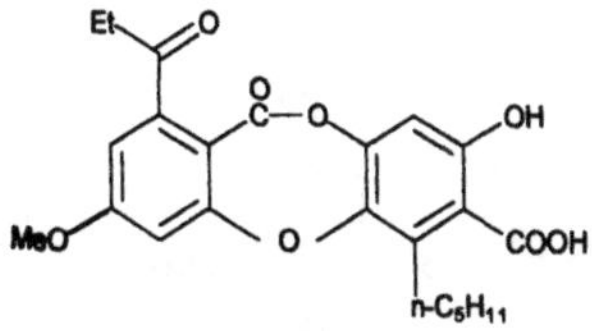

354

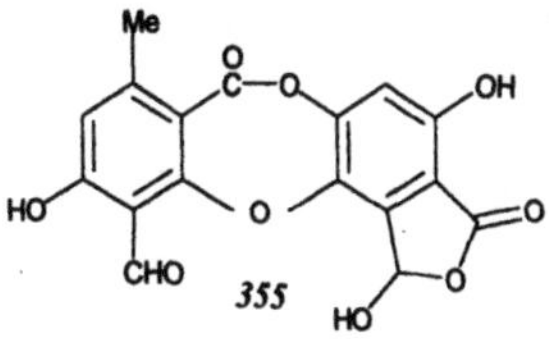

355

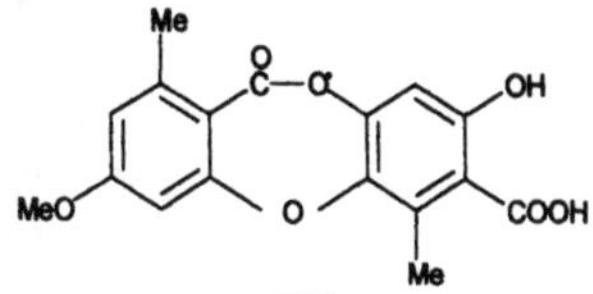

356

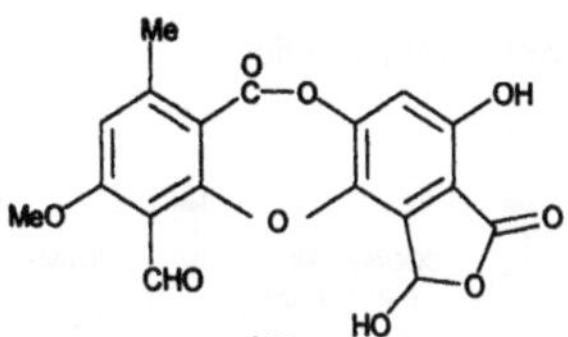

357

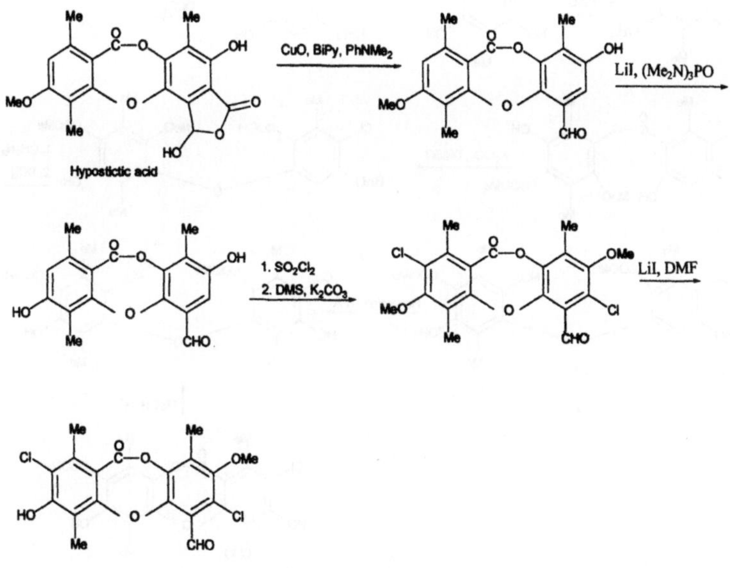

299 : Allorhizin

Scheme 112. Synthesis of allorhizin

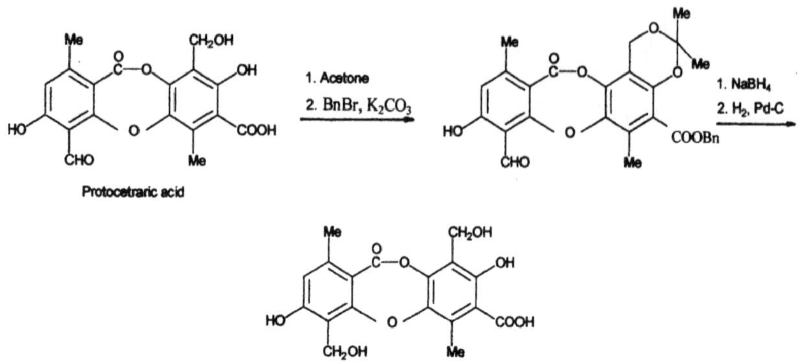

303 : Conhypoprotocetraric acid

Scheme 113. Synthesis of conhypoprotocetraric acid

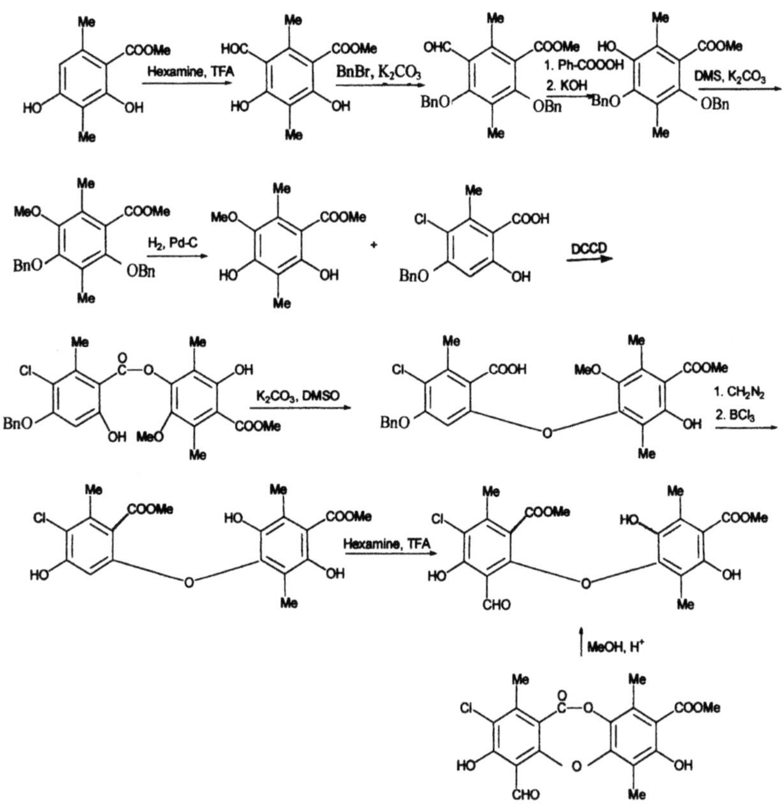

308 : Cyclographin

Scheme 114. Synthesis of the methanolysis product of cyclographin

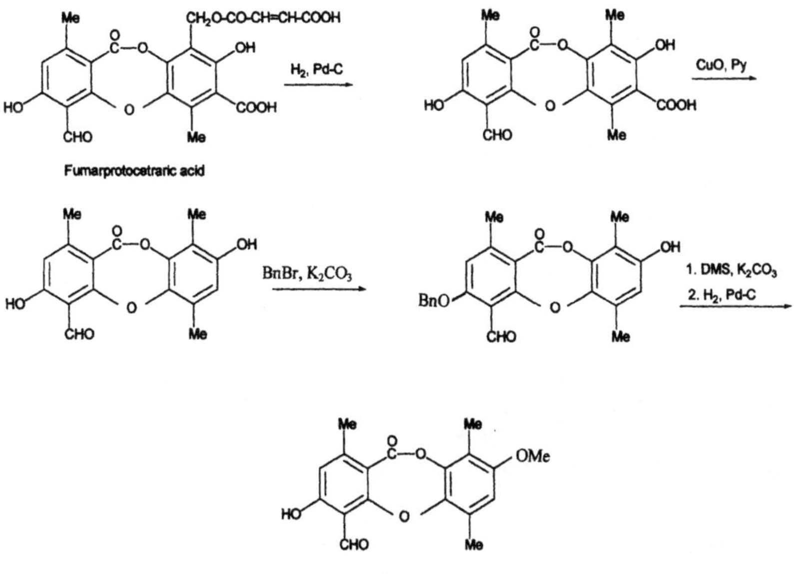

311 : Dechloropannarin

Scheme 115. Synthesis of dechloropannarin

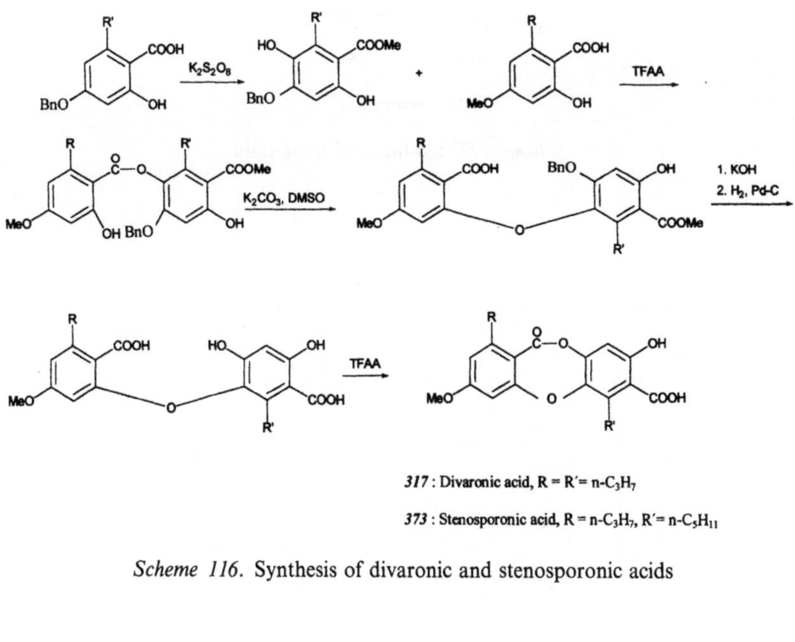

317 : Divaronic acid, R = R′= n-C_3H_7

373 : Stenosporonic acid, R = n-C_3H_7, R′= n-C_5H_{11}

Scheme 116. Synthesis of divaronic and stenosporonic acids

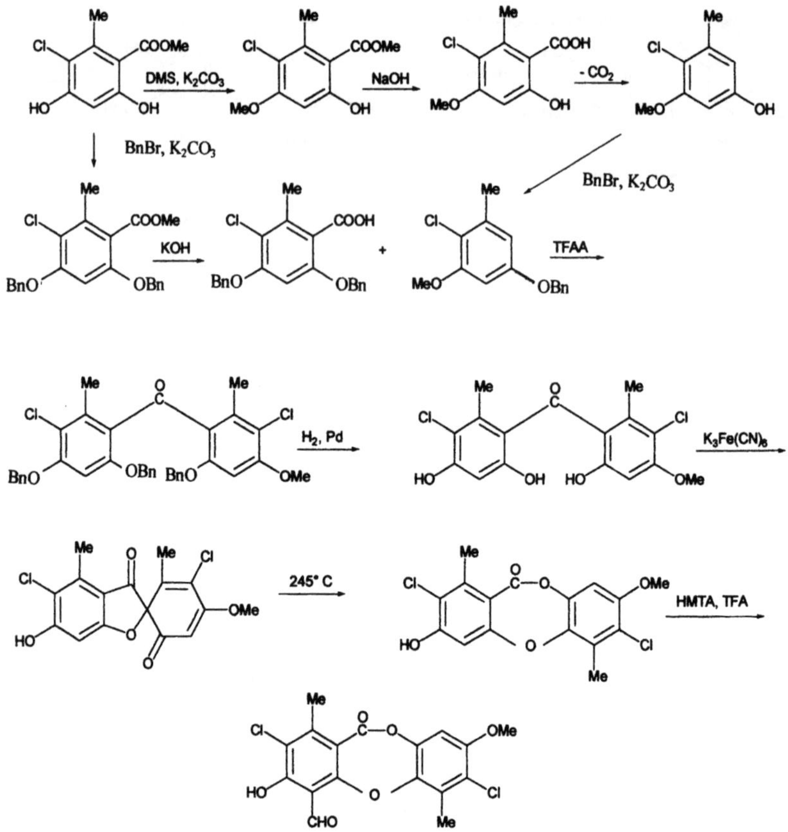

318 : Eriodermin

Scheme 117. Synthesis of eriodermin

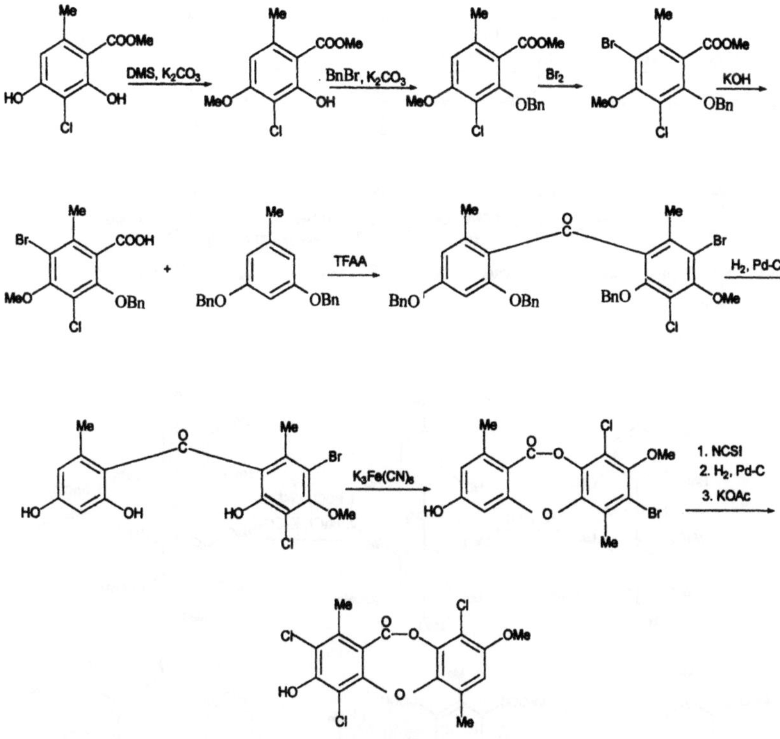

320: Fulgidin

Scheme 118. Synthesis of fulgidin

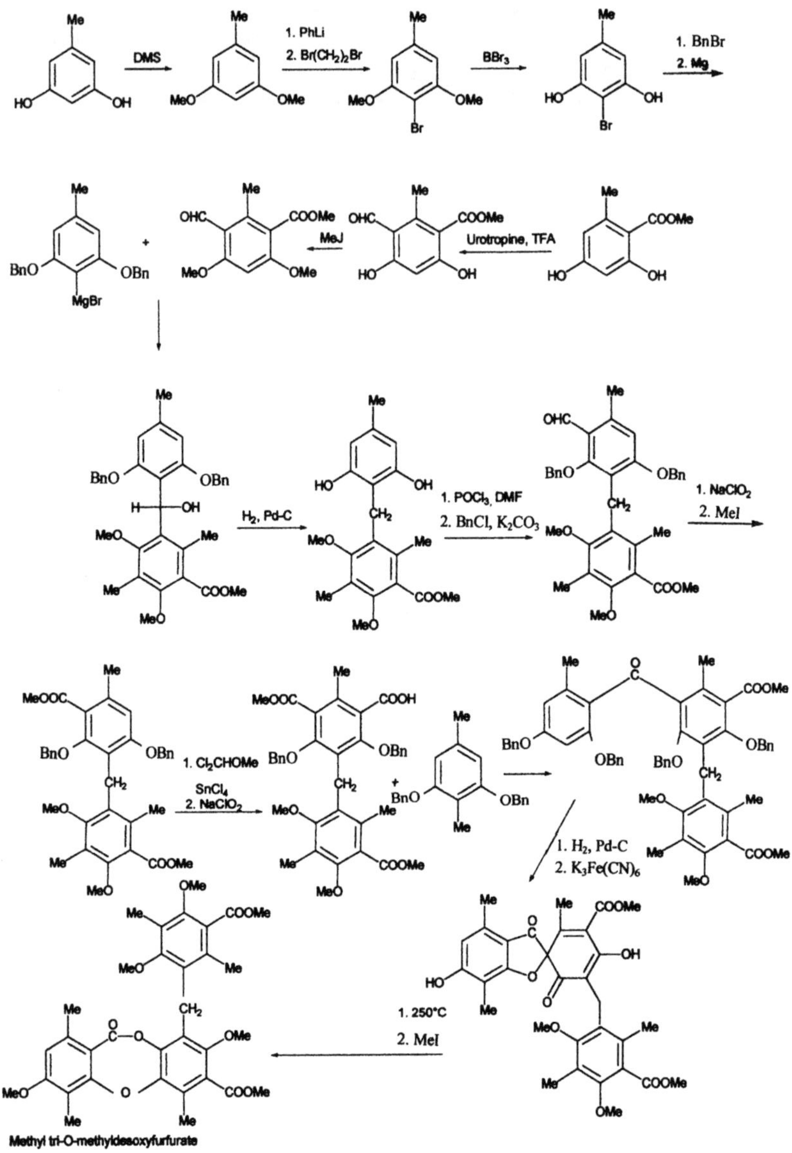

Scheme 119. Synthesis of methyl tri-*O*-methyldesoxyfurfurate

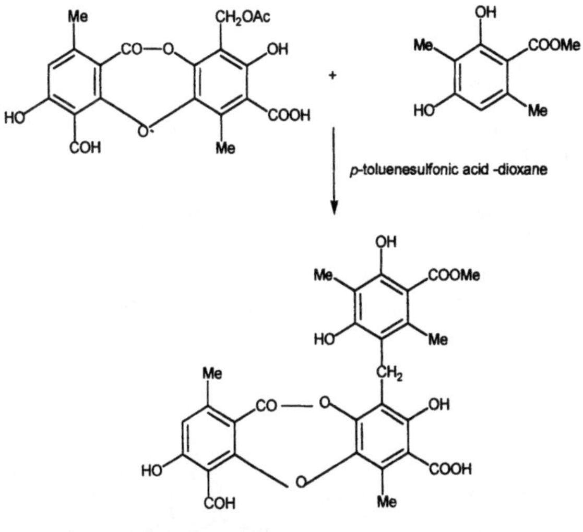

321 : Furfuric acid

Scheme 120. Synthesis of furfuric acid

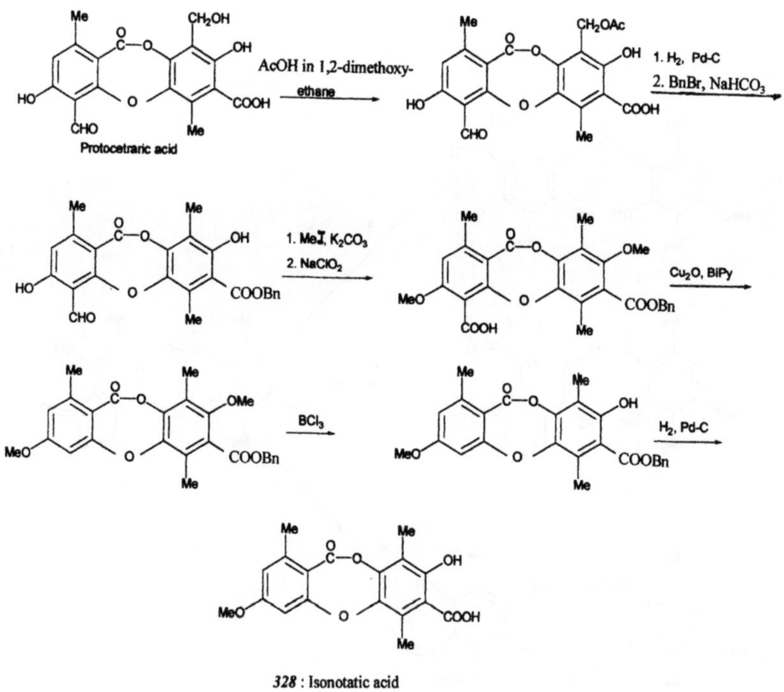

328 : Isonotatic acid

Scheme 121. Synthesis of isonotatic acid

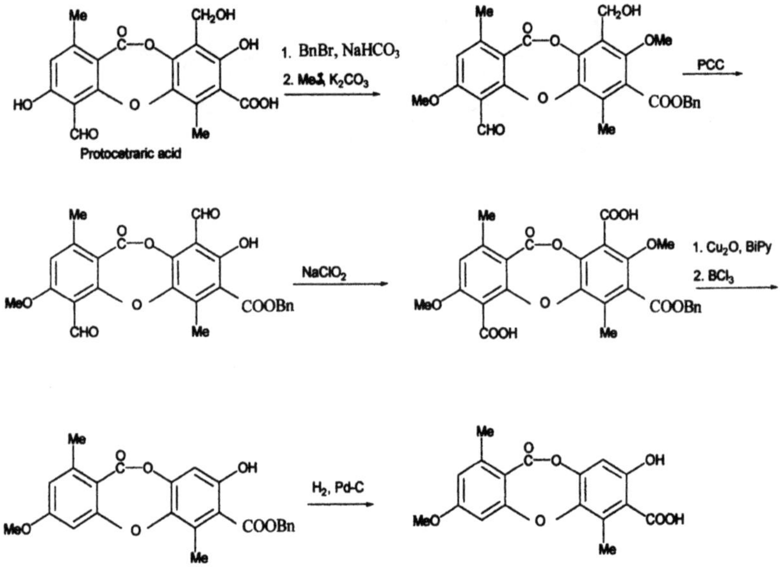

356 : Subnotatic acid

Scheme 122. Synthesis of subnotatic acid

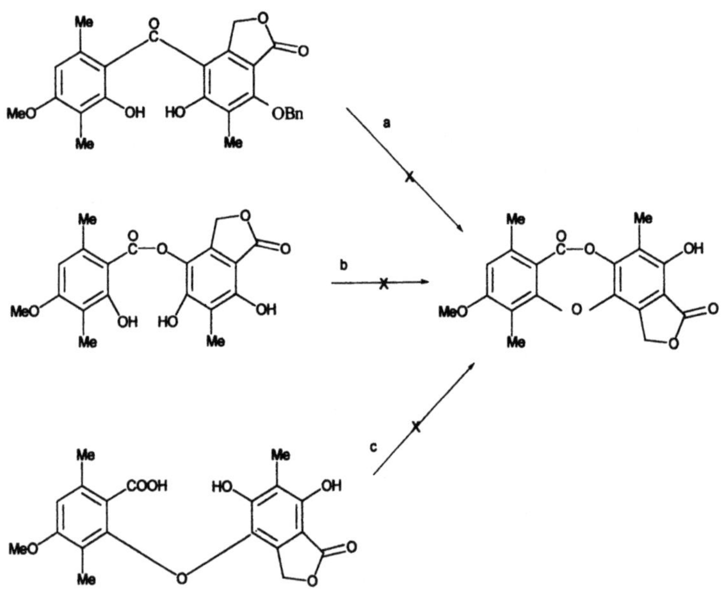

Scheme 123. Attempts to synthesize a depsidone with a lactole ring

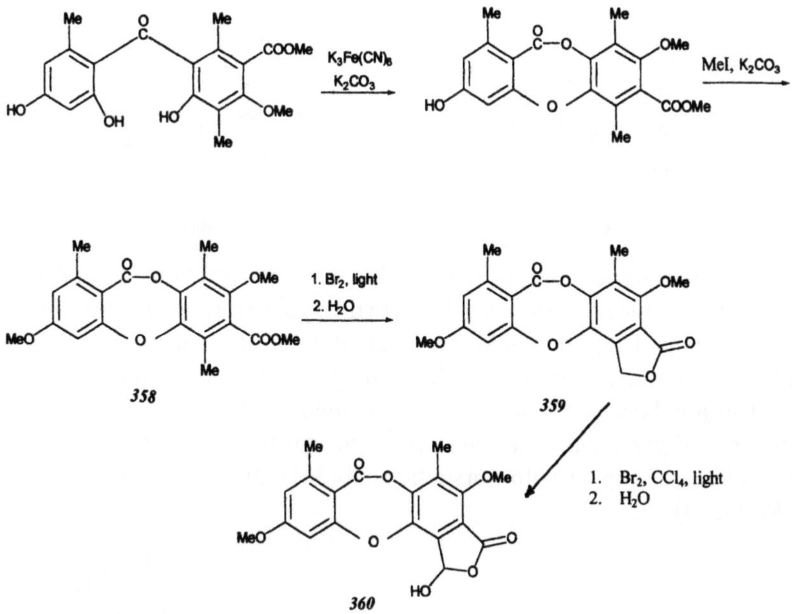

Scheme 124. Synthesis of the depsidones **359** and **360**

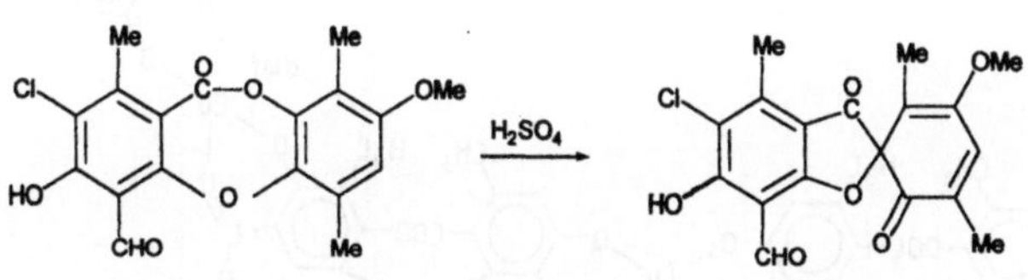

361 *362*

Scheme 125. Rearrangement of pannarin to isopannarin

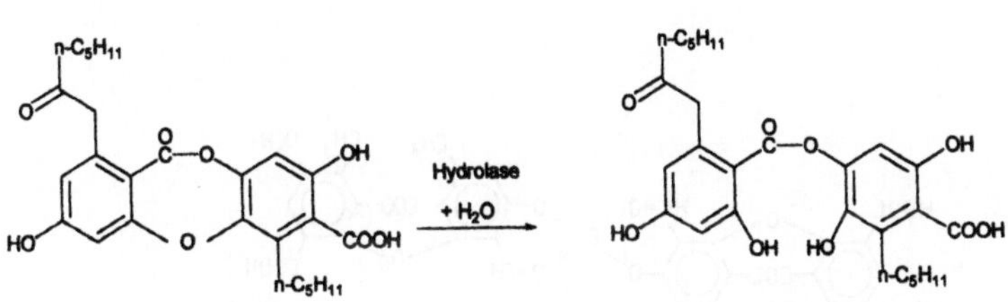

365 : Physodic acid *366* : 5′-Hydroxyolivetoric acid

Scheme 126. Enzymatic conversion of physodic acid to 5′-hydroxyolivetoric acid

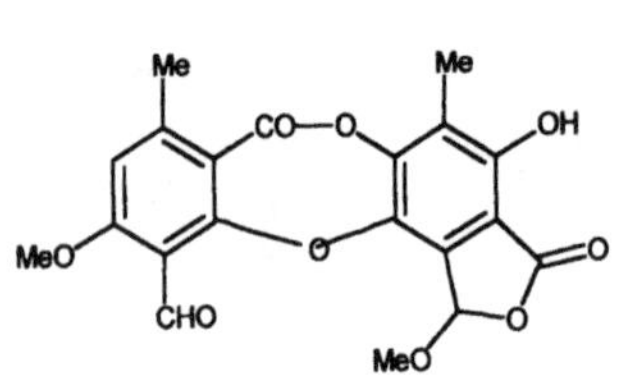

363

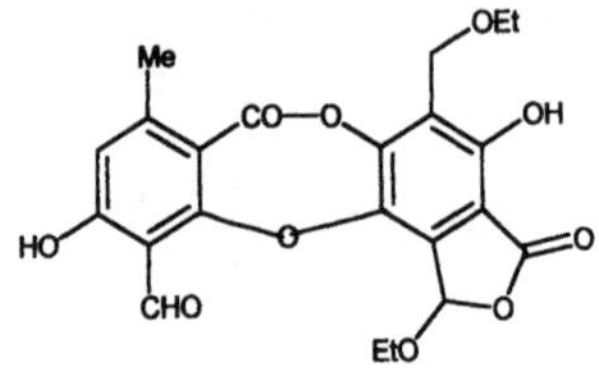

364

The depsidone physodic acid (**365**) can be converted by a hydrolase from *Pseudevernia furfuracea* to 5′-hydroxyolivetoric acid (**366**) (*315*) (see Scheme 126). It would be of great interest to look for further hydrolases able to cleave the ether linkage of other depsidones.

Certain lichens (*Acarospora smaragdula*, *Lecidea lactea*, *Lecidella bullata*, *Tephromela testaceoatra*) from cupriferous substrata form complexes of copper with norstictic acid (**367**) (*593*) and psoromic acid (**368**) (*594*).

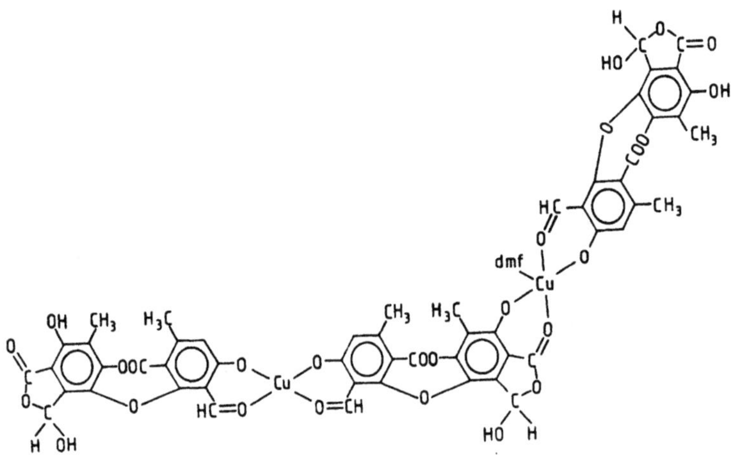

367 (dmf = dimethylformamide)

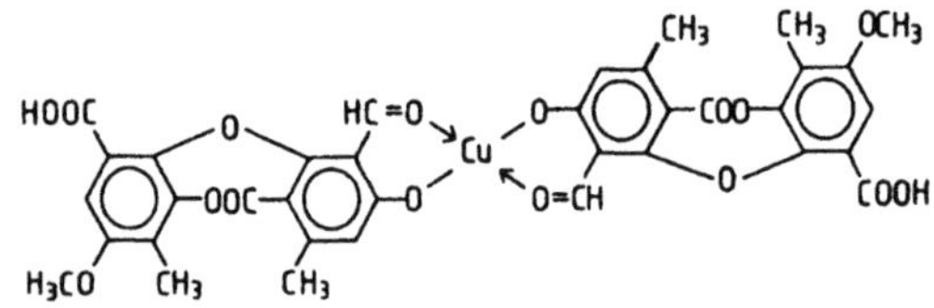

368

Aurantiacin A (**369**) and B (**370**) are depsidones from the fungus *Preussia aurantiaca* growing on a mangrove species (*577*). Three depsidones, emeguisin A (**371**), emeguisin B (**372**) and emeguisin C (**373**) were isolated from the mycelium of the fungus *Emericella unguis* (*450*).

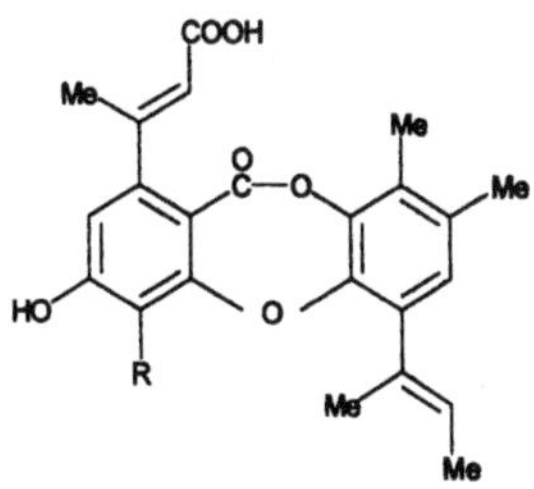

369 : Aurantiacin A , R = CH_2OH

370 : Aurantiacin B , R = COH

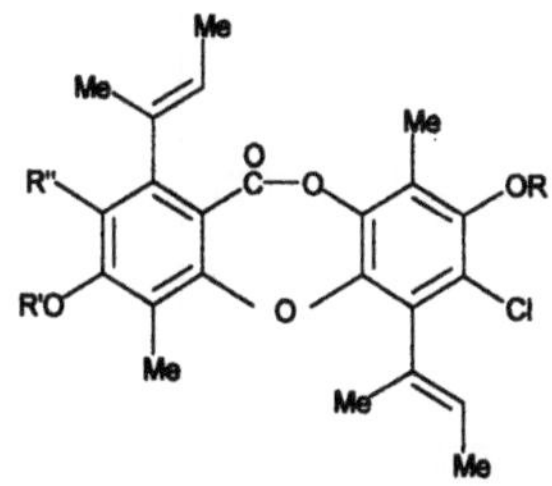

	R	R′	R′′
371 : Emeguisin A :	H	H	H
372 : Emeguisin B :	H	Me	H
373 : Emeguisin C :	Me	H	Cl

4.5.8. Depsones

New depsones are summarized in Table 6. They differ only in the length of the *n*-alkyl side chains. Their syntheses are shown in Schemes 127 and 128.

Table 6. Depsones from lichens

Compound	Formula No.	M.p. (°C)	Lichen	Synth. Scheme	Ref.
Hyperpicrolichenic acid	**374**	95–97	*Pertusaria truncata*	127	(*257*)
Isohyperpicrolichenic acid	**375**	107–107.5	*Pertusaria truncata*	127	(*257*)
Isomegapicrolichenic acid	**376**	145	*Pertusaria truncata*	127	(*257*)
Isosubpicrolichenic acid	**377**		*Pertusaria truncata*	127	(*257*)
Megapicrolichenic acid	**378**	116–117	*Pertusaria truncata*	127	(*257*)
Subpicrolichenic acid	**379**	164–165	*Pertusaria amara*	128	(*197*)
Superpicrolichenic acid	**380**	108–110	*Pertusaria truncata*	128	(*197*)

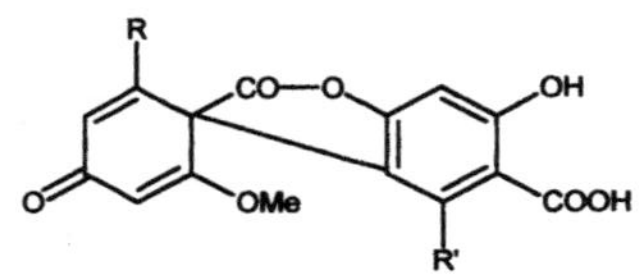

	R	R'
374 :	n-C_5H_{11}	n-C_7H_{15}
375 :	n-C_7H_{15}	n-C_5H_{11}
376 :	n-C_9H_{19}	n-C_7H_{15}
377 :	n-C_5H_{11}	n-C_3H_7
378 :	n-C_7H_{15}	n-C_9H_{19}
379 :	n-C_3H_7	n-C_5H_{11}
380 :	n-C_7H_{15}	n-C_7H_{15}

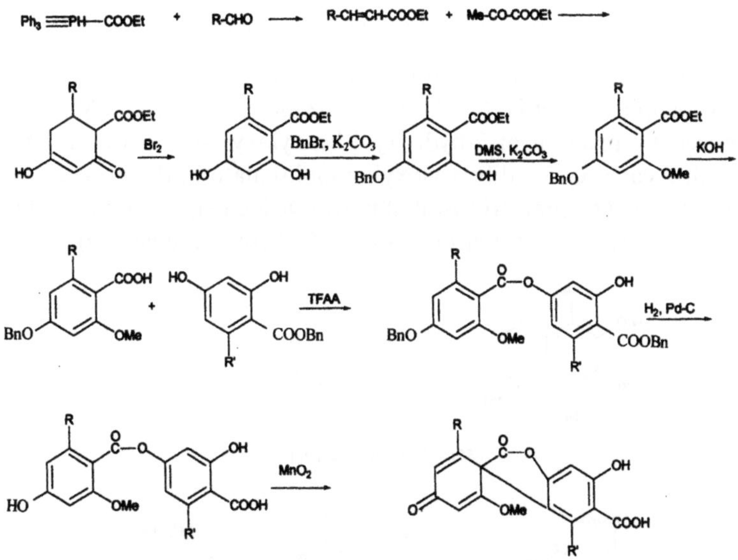

	R	R'
374 : Hyperpicrolichenic acid	n-C_5H_{11}	n-C_7H_{15}
375: Isohyperpicrolichenic acid	n-C_7H_{15}	n-C_5H_{11}
376 : Isomegapicrolichenic acid	n-C_9H_{19}	n-C_7H_{15}
377 : Isosubpicrolichenic acid	n-C_5H_{11}	n-C_3H_7
378 : Megapicrolichenic acid	n-C_7H_{15}	n-C_9H_{19}

Scheme 127. Synthesis of hyperpicrolichenic, isohyperpicrolichenic, isomegapicrolichenic, isosubpicrolichenic, and megapicrolichenic acids

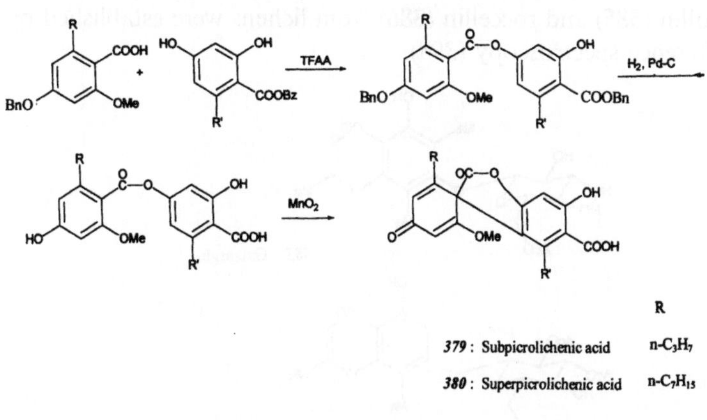

	R	R'
379 : Subpicrolichenic acid	n-C_3H_7	n-C_5H_{11}
380 : Superpicrolichenic acid	n-C_7H_{15}	n-C_7H_{15}

Scheme 128. Synthesis of subpicrolichenic and superpicrolichenic acids

4.5.9. Chromanes and Chromones

The only chromane known from lichens (*Erioderma phaeorhizum*) is acetyl-α-tocopherol (**381**) (*348*). The chromone siphulin (**382**) was synthesized by reaction of 2-hydroxy-4-benzyloxy-6-*n*-heptylacetophenone with 2-carboxy-3,5-dibenzyloxyhomophthalate in the presence of NaH to give tri-*O*-benzylprotosiphulin; this yielded siphulin after acid-induced dehydration and hydrogenolysis (*461*) (see Scheme 129).

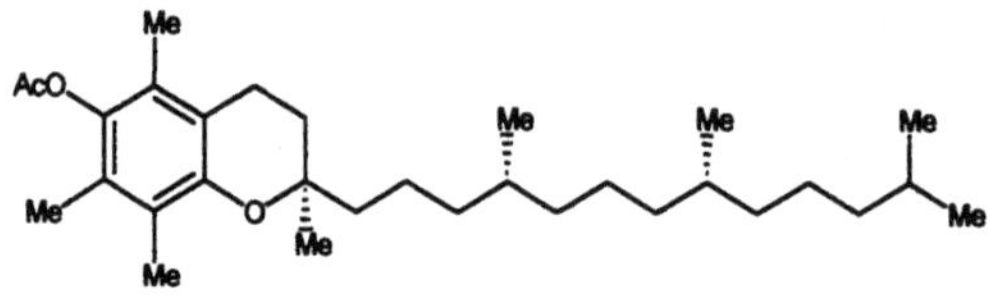

381: Acetyl-α-tocopherol

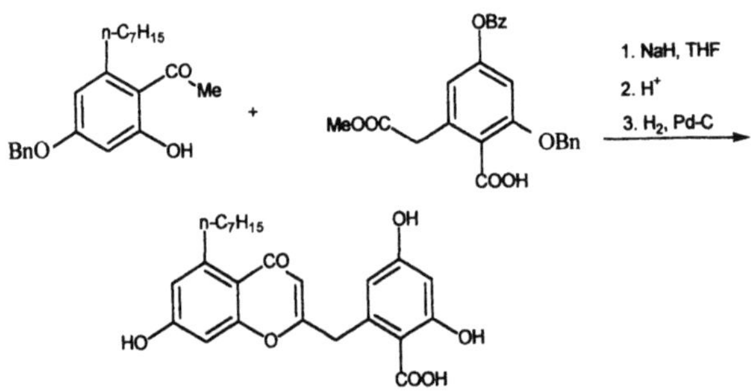

382: Siphulin

Scheme 129. Synthesis of siphulin

The final structures of the chromones galapagin (**383**), lobodirin (**384**), mollin (**385**) and roccellin (**386**) from lichens were established by NOE difference spectroscopy (*398*).

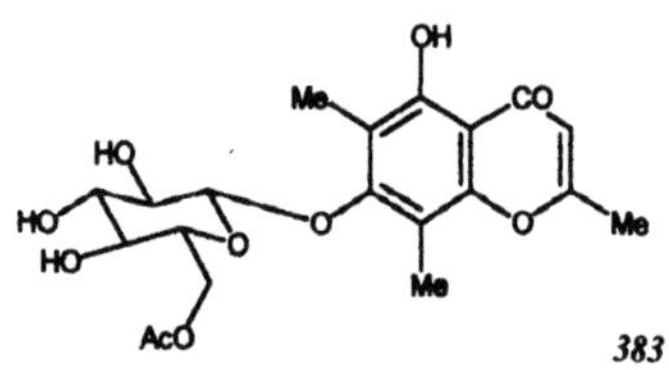

383: Galapagin

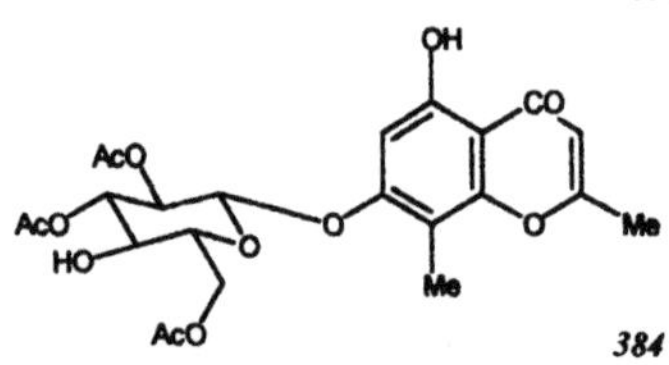

384: Lobodirin

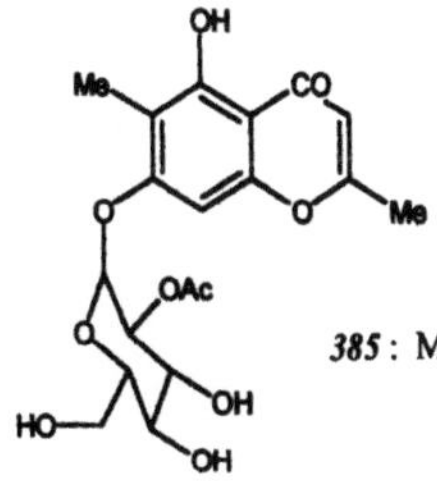

385 : Mollin

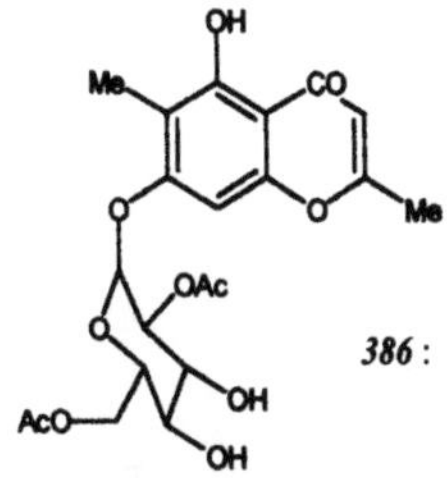

386 : Roccellin

4.5.10. Xanthones

Numerous xanthones have been isolated from lichens structurally elucidated and synthesized by Elix and co-workers and other authors since 1984; these are summarized in Table 7. Formulae **387**–**464** are listed separately.

2,2′,4,4′,6′-Pentahydroxy-6-methylbenzophenone (**465**) has been postulated as a precursor of xanthones. It was synthesized by Sundholm (*682*) and found to undergo facile dehydration to norlichexanthone (see Scheme 139). Base-catalyzed aldol-type cyclization of 1-(2,4,6-trihydroxyphenyl)octane-1,3,5,7-tetraone gave a benzophenone which underwent cyclodehydration to form norlichexanthone (*630*) (see Scheme 140).

Table 7. Xanthones from lichens

Compound	Formula No.	M.p. (°C)	Lichen	Synth. Scheme	Ref.
Aotearone: see 5,7-Dichloro-3-*O*-methylnorlichexanthone					
Arthothelin (2,4,5-Trichloro-norlichexanthone)	**387**	268–269	*Lecanora broccha*		
		283–285	*Lecidella asema*, *L. subalpicida*		(*198*)
			Dimelaena cf. *australiensis*, *Lecanora sulphurata*, *Lecidella meiococca*, *L. vorax*, *Micarea isabellina*, *Pertusaria pycnothelia*		(*201*)
			Byssoloma polychromum, *Sporopodium antonianum*, *S. leprieurii*		(*242*)
Asemone: see 4,5,7-Trichloro-3-*O*-methylnorlichexanthone					
Capistratone: see 2,5,7-Trichloro-3-*O*-methylnorlichexanthone					
2-Chlorolichexanthone	**388**	222–223	*Lecanora* spec., *Pertusaria cicatricosa*		(*201*)
4-Chlorolichexanthone	**389**	199–200	*Sporopodium citrinum*		(*242*)
5-Chlorolichexanthone	**390**	156	*Lecanora contractula*		(*201*)
				By methylation of 5-chloro-6-*O*-methylnorlichexanthone with DMS-K_2CO_3 and subsequent demethylation with BCl_3	(*202*)

7-Chlorolichexanthone	**391**	320, dec.	not yet found in lichens	By methylation of 7-chloro-6-*O*-methylnorlichexanthone with diazomethane	(*202*)
2-Chloro-3-*O*-methylnorliche-xanthone	**392**	325		By methylation of 2-chloro-6-*O*-benzylnorlichexanthone with DMS-K_2CO_3 and subsequent debenzylation with BCl_3	(*202*)
2-Chloro-6-*O*-methylnorliche-xanthone	**393**	280–283	*Lecanora salina*, *Pertusaria cicatricosa*		(*201*)
4-Chloro-3-*O*-methylnorliche-xanthone (Griseoxanthone B)	**394**	249–250	*Sporopodium citrinum*		(*242*)
				Synthesis	(*680*)
4-Chloro-6-*O*-methylnorliche-xanthone	**395**	232–235	*Sporopodium citrinum*		(*242*)
5-Chloro-3-*O*-methylnorliche-xanthone (Vinetorin)	**396**	yellowish needles, 256–256.5	*Lecanora vinetorum*		(*201*)
			Byssoloma polychromum, *B. subdiscordans*, *B. tricholomum*, *Sporopodium citrinum*		(*242*)
				130	(*681*)
5-Chloro-6-*O*-methylnorliche-xanthone	**397**		*Lecanora contractula*		(*201*)
7-Chloro-3-*O*-methylnorliche-xanthone	**398**	283–284			(*681*)
				Synthesis	(*680*)
7-Chloro-6-*O*-methylnorliche-xanthone	**399**	pale yellow crystals 310–312	*Lecanora populicola*, *L. alina*		(*201*)
				By condensation of methyl 3-chloro-6-hydroxy-4-methoxy-2-methylbenzonate and phloroglucinol	(*202*)

Table 7 (continued)

Compound	Formula No.	M.p. (°C)	Lichen	Synth. Scheme	Ref.
2-Chloronorlichexanthone	**400**	310–335, dec. 313–314.5	*Lecanora populicola*		(*201*, *202*)
				Synthesis	(*680*)
4-Chloronorlichexanthone		yellow needles, 296–298		Synthesis	(*680*)
5-Chloronorlichexanthone	**401**	yellow needles, 304–305	*Lecidella elaeochroma*		(*467*)
				Synthesis	(*680*)
7-Chloronorlichexanthone	**402**	285–286	*Lecanora populicola*		(*201*)
				Synthesis	(*680*)
Chlorovinetorin	**403**		*Lecanora dispersa*		(*497*)
2-Dechloro-8-*O*-methylthiomelin	**404**	224–225	*Rinodina thiomela*	131	(*215*)
4-Dechloro-8-*O*-methylthiomelin	**405**	35–38	*Rinodina thiomela*	132	(*215*)
2-Dechlorothiomelin	**406**	193–194, dec.	*Rinodina thiomela*	131	(*215*)
4-Dechlorothiomelin	**407**	224–227, 241–243	*Rinodina thiomela*	132	(*215*)
Demethylchodatin	**408**	yellow crystals, 284–286	*Lecanora pachysoma*	X-ray analysis of triacetyldemethylchodatin	(*254*)
2,4-Dichloro-1-hydroxy-7-methoxy-6,8-dimethylxanthone	**409**	pale yellow crystals, 157–158	*Rinodina thiomela*	133	(*215*)
2,4-Dichlorolichexanthone	**410**	184–186	*Dimelaena* cf. *australiensis*, *Pertusaria cicatricosa*		(*201*)
2,7-Dichlorolichexanthone	**411**	pale yellow crystals, 290–293	*Lecanora behringii*, *L. populicola*, *L. salina*		(*201*)
4,5-Dichlorolichexanthone	**412**	yellow crystals 283–283.5,	*Dimelaena* cf. *australiensis*, *Pertusaria cicatricosa*		(*201*)
		285–286.5	*Sporopodium citrinum*		(*242*)

4,7-Dichlorolichexanthone	**413**	274	not yet found in lichens	By brief methylation of 4,7-dichloro-3-*O*-methylnorlichexanthone with CH_2N_2	(*202*)
5,7-Dichlorolichexanthone	**414**	pale yellow crystals 202–205	*Sporopodium vezdeanum*		(*203*)
2,4-Dichloro-3-*O*-methyl-norlichexanthone	**415**		not yet found in lichens	By methylation of 6-benzyloxy-2,4-dichloronorlichexanthone with DMS-K_2CO_3 and subsequent debenzylation with BCl_3	(*202*)
2,5-Dichloro-3-*O*-methyl-norlichexanthone	**416**	276–268, 296–297			(*445*)
2,5-Dichloro-6-*O*-methyl-norlichexanthone	**417**	yellow crystals, 255-257	*Dimelaena* cf. *australiensis* *Lecanora contractula* *Pertusaria cicatricosa*		(*201*)
				By condensation of 2-benzyloxy-3-chloro-4-methoxy-6-methylbenzoic acid and 1,3,5-tribenzyloxy-2-chlorobenzene with TFAA and subsequent debenzylation with BCl_3	(*196*)
2,7-Dichloro-3-*O*-methyl-norlichexanthone	**418**	pale yellow crystals, 262–264, 297–298	*Lecanora behringii*, *L. salina*	By methylation of 6-benzyloxy-2,7-dichlorolichexanthone with DMS-K_2CO_3 and subsequent debenzylation with BCl_3	(*202*)
2,7-Dichloro-6-*O*-methyl-norlichexanthone	**419**	pale yellow needles, 309–311	*Lecanora behringii*		(*201*)
				By condensation of ethyl 3-chloro-6-hydroxy-4-methoxy-2-methylbenzoate and 2-chlorophloroglucinol	(*202*)

Table 7 (continued)

Compound	Formula No.	M.p. (°C)	Lichen	Synth. Scheme	Ref.
4,5-Dichloro-3-*O*-methyl-norlichexanthone	**420**	yellow-needles, 255–257	*Sporopodium citrinum*		(*242*)
4,5-Dichloro-6-*O*-methyl-norlichexanthone	**421**	280–282	*Dimelaena citrinum*		(*201*)
			Sporopodium citrinum		(*242*)
4,7-Dichloro-6-*O*-methyl-norlichexanthone	**422**	yellow prisms, 286–288	not yet found in lichens	By condensation of 3-chloro-6-hydroxy-4-methoxy-2-methyl-benzoate and 2-chlorophloroglucinol	(*202*)
5,7-Dichloro-3-*O*-methyl-norlichexanthone	**423**	pale yellow crystals, 248–250		134	(*231*)
			Lecanora broccha, L. vinetorum, Lecidella meiococca, L. vorax, Byssoloma subdiscordans, B. tricholomum, Sporopodium, phyllocharis		(*201*)
5,7-Dichloro-6-*O*-methylnor-lichexanthone	**424**	yellow rods, 275–278	*Byssoloma subdiscordans*		(*242*)
2,4-Dichloronorlichexanthone	**425**	yellow prisms, 285–287	*Lecidella vorax*		(*201*)
2,5-Dichloronorlichexanthone	**426**	yellow needles, 245–247	*Buellia* spec.,		(*198*)
			Lecanora broccha, Lecidella meiococca, L. vorax		(*201*)
2,7-Dichloronorlichexanthone	**427**	yellow crystals 298–299	*Lecanora broccha*		(*198*)

			Lecanora behringii, *L. populicola*, *L. salina*, *Lecidella meiococca*		(*201*)
				Synthesis	(*680*)
4,5-Dichloronorlichexanthone	**428**	yellow prisms 291–293	*Lecanora broccha*, *Lecidella asema*, *L. subalpicida*,		(*198*)
			L. vorax, *Micarea isabellina*, *Pertusaria pycnothelia*		(*201*)
4,7-Dichloronorlichexanthone	**429**	yellow crystals	*Lecidella asema* *L. meiococca*		(*198*)
		290–293,			(*201*)
		297–298		Synthesis	(*680*)
5,7-Dichloronorlichexanthone	**430**	yellow crystals, 285–290	*Lecanora broccha*, *Lecidella vorax*		(*201*)
1,7-Dihydroxy-2,4-dichloro-6,8-dimethylxanthone	**431**	pale yellow crystals 188–190	*Rinodina thiomela*	133	(*215*)
1,8-Dihydroxy-3,6-dimethoxy-xanthone	**432**	pale yellow crystals 193–194	*Diploschistes* spec.	135	(*217*)
1,5-Dimethoxy-4-chloro-6-methyl-8-hydroxyxanthone	**433**	yellow crystals 222–224	*Rinodina thiomela*	131	(*215*)
3,6-Di-*O*-methylthiophanic acid	**434**	yellow crystals, >315, dec.		By permethylation of 6-*O*-methlythio-phanic acid with DMS-K_2CO_3 and subsequent demethylation with BCl_3	(*202*)
Ergochrome **BB** (Secalonic Acid B)	**435**	254–256, dec. $[\alpha]_D + 156$ ($CHCL_3$)	*Diploicia canescens* ssp. *australasica*		(*229*)
Granulosin: see 6-O-Methyl-arthothelin					
Isoarthothelin: (2,5,7-Tri-chloronorlichexanthone)	**436**	yellow needles 253–255	*Lecanora broccha*, *Buellia* spec.	136	(*232*)

Table 7 (continued)

Compound	Formula No.	M.p. (°C)	Lichen	Synth. Scheme	Ref.
Lichexanthone	**437**	yellowish needles 189–190	*Lecidella patavina*, *L. stigmatea*		(*464*)
1-Methoxy-4-chloro-5,8-dihydroxy-6-methylxanthone	**438**	yellow crystals 194–196	*Rinodina thiomela*	131	(*215*)
6-*O*-Methylarthothelin	**439**	yellow needles 282–284	*Dimelaena* cf. *australiensis*, *Micarea isabellina*, *Pertusaria pycnothelia*		(*201*)
			Buellia granulosa, *Lecidella elaeochromoides*		(*464*)
				By methylation of arthothelin with DMS-K_2CO_3	(*195*)
3-*O*-Methylnorlichexanthone (Griseoxanthone C)	**440**	253–255	*Lecanora vinetorum*		(*201*)
8-*O*-Methylthiomelin	**441**	yellow threads, 219	*Rinodina thiomela*		(*215*, *218*)
3-*O*-Methylthiophanic acid	**442**	yellow needles 290–222	*Lecidella meiococca*	By chlorination of 3-*O*-methylnorlichexanthone with SO_2Cl_2	(*202*)
6-*O*-Methylthiophanic acid	**443**	yellow needles 250–252	*Micarea isabellina*		(*201*)
Northiomelin (1,5,8-Trihydroxy-2,4-dichloro-6-methylxanthone)	**444**	yellow crystals 212–213	*Rinodina thiomela*	By demethylation of thiomelin with BBr_3	(*215*)
Thiomelin	**445**	yellow threads 185	*Rinodina lepida*, *R. thiomela*	137	(*252*)
Thiophanic acid	**446**	yellow needles 242–243	*Dimelaena* spec., *Lecanora sulphurata*, *Lecidella meiocarpa*, *L. vorax*, *Micarea isabellina*, *Pertusaria*, *pycnothelia*,		(*201*)

			Byssoloma polychromum, *Sporopodium antonianum*, *S. leprieurii*		(*242*)
			Lecidella asema, *L. elaeochroma*, *L. flavisorediata*, *L. leuckertiana*, *L. meiococca*		(*467*)
Thiophaninic acid	**447**	yellow prisms 278–279	*Dimelaena* spec.		(*196*)
			D. australiensis		(*201*)
Thuringione	**448**	yellow needles 278–279	*Byssoloma polychromum*		(*242*)
2,4,5-Trichlorolichexanthone	**449**	yellowish needles 225–227	*Dimelaena* cf. *australiensis* *Pertusaria cicatricose*		(*201*)
2,4,7-Trichlorolichexanthone	**450**	pale yellow crystals 254–256, dec.		By permethylation of 2,4,7-trichloro-3-*O*-methylnorlichexanthone with DMS-K_2CO_3 and subsequent demethylation with BCl_3	(*202*)
2,5,7-Trichlorolichexanthone	**451**	pale yellow crystals 245	*Dimelaena* cf. *australiensis*, *Lecanora broccha*		(*201*)
4,5,7-Trichlorolichexanthone	**452**	pale yellow crystals 205–210	not yet found in lichens	By methylation of 4,5,7-trichloro-6-*O*-methylnorlichexanthone with CH_2N_2	(*202*)
2,4,7-Trichloro-3-*O*-methylnorlichexanthone	**453**	yellow crystals 257–259, dec.	not yet found in lichens		(*445*)
2,4,7-Trichloro-6-*O*-methylnorlichexanthone	**454**	yellow crystals >310, dec.	not yet found in lichens	By chlorination of 7-chloro-6-*O*-methylnorlichexanthone with SO_2Cl_2	(*202*)
2,5,7-Trichloro-3-*O*-methylnorlichexanthone	**455**	pale yellow crystals 259–261	*Lecanora broccha*, *Lecidea asema*, *L. subalpicida*,		(*198*)
			L. behringii, *Sporopodium vezdeanum*,		(*201*)

Table 7 (continued)

Compound	Formula No.	M.p. (°C)	Lichen	Synth. Scheme	Ref.
			S. xantholeucum, *Byssoloma polychromum*, *B. subdiscordans*, *B. tricholomum*, *Sporopodium phyllocharis*		(*203*) (*242*)
2,5,7-Trichloro-6-*O*-methylnorlichexanthone	**456**	yellow needles 249–253		By condensation of ethyl 3-chloro-6-hydroxy-4-methoxy-2-methylbenzoate and 2-chlorophloroglucinol	(*202*)
4,5,7-Trichloro-3-*O*-methylnorlichexanthone (3-*O*-Methylasemone)	**457**	yellow needles 285–287, 294–295	*Lecanora broccha*, *Lecidella meiocarpa*,	Synthesis	(*201*) (*680*)
			Byssoloma subdiscordans, *B. tricholomum*		(*231*, *242*)
4,5,7-Trichloro-6-*O*-methylnorlichexanthone (6-*O*-Methylasemone)	**458**	yellow needles 298–300, dec.	*Pertusaria pycnothelia*		(*201*)
				138	(*233*)
2,4,7-Trichloronorlichexanthone	**459**	yellow crystals 228–229	*Lecanora sulphurata*		(*201*)
4,5,7-Trichloronorlichexanthone	**460**	yellow needles 308–309, dec.	*Micarea austroternaria* var. *isabellina*		(*233*)
1,5,8-Trihydroxy-4-chloro-6-methylxanthone	**461**	yellow crystals 212–214	*Rinodina thiomela*	131	(*215*)
1,5,8-Trimethoxy-2-chloro-6-methylxanthone	**462**	pale yellow gum	*Rinodina thiomela*		(*215*)
1,5,8-Trimethoxy-4-chloro-6-methylxanthone	**463**	yellow crystals 197–199	*Rinodina thiomela*	131	(*215*)
1,3,6-Tri-*O*-methylarthothelin	**464**	yellow needles 200–202		By methylation of arthothelin with DMS-K_2CO_3	(*195*)

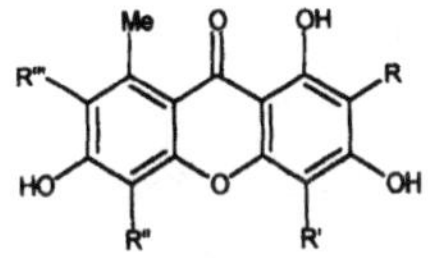

	R	R'	R''	R'''
387:	Cl	Cl	Cl	H
400:	Cl	H	H	H
401:	H	H	Cl	H
402:	H	H	H	Cl
408:	Cl	OMe	Cl	Cl
425:	Cl	Cl	H	H
426:	Cl	H	Cl	H
427:	Cl	H	H	Cl
428:	H	Cl	Cl	H
429:	H	Cl	H	Cl
430:	H	H	Cl	Cl
436:	Cl	H	Cl	Cl
446:	Cl	Cl	Cl	Cl
459:	Cl	Cl	H	Cl
460:	H	Cl	Cl	Cl

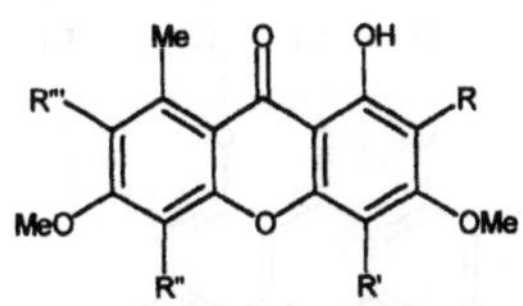

	R	R'	R''	R'''
388:	Cl	H	H	H
389:	H	Cl	H	H
390:	H	H	Cl	H
391:	H	H	H	Cl
410:	Cl	Cl	H	H
411:	Cl	H	H	Cl
412:	H	Cl	Cl	H
413:	H	Cl	H	Cl
414:	H	H	Cl	Cl
434:	Cl	Cl	Cl	Cl
437:	H	H	H	H
449:	Cl	Cl	Cl	H
450:	Cl	Cl	H	Cl
451:	Cl	H	Cl	Cl
452:	H	Cl	Cl	Cl

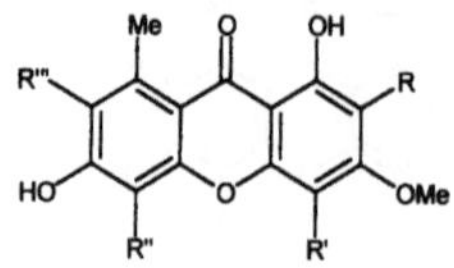

	R	R'	R''	R'''
392:	Cl	H	H	H
394:	H	Cl	H	H
396:	H	H	Cl	H
398:	H	H	H	Cl
403:	R'' = Cl, R = R' = R''' = H, H, Cl			
415:	Cl	Cl	H	H
416:	Cl	H	Cl	H
418:	Cl	H	H	Cl
420:	H	Cl	Cl	H
423:	H	H	Cl	Cl
440:	H	H	H	H
442:	Cl	Cl	Cl	Cl
448:	Cl	Cl	Cl	H
453:	Cl	Cl	H	Cl
455:	Cl	H	Cl	Cl
457:	H	Cl	Cl	Cl

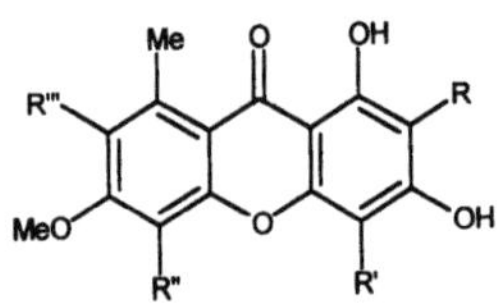

	R	R'	R''	R'''
393:	Cl	H	H	H
395:	H	Cl	H	H
397:	H	H	Cl	H
399:	H	H	H	Cl
417:	Cl	H	Cl	H
419:	Cl	H	H	Cl
421:	H	Cl	Cl	H
422:	H	Cl	H	Cl
424:	H	H	Cl	Cl
439:	Cl	Cl	Cl	H
443:	Cl	Cl	Cl	Cl
447:	Cl	Cl	H	H
454:	Cl	Cl	H	Cl
456:	Cl	H	Cl	Cl
458:	H	Cl	Cl	Cl

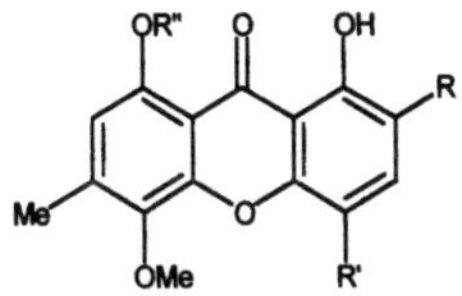

	R	R′	R″
404 :	H	Cl	Me
405 :	Cl	H	Me
406 :	H	Cl	H
407 :	Cl	H	H
441 :	Cl	Cl	Me
445 :	Cl	Cl	H

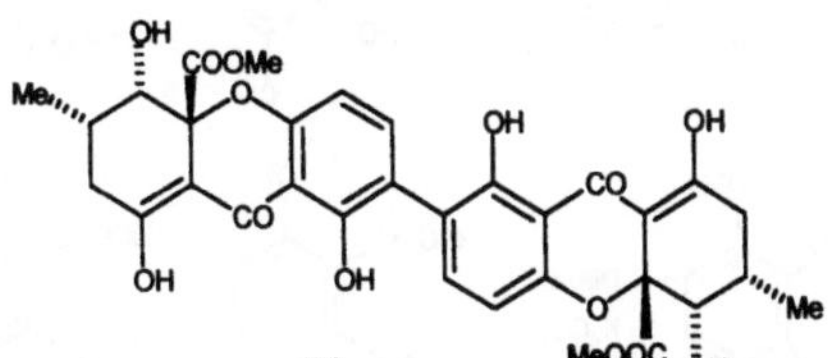

435

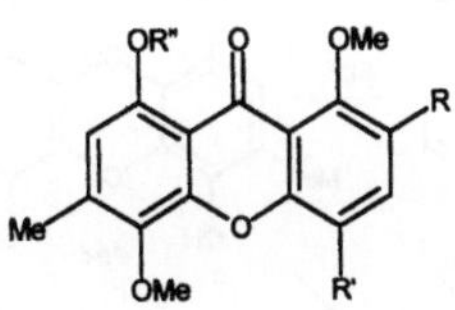

	R	R′	R″
433 :	H	Cl	H
462 :	Cl	H	Me
463 :	H	Cl	Me

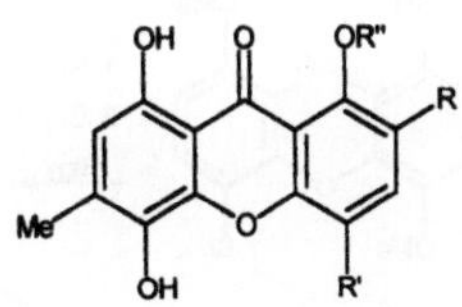

	R	R′	R″
438 :	H	Cl	Me
444 :	Cl	Cl	H
461 :	H	Cl	H

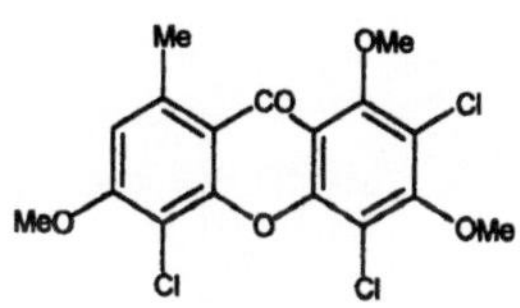

464

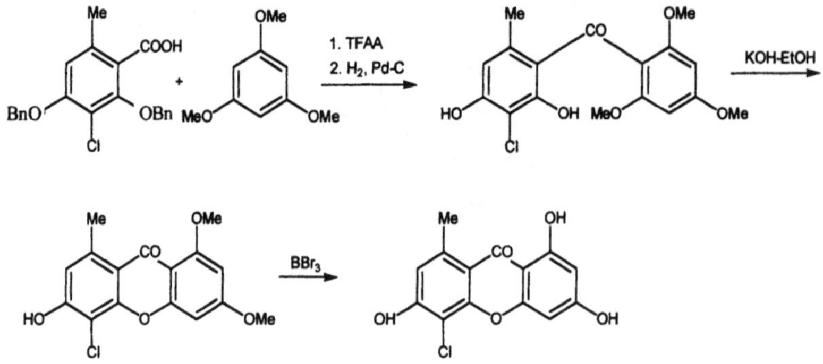

396 : Vinetorin

Scheme 130. Synthesis of vinetorin

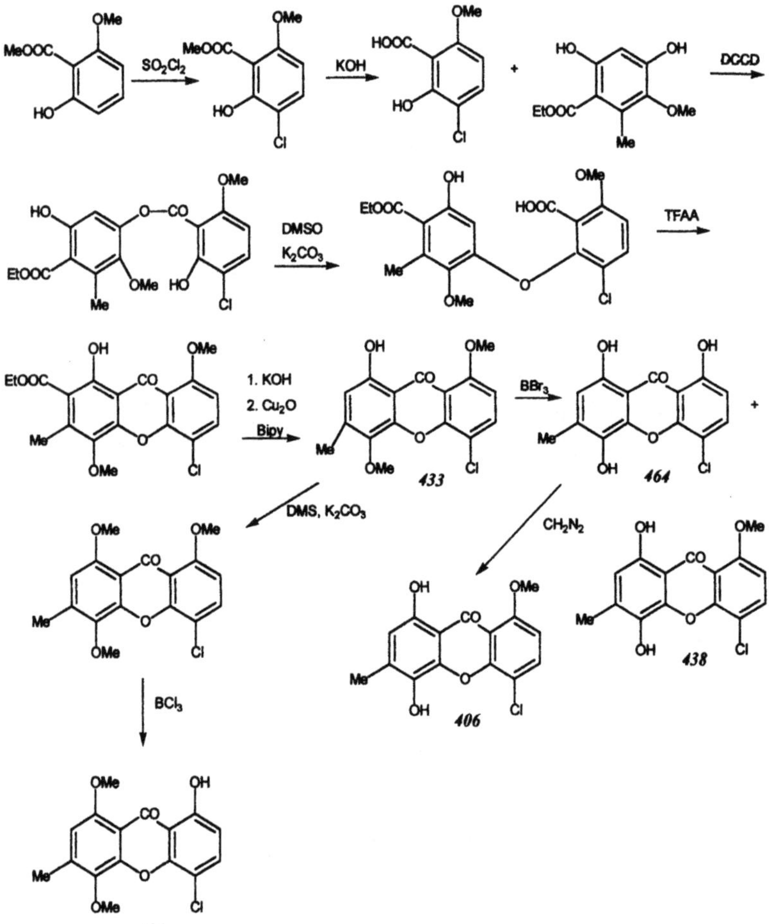

Scheme 131. Synthesis of 2-dechloro-8-*O*-methylthiomelin and related compounds

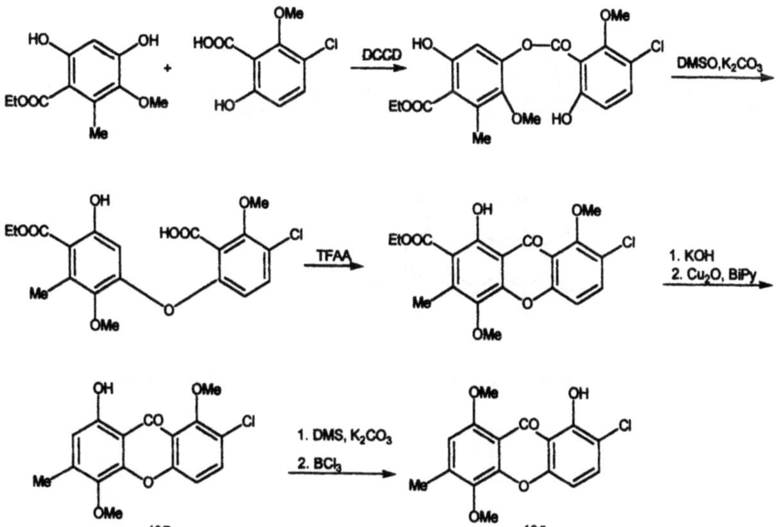

407 405

Scheme 132. Synthesis of 4-dechloro-8-*O*-methylthiomelin and 4-dechlorothiomelin

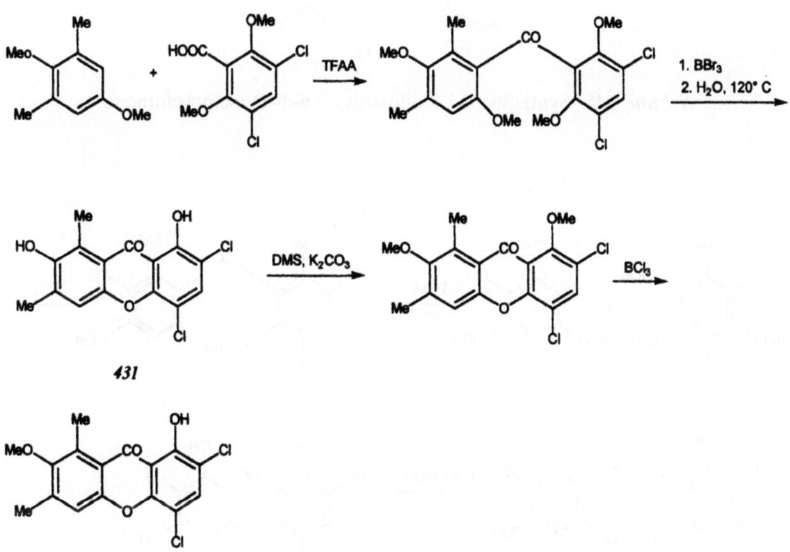

431

409

Scheme 133. Synthesis of 2,4-dichloro-1-hydroxy-7-methoxy-6,8-dimethylxanthone

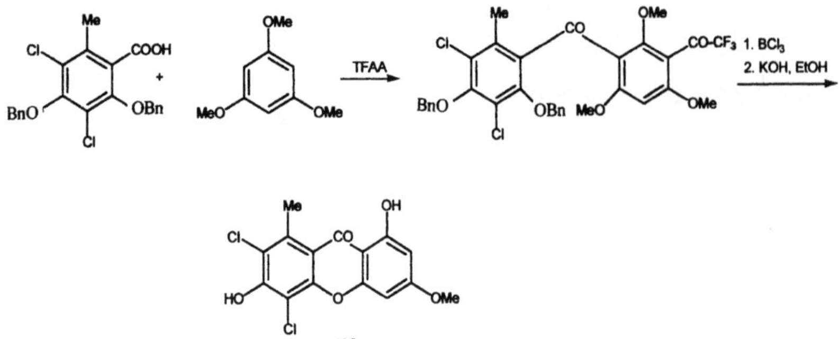

423

Scheme 134. Synthesis of 5,7-dichloro-3-*O*-methylnorlichexanthone

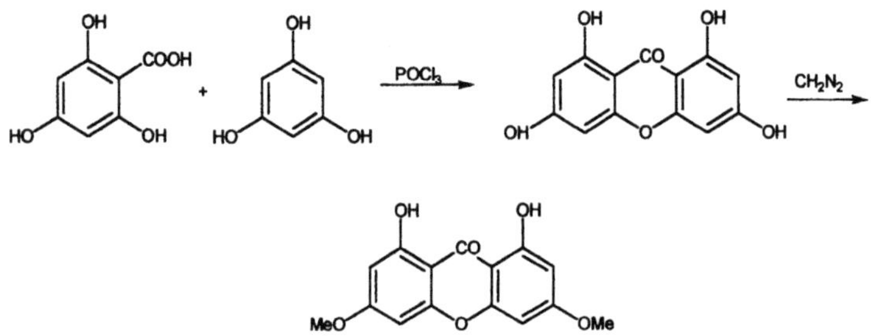

432

Scheme 135. Synthesis of 1,8-dihydroxy-3,6-di-*O*-methylxanthone

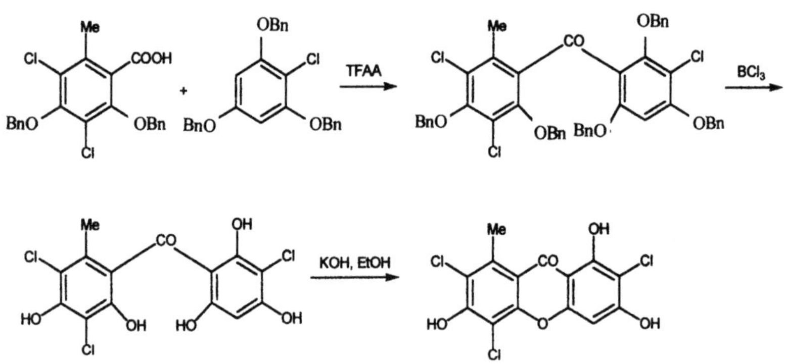

Scheme 136. Synthesis of isoarthothelin

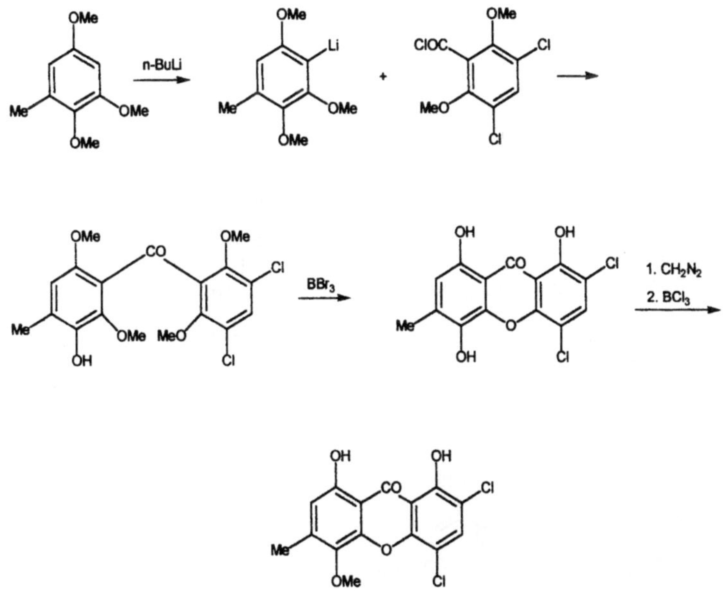

445

Scheme 137. Synthesis of thiomelin

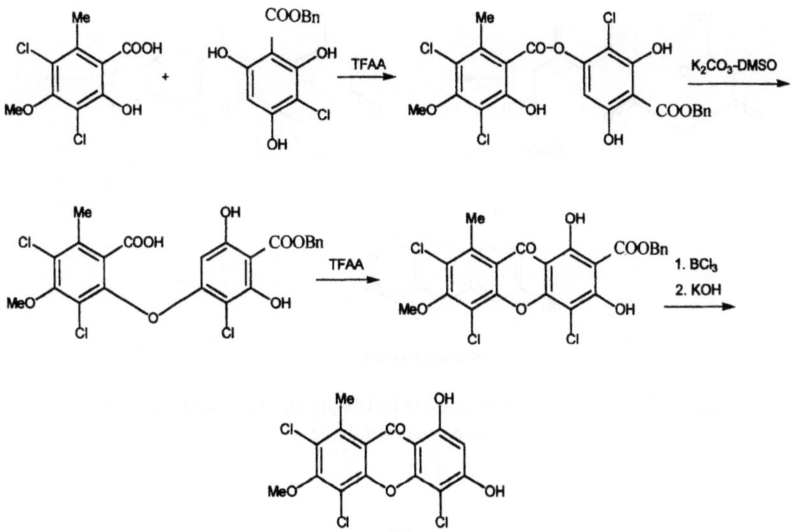

458

Scheme 138. Synthesis of 4,5,7-trichloro-6-*O*-methylnorlichexanthone

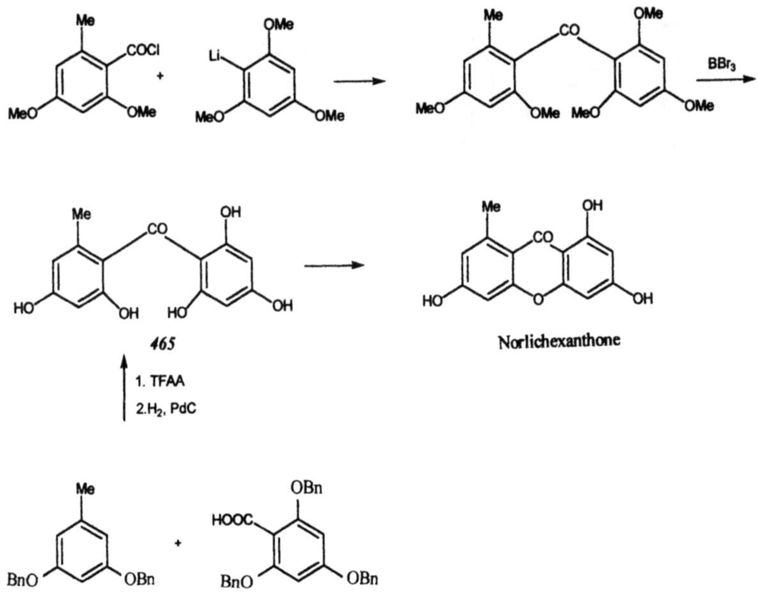

Scheme 139. Dehydration of 2,2′,4,4′,6′-pentahydroxybenzophenone to norlichexanthone

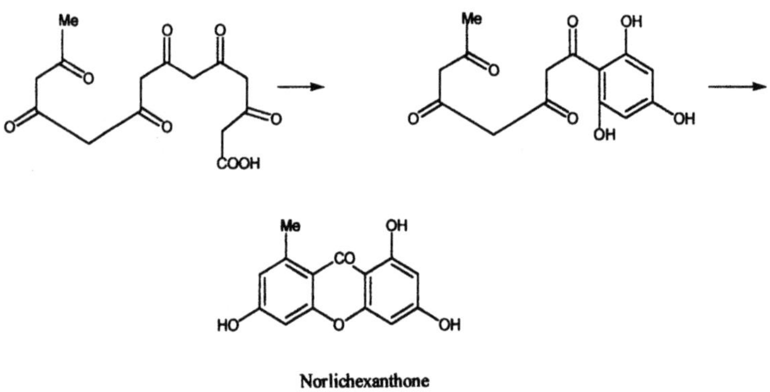

Scheme 140. Cyclization of 1-(2,4,7-trihydroxyphenyl)-octane-1,3,5,7-tetraone to norlichexanthone

4.5.11. Benzofurans

1,3-Dihydroxy-6-(1′-hexanoyl)-benzofuran (**466**) is the only known benzofuran from a lichen (*Letharia vulpina*) (*685*).

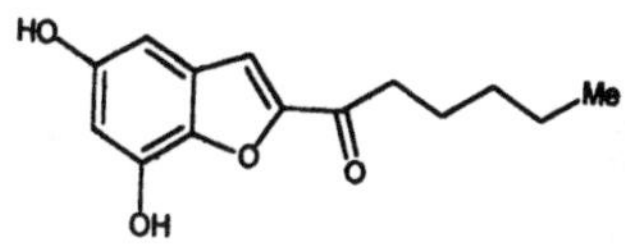

466

4.5.12. Dibenzofurans

Structure, occurrence and synthesis of new dibenzofurans are summarized in Table 8.

Shibata and Iitaka (*647*) confirmed the structure of didymic acid by an X-ray analysis of its oxidation product **491** (see Scheme 151).

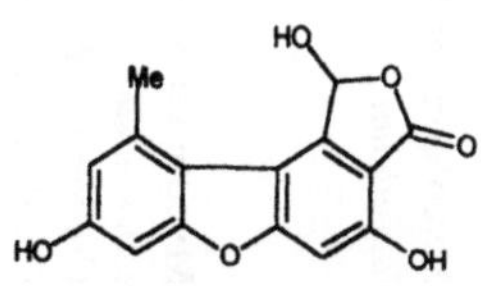

467

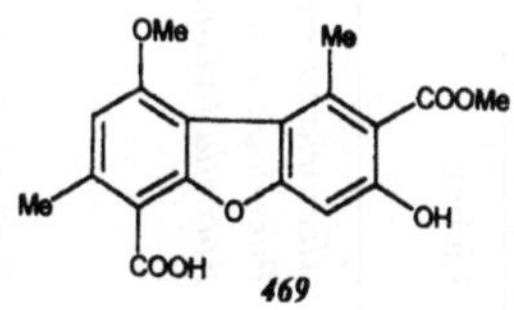

469

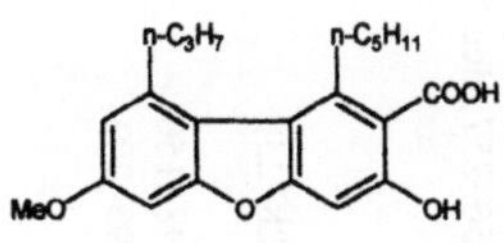

470

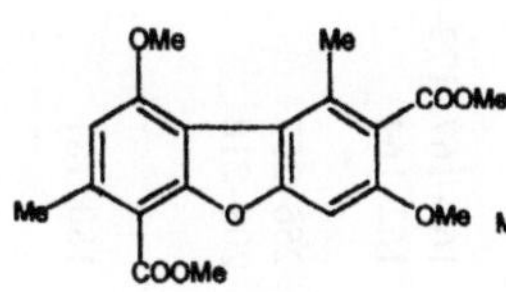

471

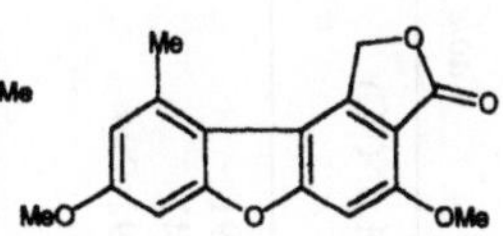

472

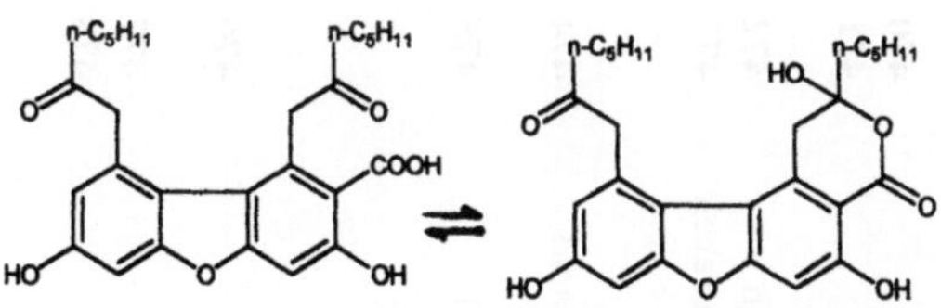

473

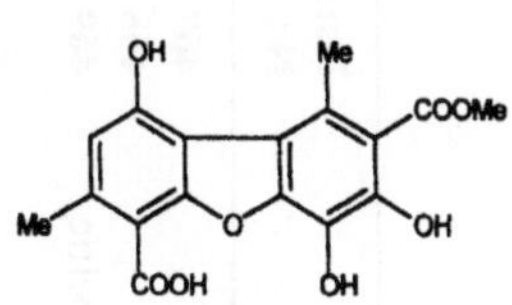

474

Table 8. Dibenzofurans from lichens

Compound	Formula No.	M.p. (°C)	Lichen	Synth. Scheme	Ref.
Alectosarmentin	**467**	291	*Alectoria sarmentosa*		(*327*)
Ascomatic acid	**468**	242	*Bunodophoron patagonicum*	141	(*260*)
3-*O*-Demethylschizopeltic acid	**469**	292–294	*Roccella hypomecha*	Synthesis by demethylation of schizopeltic acid with BCl_3	(*392*)
Didymic acid	**470**	165–167, 172–173		142	(*61*)
Dimethyl di-*O*-methylpannarate	**471**	162–164	*Roccellina luteola*		(*300*)
Di-*O*-methylstrepsilin	**472**	256–257		143	(*325*)
Haemophaein (Furfuraceic acid)	**473**	208–210	*Phyllopsora haemophaea*		(*369*)
			Phyllopsora furfuracea		(*259*)
4-Hydroxypannaric acid 9-methyl ester	**474**	180–181	*Leproloma diffusum*	144	(*249*)
Hypostrepsilalic acid	**475**	yellow crystals 256–257, dec.	Mycobiont of *Stereocaulon japonicum*	Methyl di-*O*-methylhypostrepsilalate gave on Clemmensen reduction methyl di-*O*-methylhypostrepsilate	(*550*)
Hypostrepsilic acid	**476**	228–230	Mycobiont of *Evernia esorediosa*	X-ray analysis of methyl di-*O*-methylhypostrepsilate	(*468, 548*)
Isodidymic acid	**477**	148–150	*Cladonia didyma*	145	(*69*)
Isoschizopeltic acid	**478**	199–201	*Combea californica*		(*247*)

Isostrepsilic acid	**479**	210–220	Mycobiont of *Usnea orientalis*		(*468*)
Melacarpic acid	**480**	148–149	*Neophyllis melacarpa*	146	(*60*)
Methyl ascomatate	**481**	154	*Bunodophoron patagonicum*	141	(*260*)
Methyl di-*O*-methylporphyrilate	**482**	280–282		147	(*632*)
Methyl 6-*O*-methylnorascomatate	**483**	147–149	*Bunodophoron patagonicum*	141	(*260*)
6-*O*-Methylnorascomatic acid	**484**	204	*Bunodophoron patagonicum*	141	(*260*)
3-*O*-Methylpannaric acid	**485**	164–166	*Combea californica*	148	(*247*)
Norascomatic acid	**486**	227–229	*Bunodophoron patagonicum*	141	(*260*)
Norschizopeltic acid	**487**	Structure not yet established	*Roccellina luteola*		(*300*)
Pannaric acid 9-methylester	**488**	218–220	*Roccella capensis*	This compound was erroneously assigned the structure 3-*O*-methylpannaric acid; its structure was revised by Elix *et al.* (*248*)	(*397*)
Pannaric acid 10-methylester	**489**	pale yellow crystals 226–228		149	(*253*)
Subdidymic acid	**490**	158.5–160		150	(*237*)

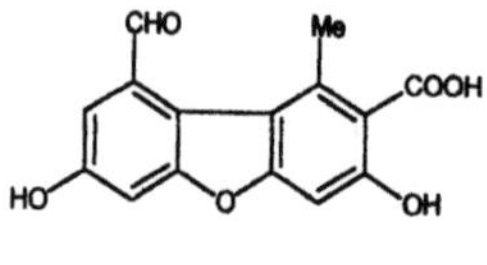

475

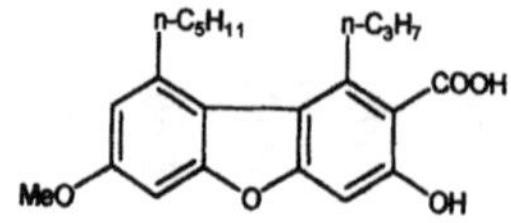

477

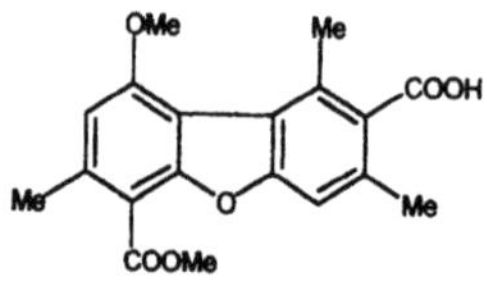

478

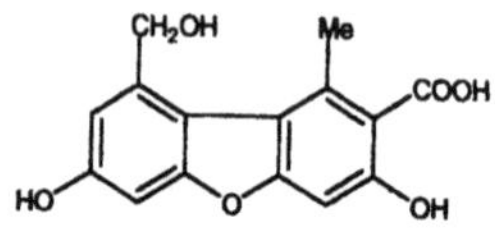

479

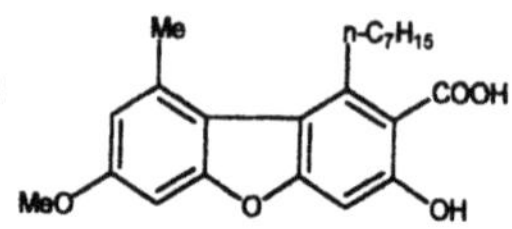

480

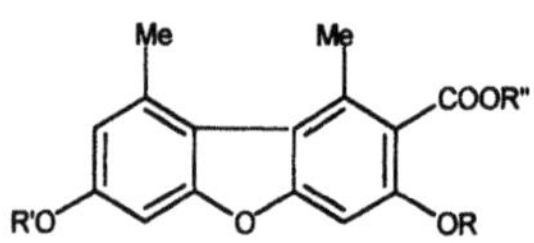

	R	R'	R''
468 :	Me	Me	H
476 :	H	H	H
481 :	Me	Me	Me
483 :	H	Me	Me
484 :	H	Me	H
486 :	H	H	H

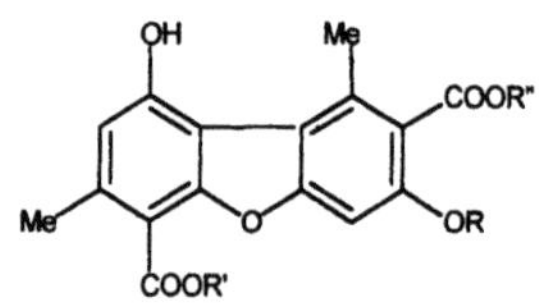

	R	R'	R''
485 :	Me	H	H
487 :	Me	H	Me
488 :	H	H	Me
489 :	H	Me	H

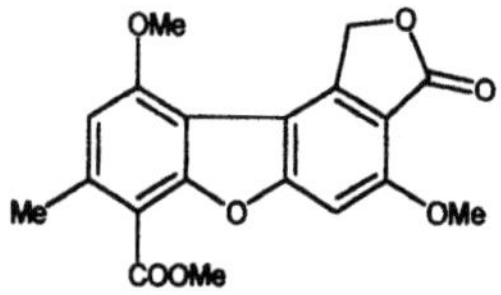

482

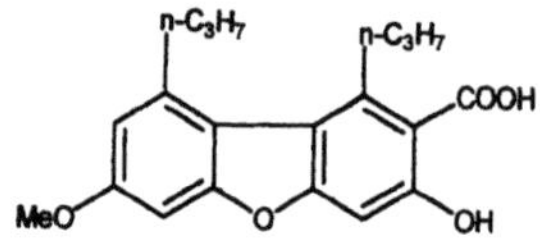

490

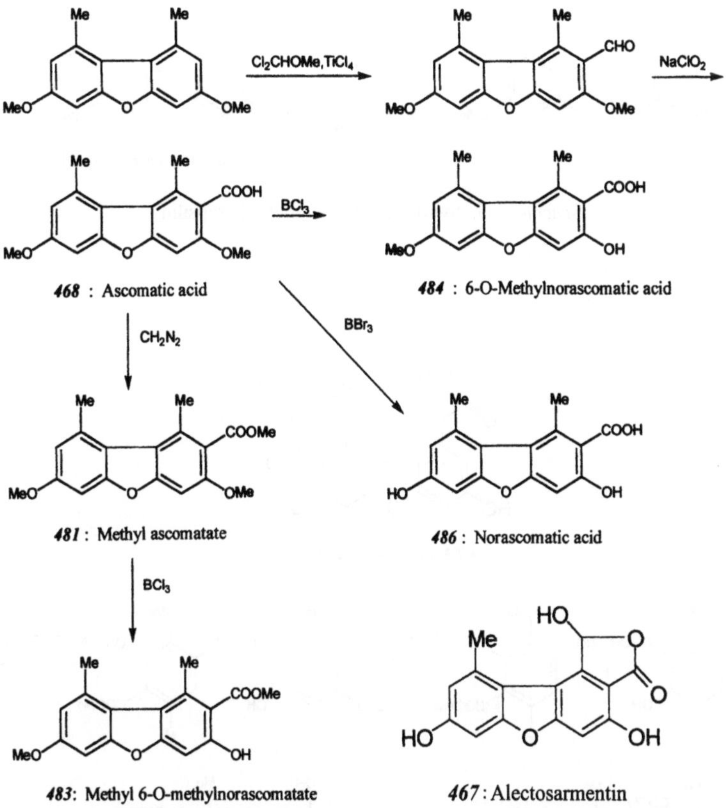

Scheme 141. Synthesis of ascomatic acid and derivatives

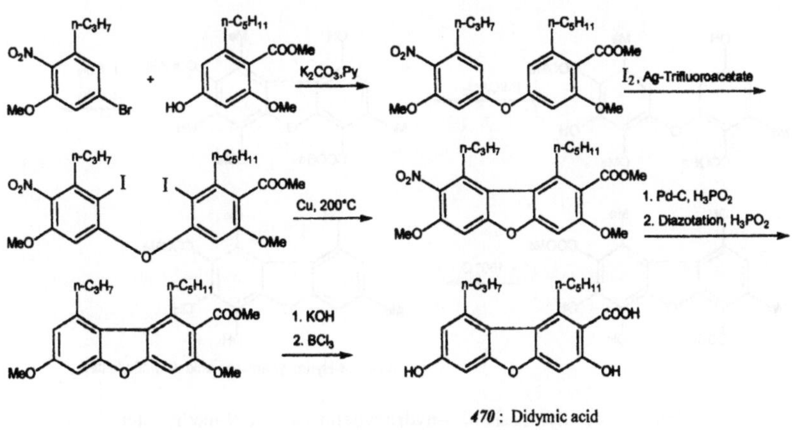

Scheme 142. Synthesis of didymic acid

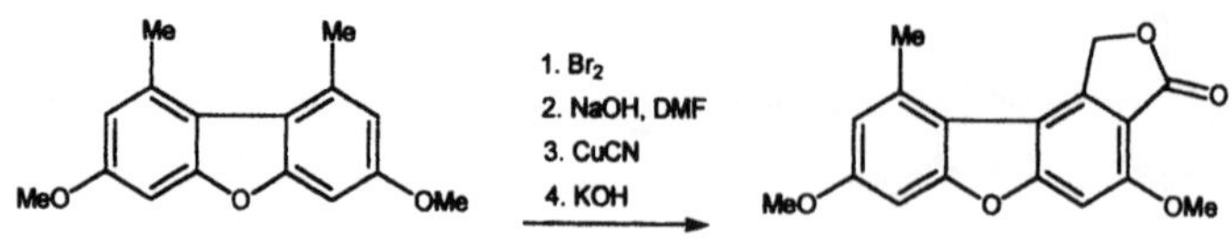

472 : Di-O-methylstrepsilin

Scheme 143. Synthesis of di-*O*-methylstrepsilin

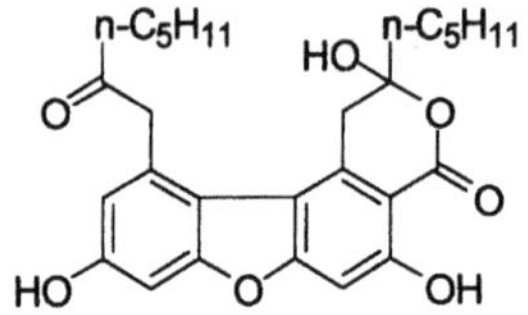

473: Haemophaein

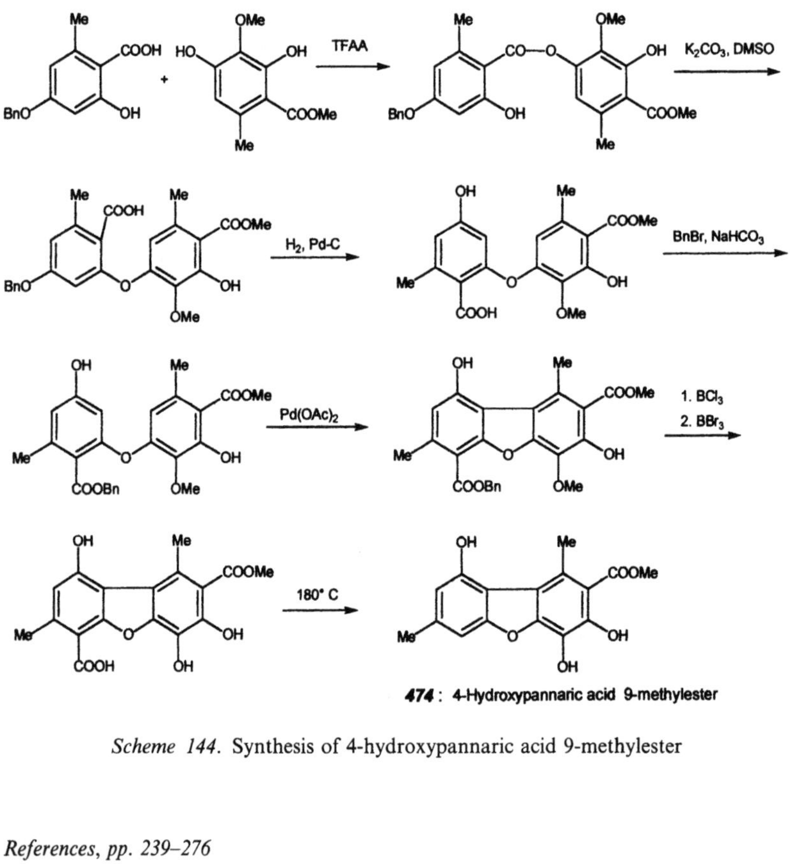

474 : 4-Hydroxypannaric acid 9-methylester

Scheme 144. Synthesis of 4-hydroxypannaric acid 9-methylester

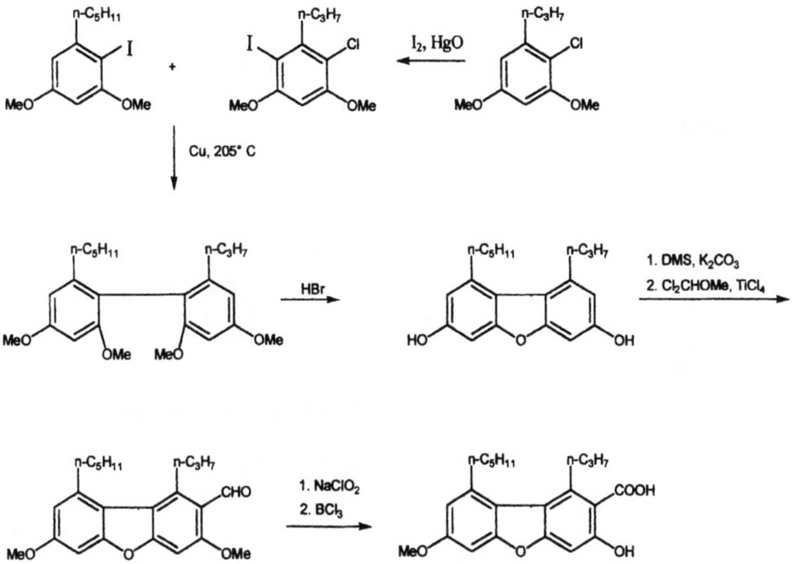

477: Isodidymic acid

Scheme 145. Synthesis of isodidymic acid

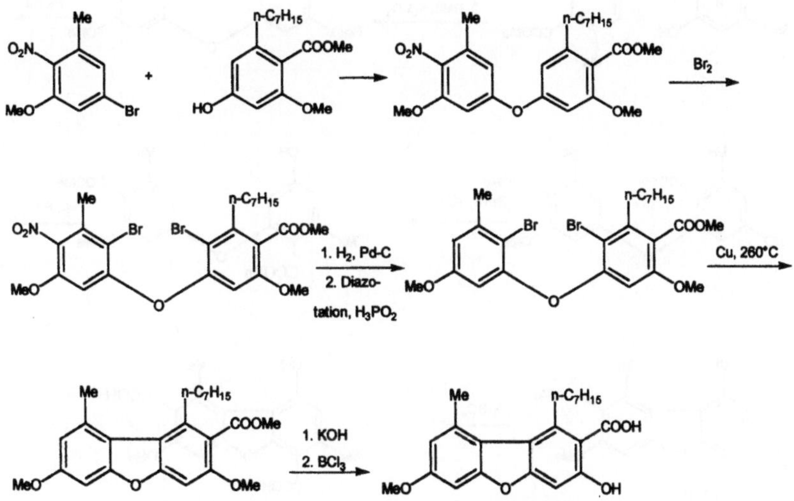

480: Melacarpic acid

Scheme 146. Synthesis of melacarpic acid

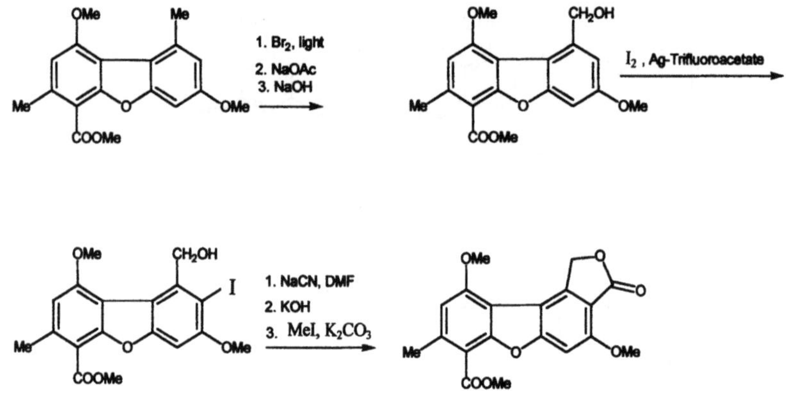

482 : Methyl di-O-methylporphyrilate

Scheme 147. Synthesis of methyl di-O-methylporphyrilate

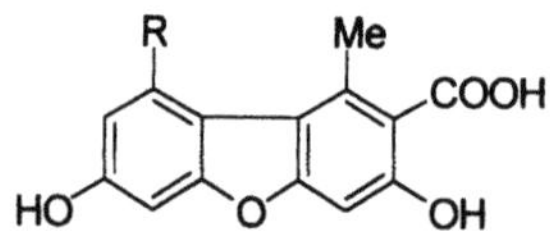

475: R=CHO

479: R=Me

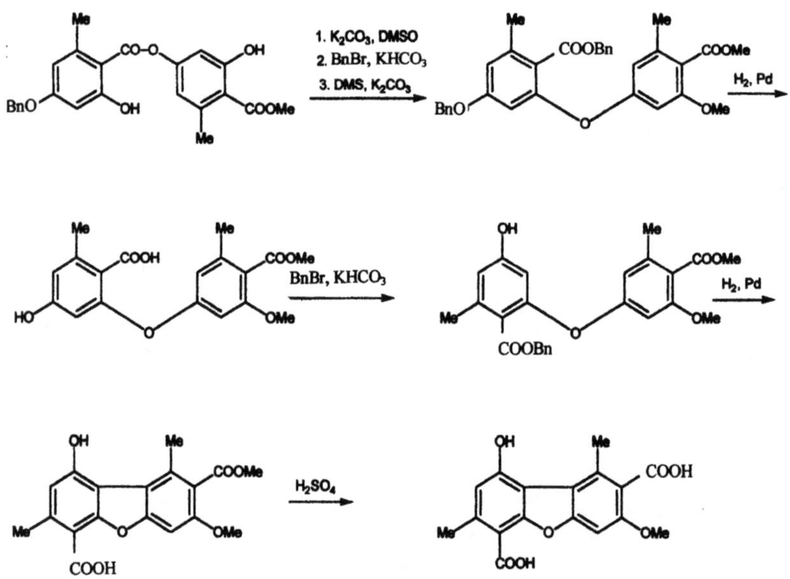

485 : 3-O-Methylpannaric acid

Scheme 148. Synthesis of 3-O-methylpannaric acid

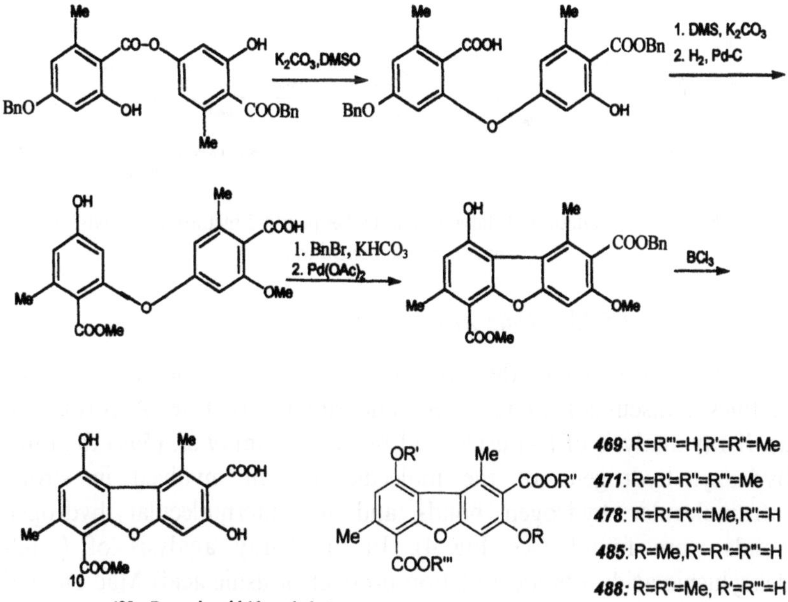

489 : Pannaric acid 10-methylester

Scheme 149. Synthesis of pannaric acid 10-methylester

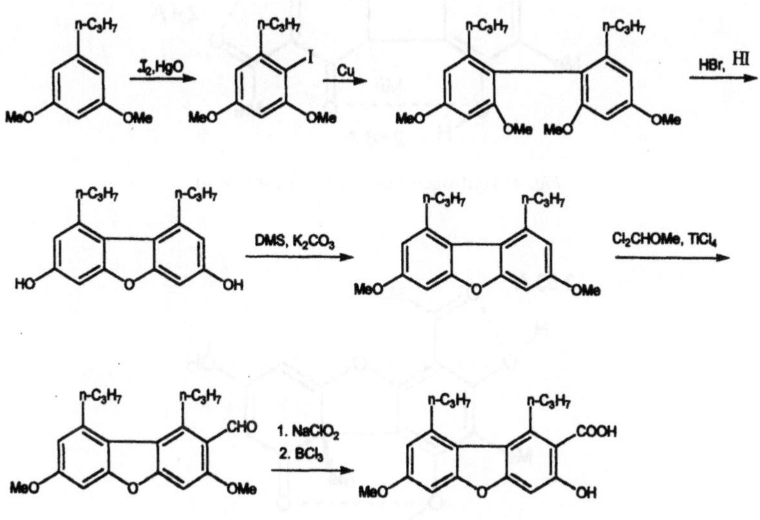

490 : Subdidymic acid

Scheme 150. Synthesis of subdidymic acid

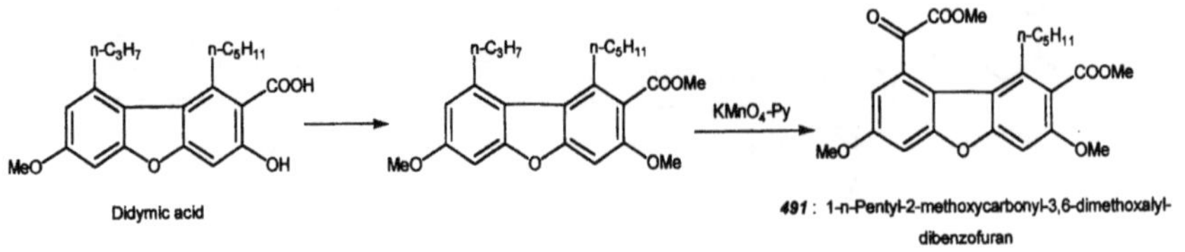

Scheme 151. Oxidation of didymic acid to 1-*n*-pentyl-2-methoxycarbonyl-3,6-dimethoxalyldibenzofuran

4.5.13. *Usnic Acid and Related Compounds*

Usnic acid is one of the biologically most active compounds which justifies a discussion of its three-dimensional structure. According to an X-ray analysis of (−)-usnic acid by Norrestam *et al.* (*566*) the three hydroxyl hydrogens of the molecule are all involved in strong intramolecular hydrogen bonds and no intermolecular hydrogen bonds were found (see Fig. 4). In an X-ray analysis of (+)-2-deacylusnic acid (a biodegradation product of usnic acid) Mac Donald *et al.* (*519*) observed the O–O-distances shown in Fig. 5. Kutney and Sanchez (*481*) carried out an X-ray analysis of (−)-$\Delta^{2,11}$-enaminousnic

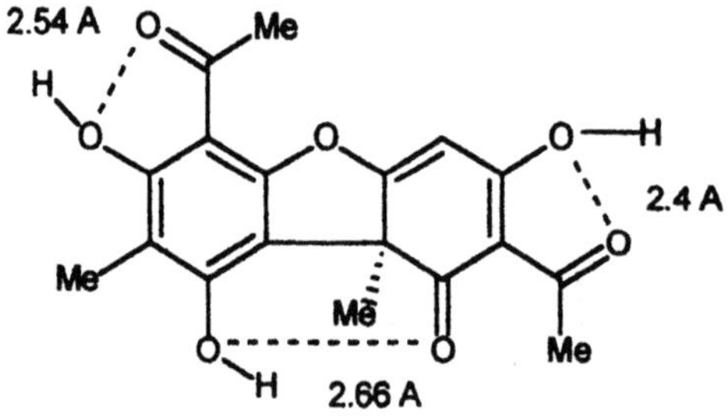

Fig. 4. Hydrogen bonds in (−)-usnic acid

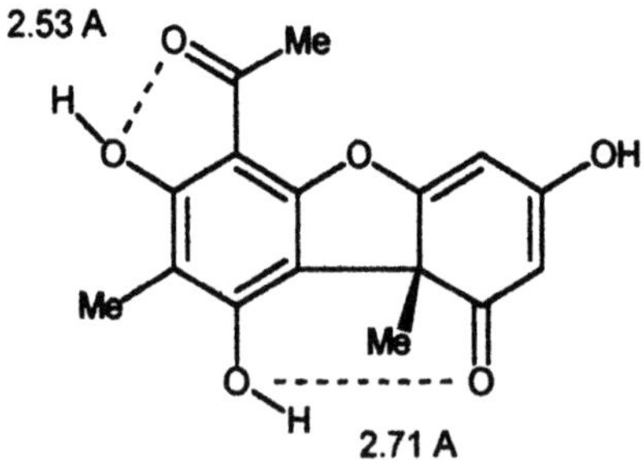

Fig. 5. O–O-Distances in (+)-2-deacylusnic acid

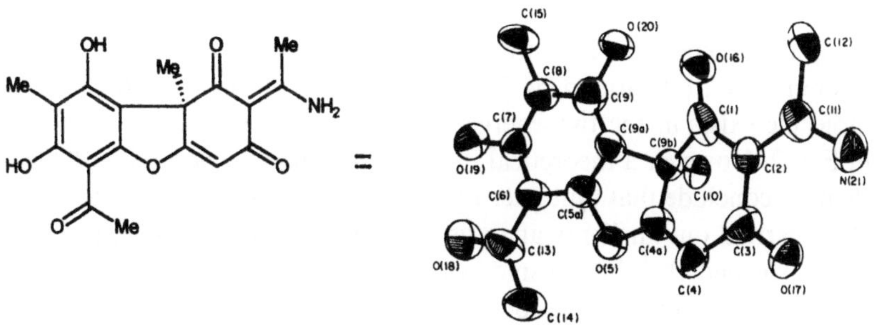

Fig. 6. Computer generated structure of (−)-$\Delta^{2,11}$-enaminousnic acid

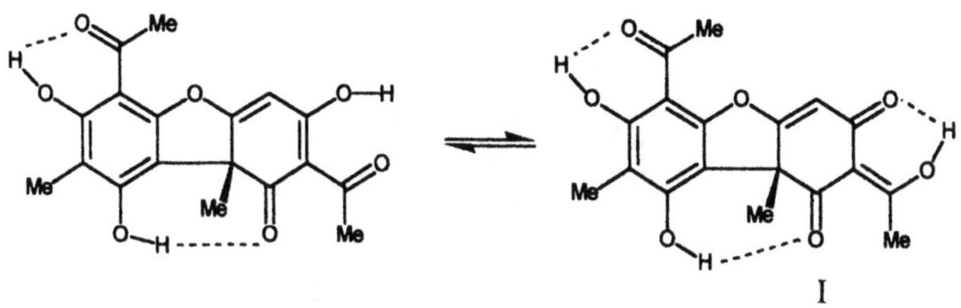

Fig. 7. Tautomeric forms of (+)-usnic acid

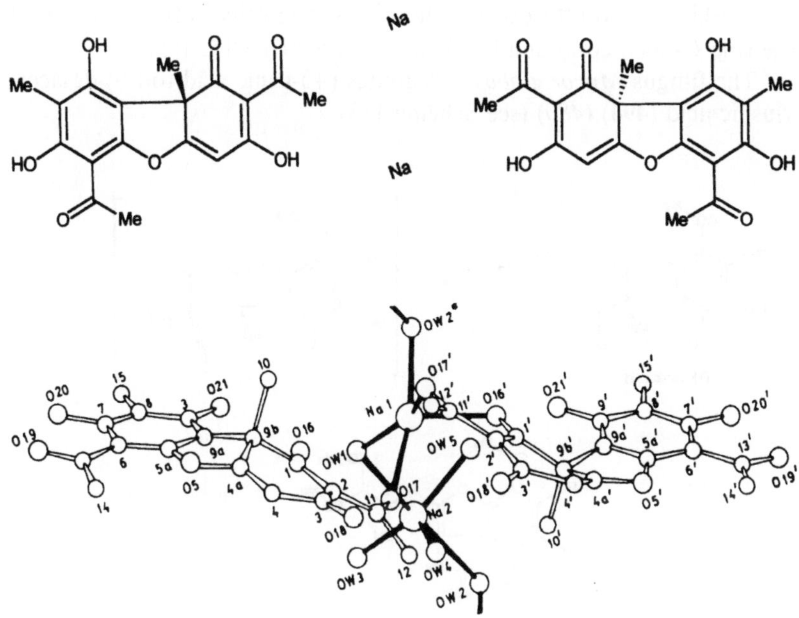

Fig. 8. Computer drawing of the structure of the sodium 2½ hydrate of (+)-usnic acid

acid which was prepared from (+)-usnic acid with ammonia. The computer generated structure of this compound is shown in Fig. 6. Behrens *et al.* (*30*) found by X-ray analysis of (+)-usnic acid that the molecule exists in the two energetically equal tautomeric forms shown in Fig. 7. However a theoretical study by Buemi and Zuccarello (*49*) led them to conclude that tautomer I predominated in solution. Ribár *et al.* (*620*) carried out an X-ray analysis of the sodium salt of (+)-usnic acid which showed that both usnic acid ligands are different as shown in Fig. 8: one is bound to the sodium 1 ion as a monodentate, the other is bidendate.

The absolute configuration of (+)-usnic acid has been determined by an X-ray analysis of the (−)-α-phenylethylamine derivative **492** (*389*).

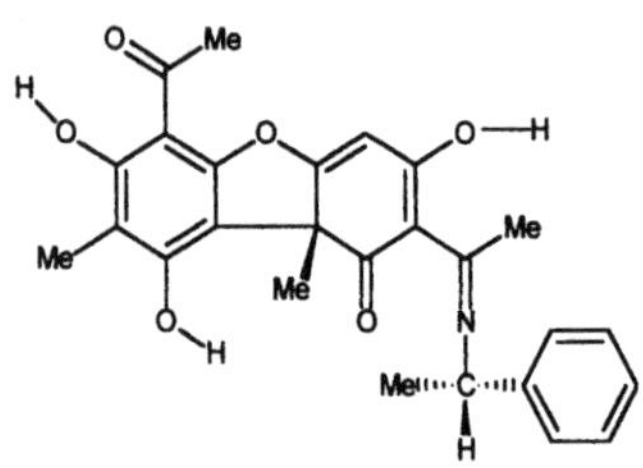

492

(+)-Usnic acid afforded on photolysis in tetrahydrofuran-methanol the ring C-open compound **493** shown in Scheme 152 (*690*).

The fungus *Mucor globosus* degrades (+)-usnic acid to (+)-desacetylusnic acid (**494**) (*480*) (see Scheme 153).

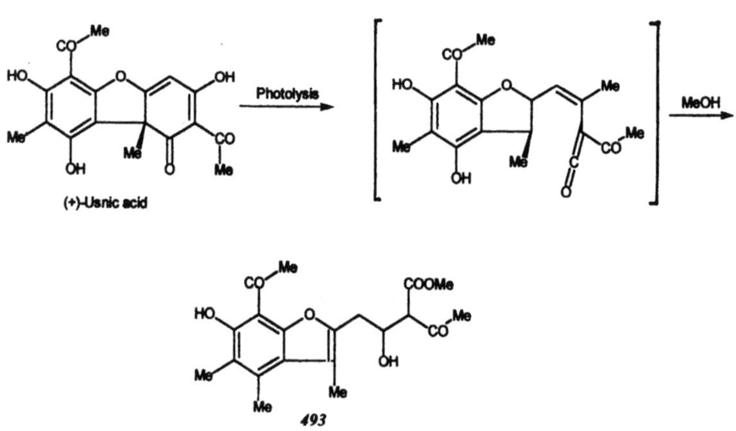

493

Scheme 152. Photolysis of (+)-usnic acid

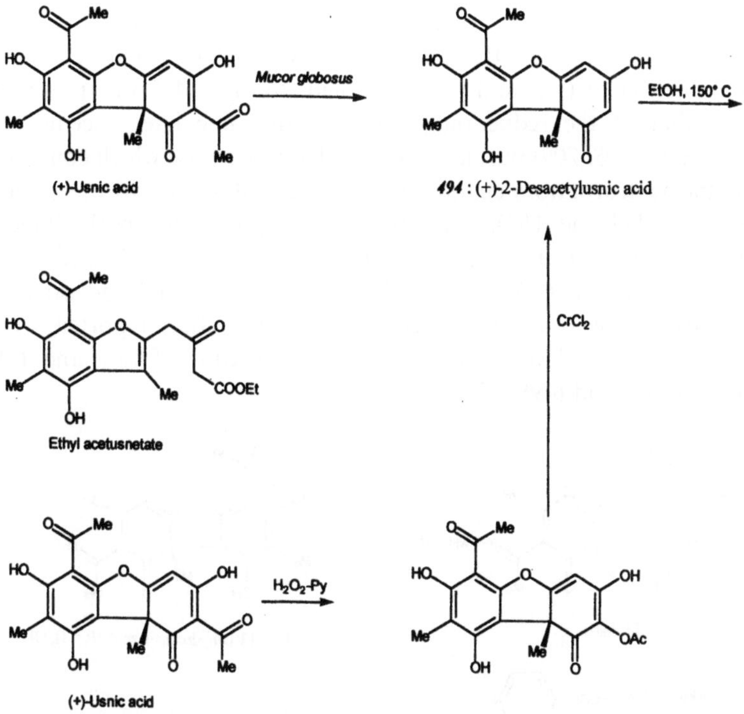

Scheme 153. Degradation of (+)-usnic acid by *Mucor globosus*

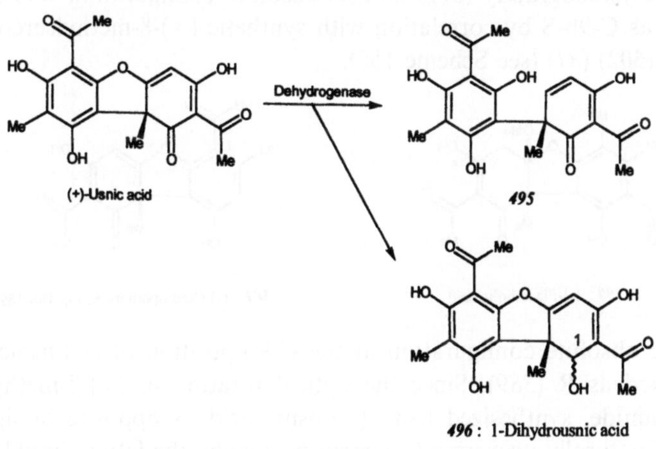

Scheme 154. Enzymatic hydrogenation of (+)- and (−)-usnic acid

Estévez *et al.* (*285*) isolated from cell-free extracts of *Evernia prunastri* the enzyme DL-usnic acid dehydrogenase which catalyzed the reduction of (+)- and (−)-usnic acid; formula **495** was presumed to be that of the reduction product. Some time later Vicente and co-workers (*708*, *709*) purified a second dehydrogenase which converted only the dextrorotatory enantiomer of usnic acid to l-dihydrousnic acid (**496**) (see Scheme 154). The structure proposals for both dihydro derivatives were primarily based on the IR spectra and need to be confirmed by NMR, MS and/or X-ray analysis. Phenyl and isonicotinoyl hydrazones of (−)-usnic acid (**497**, **498**) have been reported, while reduction with sodium borohydride gave a mixture of 13α- and 13β-dihydrousnic acid **(499)** (*585*).

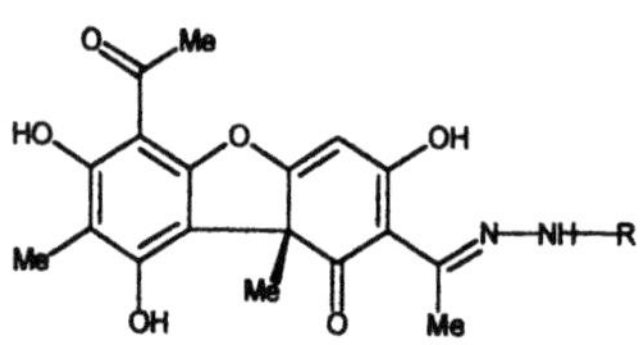

497 : R = $-C_6H_5$

498 : R = —CO—

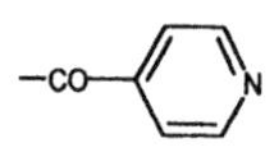

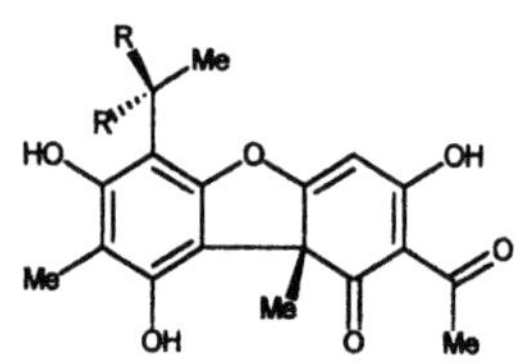

499 : 13-Dihydrousnic acid, R = H, OH

The *trans*-stereochemistry at the ring junction of (−)-placodiolic acid (**500**) from the lichen *Rhizoplaca chrysoleuca* was established by an X-ray analysis (*78*). (−)-Cercosporamide (**501**) is a phytotoxin from the fungus *Cercidosporidium henningsii*; its structure was established by X-ray crystallography (*679*) and its absolute configuration was determined as C-9b-S by correlation with synthetic (+)-8-methylcercosporamide (**502**) (*81*) (see Scheme 155).

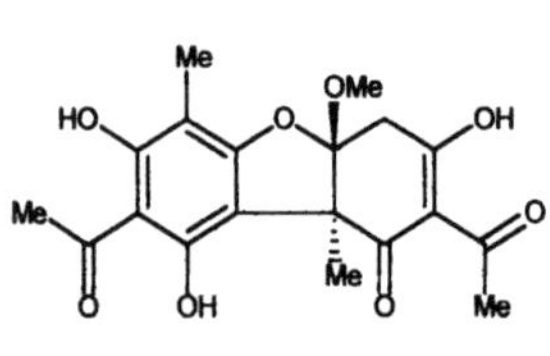

500 : (-)-Placodiolic acid

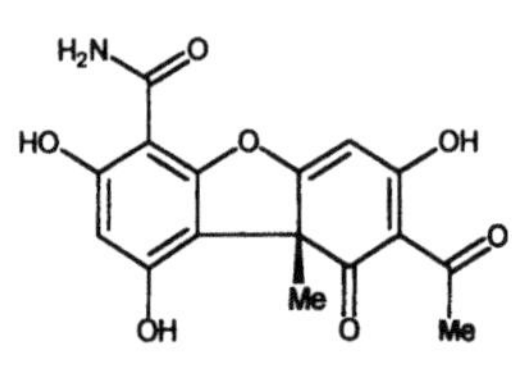

501 : (-)-Cercosporamide , mp 188-189°C

The absolute configuration at the C-9b-position of (+)-usnic acid was fixed as R (*389*). Since the optical rotation of (+)-8-methylcercosporamide, synthesized from (+)-usnic acid, is opposite in sign to that of naturally occurring (−)-cercosporamide, the latter should have the S-configuration at C-9b. The CD spectra of (−)-cercosporamide,

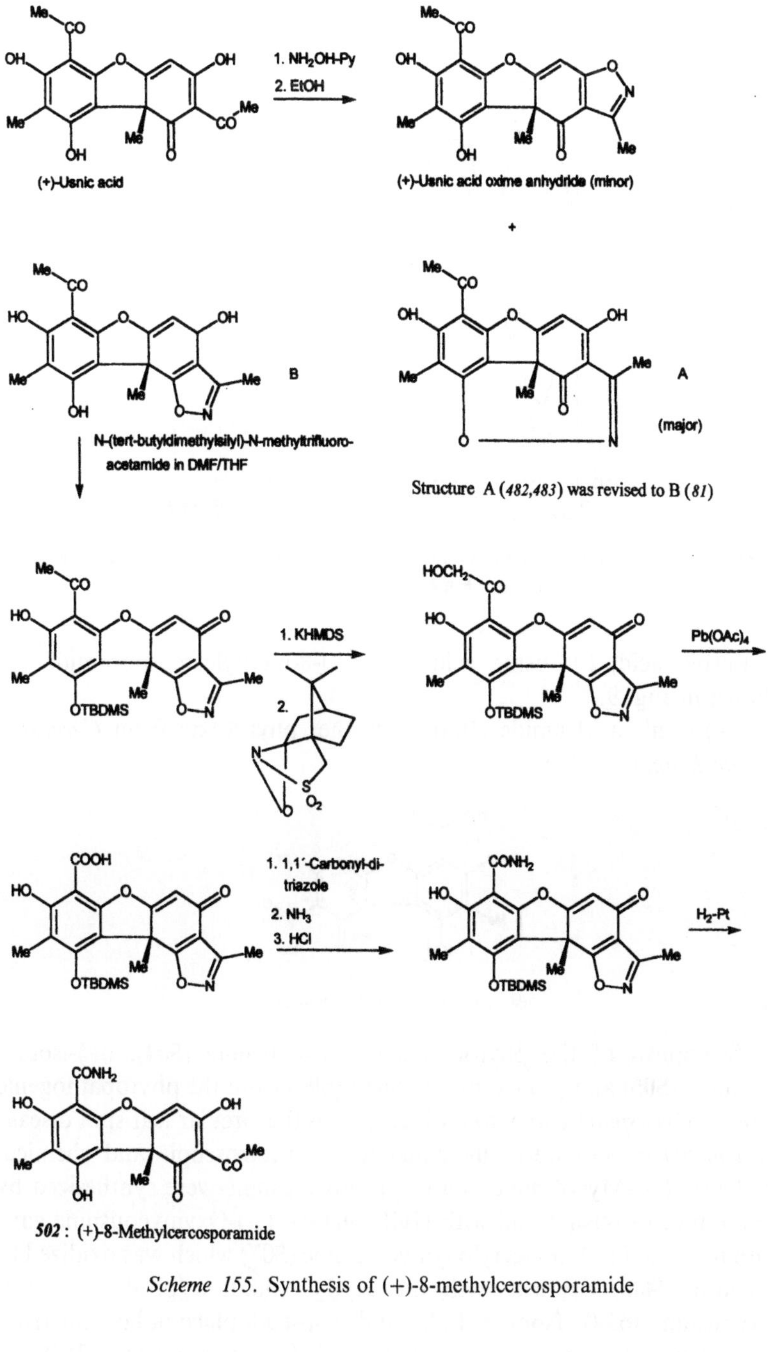

502 : (+)-8-Methylcercosporamide

Scheme 155. Synthesis of (+)-8-methylcercosporamide

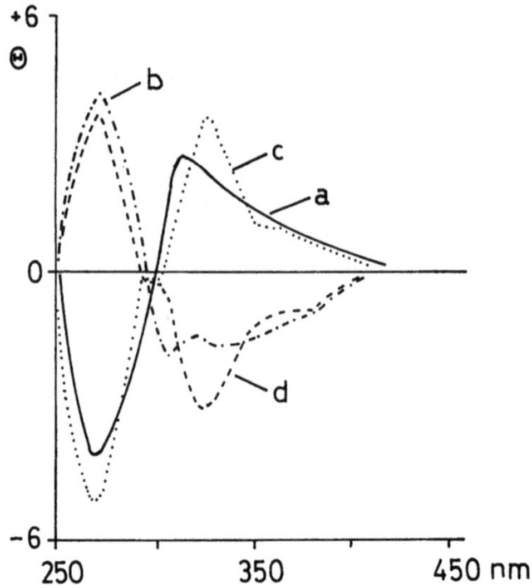

Fig. 9. CD spectra of (−)-cercosporamide (a), (−)-usnic acid (b), (+)-usnic acid (c), and (+)-8-methylcercosporamide (d)

(−)-usnic acid, (+)-usnic acid and (+)-8-methylcercosporamide are shown in Fig. 9.

(−)-Usnic acid amide (**503**) is another phytotoxin from *Cercosporidium henningsii* (*80*).

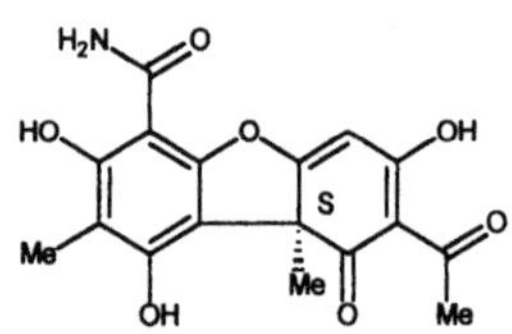

503: (-)-Usnic acid amide, yellow powder

Structures of the phytotoxins (−)-mycousnine (**504**), (+)-isomycousnine (**505**) and (+)-oxymycousnine (**506**) from the phytopathogenic fungus *Mycosphaerella nawae* which causes the circular leaf spot disease of *Diospyros kaki* were elucidated using spectroscopic and chemical evidence. (−)-Mycousnine and (+)-isomycousnine were synthesized by reaction of (+)-usnic acid with HCl–MeOH. (+)-Oxymycousnine gave with methanol (+)-desacetyloxymycousnine (**507**) which was oxidized to triketone **508**, identical with the H_2O_2 oxidation product of (−)-mycousnine (*634*). Isoplacodiolic and isopseudoplacodiolic acid from the lichens *Haematomma flexuosum* and *H. matogrossense* (*255*) are identical with mycousnine and isomycousnine respectively.

References, pp. 239–276

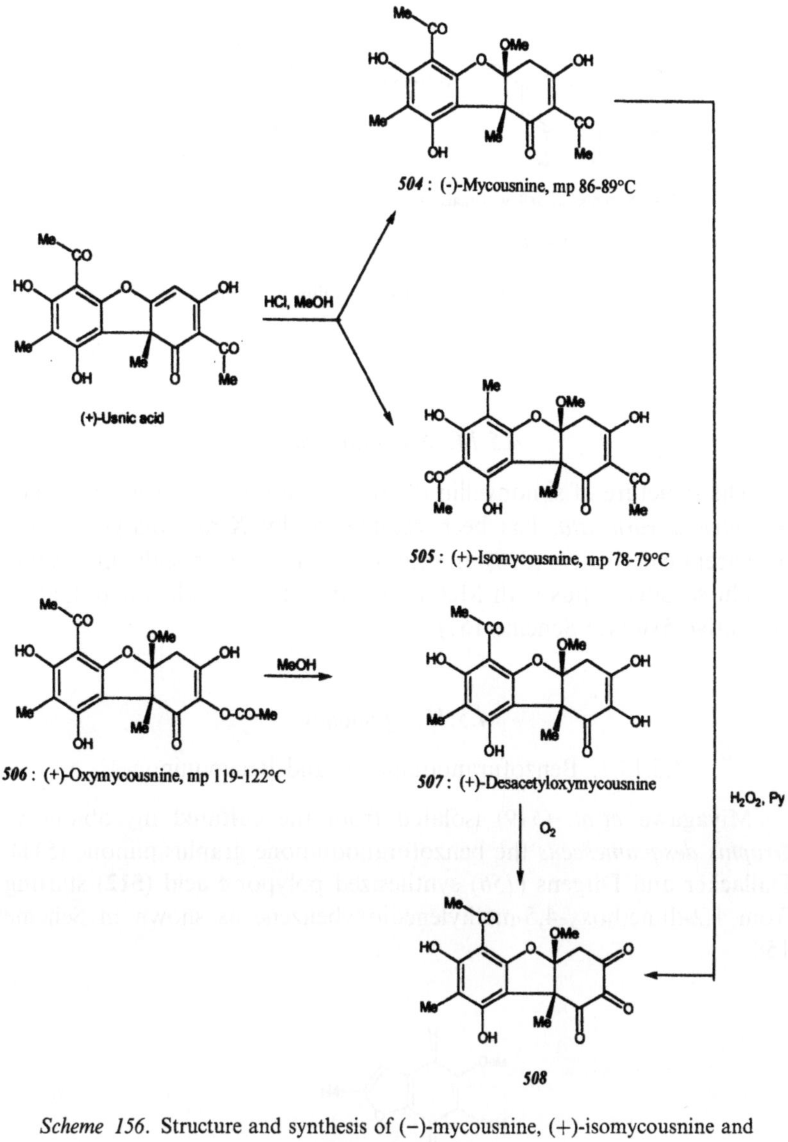

Scheme 156. Structure and synthesis of (−)-mycousnine, (+)-isomycousnine and (+)-oxymycousnine

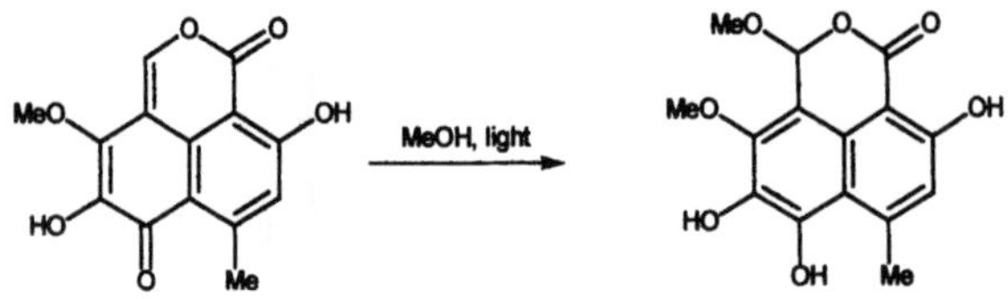

509 : Simonyellin, yellow crystals, mp 274-275°C

510

Scheme 157. Reaction of simonyellin with methanol

4.5.14. Naphthopyrans

The structure of simonyellin (**509**), a naphthopyran from the lichen *Simonyella variegata*, has been established by X-ray analysis of the tetraacetate derivative (*210*). In solution of alcohols or acids simonyellin is light sensitive; thus with MeOH it forms the colourless naphthalene derivative **510** (see Scheme 157).

4.5.15. Quinones

4.5.15.1. Benzofuranoquinones and Benzoquinones

Miyagawa *et al.* (*549*) isolated from the cultured mycobiont of *Graphis desquamescens* the benzofuranoquinone graphisquinone (**511**). Dallacker and Ditgens (*156*) synthesized polyporic acid (**512**) starting from 1,2-dimethoxy-4,5-methylenedioxybenzene as shown in Scheme 158.

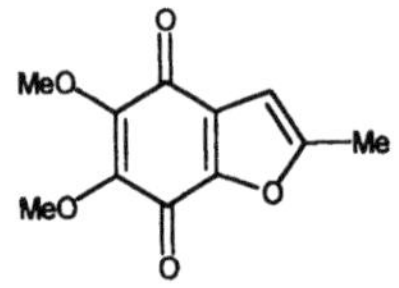

511 : Graphisquinone, red needles

mp 127-128°C

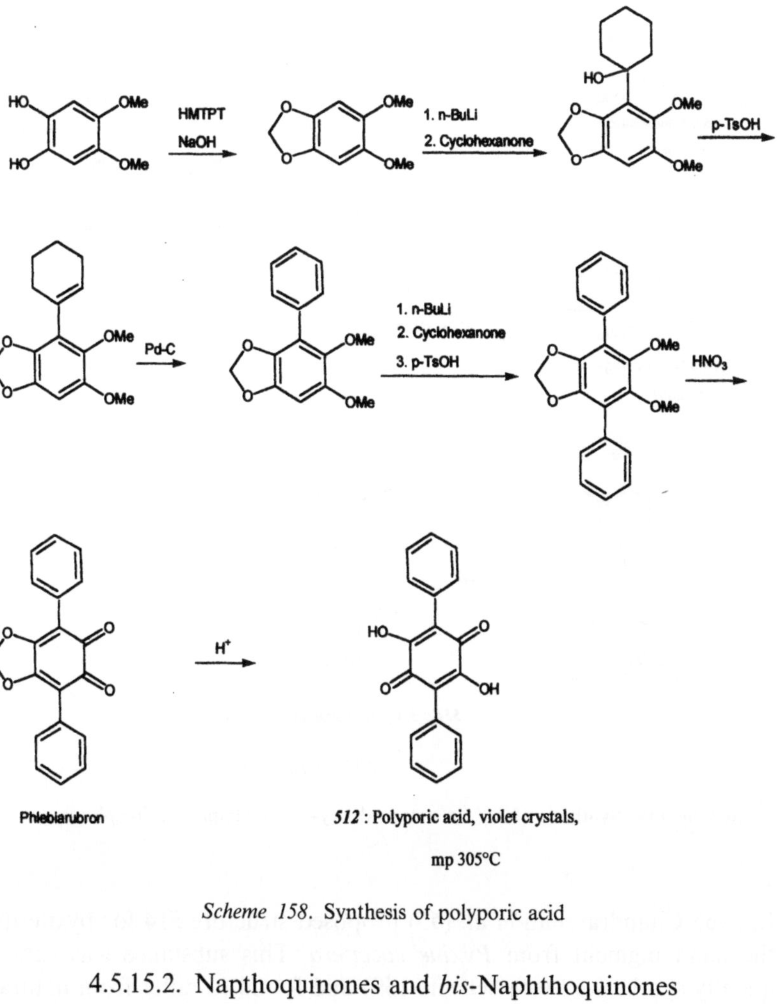

Phlebiarubron

512 : Polyporic acid, violet crystals, mp 305°C

Scheme 158. Synthesis of polyporic acid

4.5.15.2. Napthoquinones and *bis*-Naphthoquinones

The structure of the red pigment from the apotheciae of *Cladonia deformis* and *C. floerkeana*, rhodocladonic acid, was shown to be 2-acetyl-3,4,6-trihydroxy-7-methoxynaphtho[2,3-b]furan-5,8-dione (**513**) (*540*). Lichenicolous mites trigger the production of rhodocladonic acid in *Cladonia bacilliformis* and *C. norvegica* (*697*).

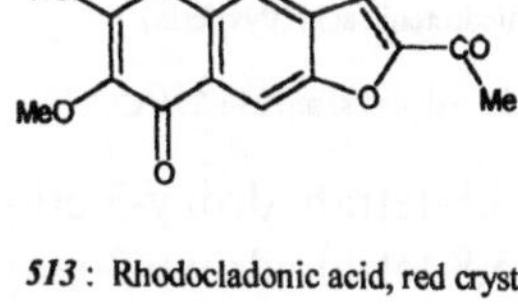

513 : Rhodocladonic acid, red crystals, mp > 360°C

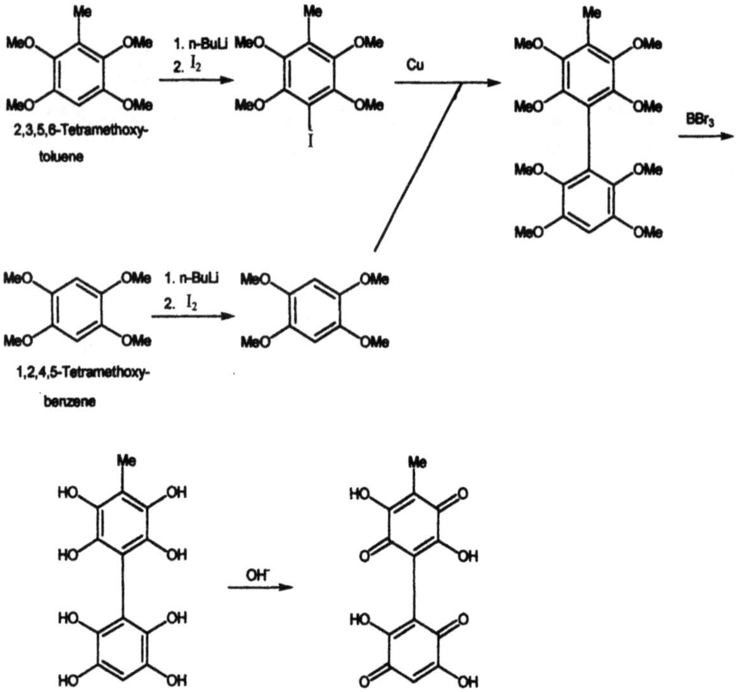

514: 3,3′,6,6′-Tetrahydroxy-4-methyl-biphenyl-di-*p*,*p*′-quinone

Scheme 159. Synthesis of 3,3′,6,6′-tetrahydroxy-4-methylbiphenyl-di-*p*,*p*′-quinone

In 1965 Chandrasenan *et al.* (*65*) proposed structure **514** for pyxiferin, the main pigment from *Pyxine coccifera.* This substance was subsequently synthesized (see Scheme 159) and found to differ from natural pyxiferin. Later on Huneck (*380*) identified pyxiferin with chiodectonic acid and proposed structure **515** for this pigment.

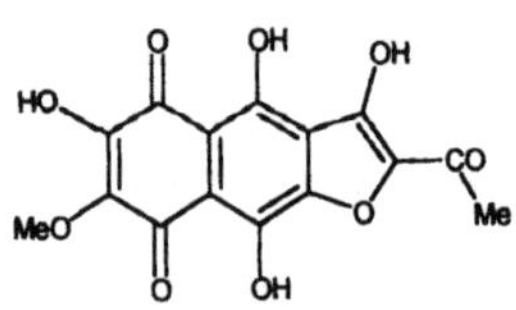

515: Chiodectonic acid (Pyxiferin)

dark red plates, mp 294-296°C

The naphthoquinones 1,4,5,8-tetrahydroxy-3-ethyl-naphthalene-2,6-dione (**516**) and 7,7′-*bis*-(1,4,5,8-tetrahydroxy-3-ethyl-naphthalene-2,6-

dione (**517**) have been found in *Cetraria cucullata* (*475*). Islandoquinone, 2-*O*-(3′-ethyl-1′,2′,3′,4′-tetrahydro-5′,7′,8′-trihydroxy-1′,2′,4′-triketo-3′-naphthyl-3-ethyl-7-hydroxynaphthazarin (**518**), is a new naphthaquinone from *Cetraria islandica* (*671*).

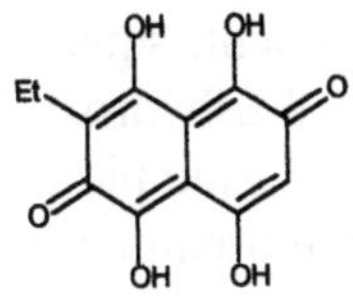

516: red crystals, mp 183-185°C

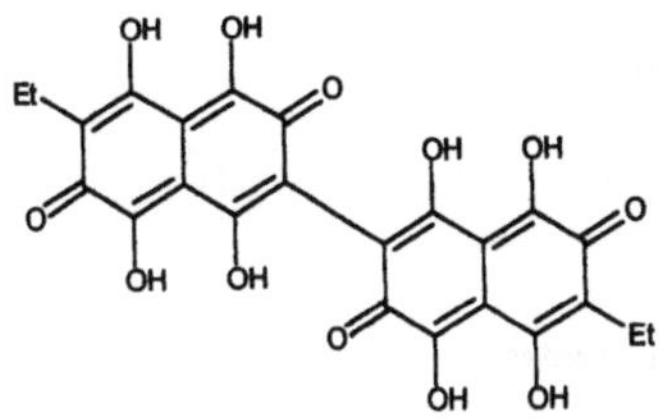

517: red crystals, mp 265°C, dec.

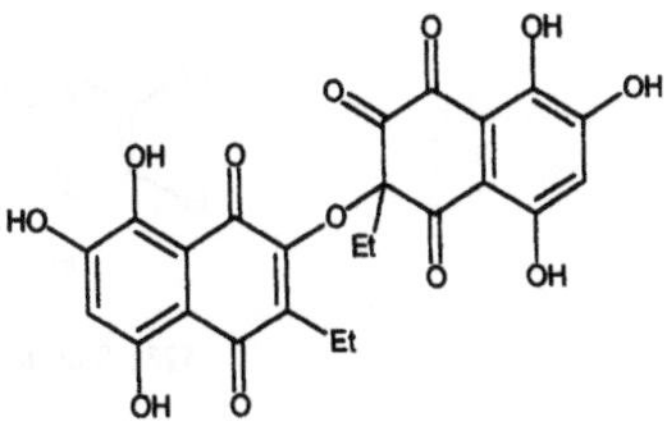

518: orange crystals, mp 243-244°C

Formula **519** was originally proposed (*47*) for the red pigment haemoventosin of the apotheciae from *Ophioparma ventosa* and subsequently revised to **520** (*536*). Its structure was finally established

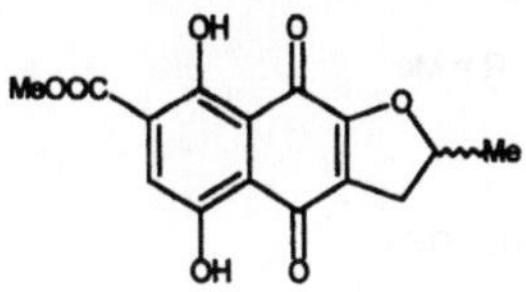

519

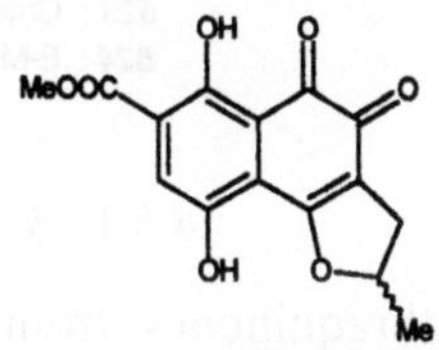

520

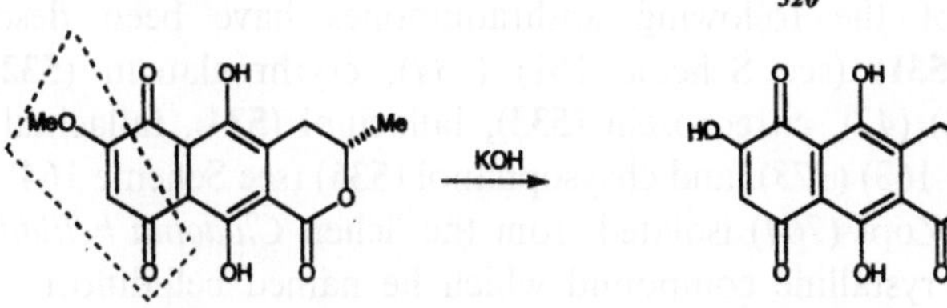

521: Haemoventosin, redbrown crystals, mp 202-204°C

Norhaemoventosin

Scheme 160. Structure and saponification of haemoventosin

as **521** on the basis of NMR studies and ORD data (*625*). The encircled part of the structure explains the facile saponification of haemoventosin to norhaemoventosin (see Scheme 160).

The lichen *Squamarina cartilaginea* is sometimes infected by a *Wardomyces* fungus and then forms deep red spots on the thallus. The structure of this pigment, squamarone, has been elucidated by NMR spectroscopy as 2,6,8-trihydroxy-3-methyl-7-ethyl-1,4-naphthaquinone (**522**). This is the first example of the formation of a phytoalexin by a lichen (*370*). The naphthaquinones cristazarin (**523**) and 6-methylcristazarin (**524**) have been isolated from cultures of the spore-derived mycobiont of *Cladonia cristatella* (*744*).

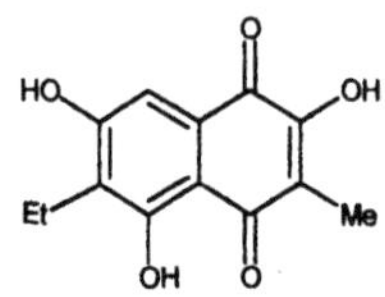

522: Squamarone, orange needles,

mp 275-277°C

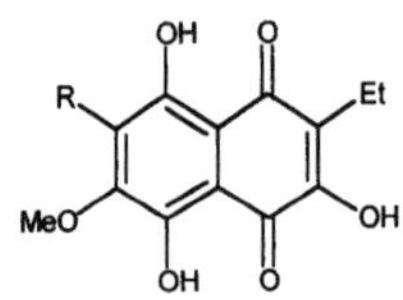

523: Cristazarin, R = H
524: 6-Methylcristazarin, R = Me

4.5.15.3. Anthraquinones

New anthraquinones from lichens are summarized in Table 9. Syntheses of the following anthraquinones have been described: islandicin (**531**) (see Scheme 161) (*734*), erythroglaucin (**532**) (see Scheme 162) (*42*), citreorosein (**533**), fallacinol (**534**), fallacinal (**535**) (see Scheme 163) (*373*), and chrysophanol (**536**) (see Scheme 164) (*560*).

In 1907 Zopf (*764*) isolated from the lichen *Cladonia bellidiflora* a red-brown crystalline compound which he named bellidiflorin. More than 80 years later Alagna *et al.* (*7*) showed by X-ray crystallography that bellidiflorin is an iron complex of the *bis*-anthraquinone graciliformin (*186*); the coordination of the ligands to the iron atom is shown in Fig. 10.

Table 9. Anthraquinones from lichens

Compound	Formula No.	M.p. (°C)	Lichen	Ref.
Asahinin	**525**	red crystals, >320	*Asahinea chrysantha*	(*547*, *672*)
Averythrin	**526**	red prisms, 229–231	*Solorina crocea*	(*48*)
Haematommone	**527**	red needles, >300, dec.	*Haematomma puniceum*	(*390*)
Nemetzone	**528**	orange red needles 289–290	*Haematomma nemetzii*	(*396*)
Russulone	**529**	red substance	*Lecidea russula*	(*528*, *532*)
1,5,6,8-Tetrahydroxy-3-methylanthraquinone	**530**	dark red crystals, >320	*Asahinea chrysantha*	(*547*)

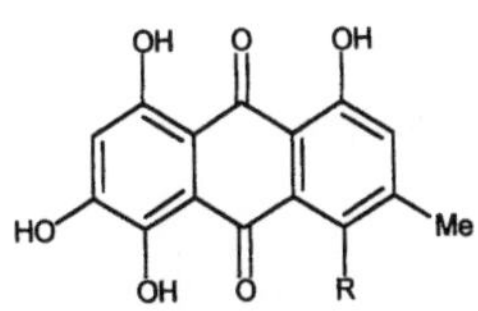

525: R = OH
530: R = H

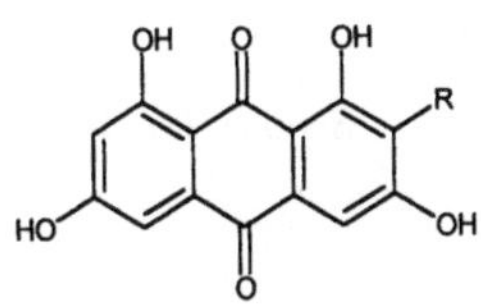

526 : R = $-CH{=}CH-(CH_2)_3-Me$
527 : R = -CO-Me

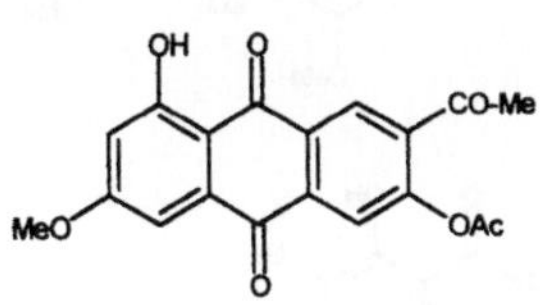

528

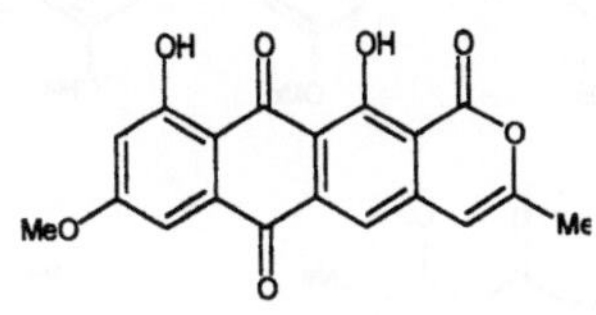

529

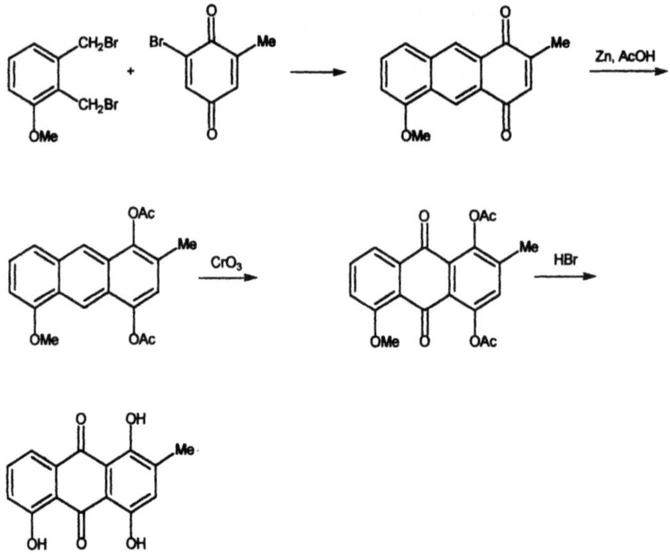

531 : Islandicin

Scheme 161. Synthesis of islandicin

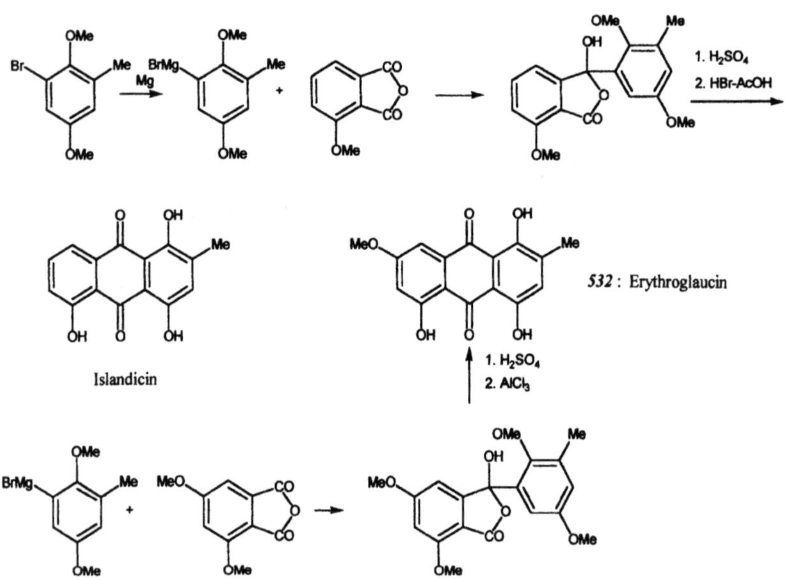

Scheme 162. Synthesis of islandicin and erythroglaucin

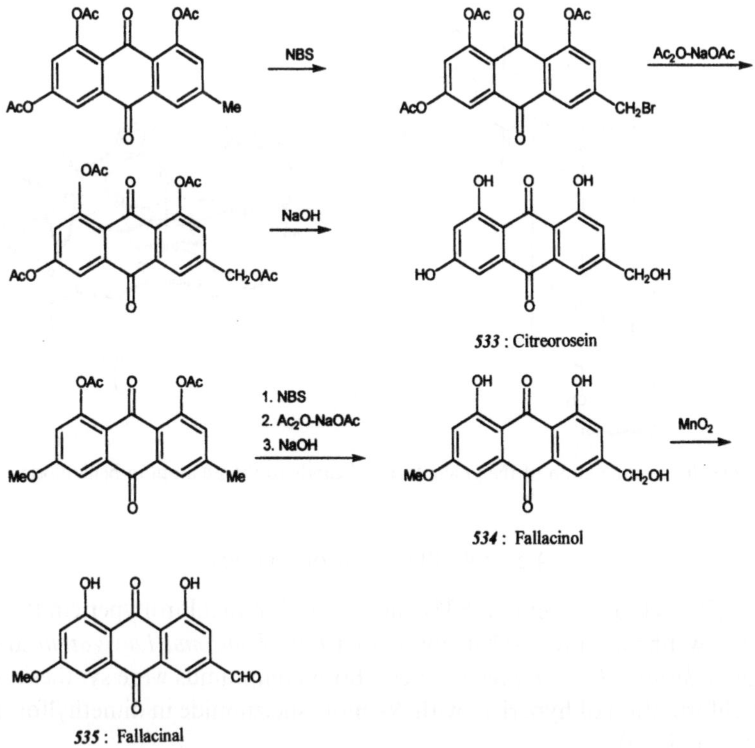

Scheme 163. Synthesis of citreorosein, fallacinol and fallacinal

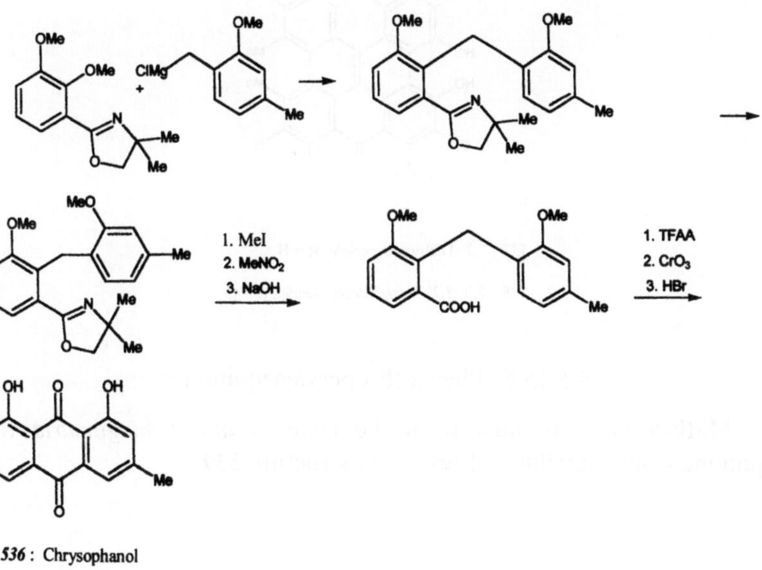

Scheme 164. Synthesis of chrysophanol

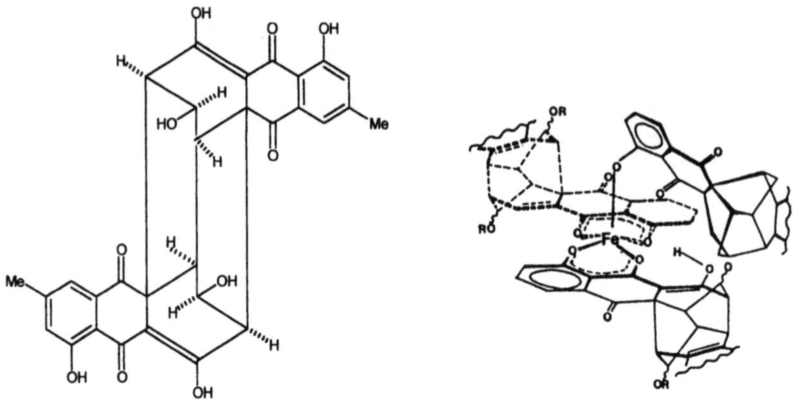

Fig. 10. Coordination of the gracilliformin ligands to the Fe atom in bellidiflorin

4.5.15.4. Phenanthroquinones

7,7′-Dichlorohypericin (**537**) and 2,2′,7,7′-tetrachlorohypericin (**538**) are new phenanthroperylenequinones from *Nephroma laevigatum* and *Heterodermia obscurata* respectively. Both compounds were synthesized by chlorination of hypericin with *N*-chlorosuccinimide in dimethylformamide (*73, 74*).

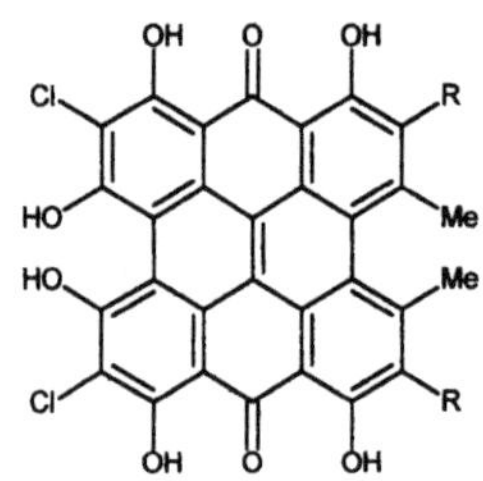

537: 7,7′-Dichlorohypericin, R = H

538: 2,2′,7,7′-Tetrachlorohypericin, R = Cl

4.5.15.5. Phenanthroperylenequinones

Mathey (*528*) isolated from the lichen *Laurera sanguinaria* the quinone isohypocrellin and proposed structure **539**.

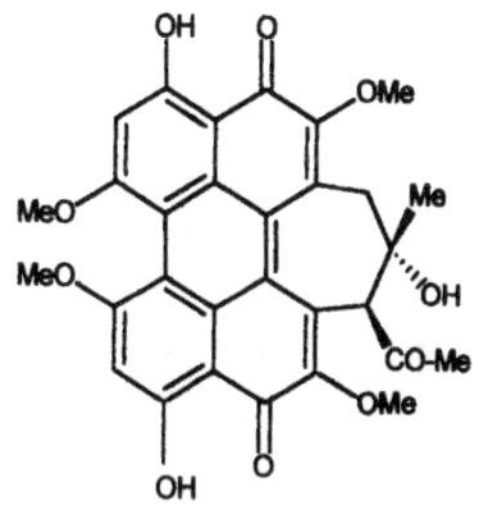

539 : Isohypocrellin

4.5.15.6. Perylenequinones

Biruloquinone is the only phenanthrenequinone known from a lichen (*Parmelia birulae*). Structure **540** was originally proposed by Krivoshchekova *et al.* (*477*) but was later revised to **541** by Arnone *et al.* (*20*) who isolated it from the fungus *Mycosphaerella rubella.*

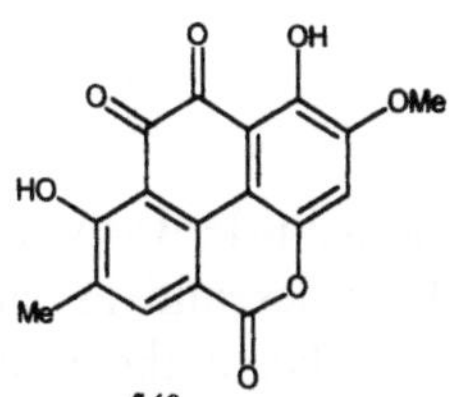

540

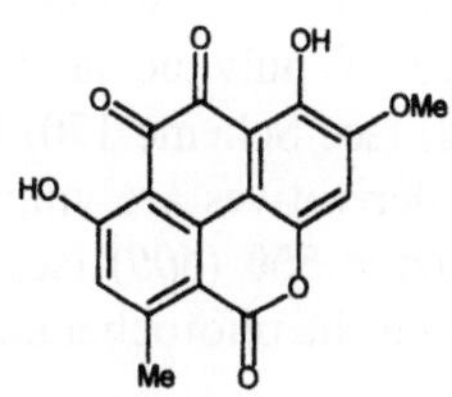

541 : Biruloquinone, dark violet crystals, mp 307-308°C

4.5.16. Pulvinic Acid Derivatives

Demethylleprapinic acid (**542**), 4-hydroxyvulpinic acid (**543**), 2-hydroxypulvinic acid dilactone (**544**) and methylcalycin (**545**) have been isolated from Himalayan *Lepraria* species and from *Letharia vulpina* respectively (*14*, *685*). The structure of vulpinic acid (**546**) has been confirmed by an X-ray analysis (*41*). More than twenty years ago Pattenden (*573*) reviewed naturally occurring 4-ylidenebutenolides in vol. 38 of this series. Since then numerous new synthetic ways to pulvinones have been developed (see Schemes 165–168).

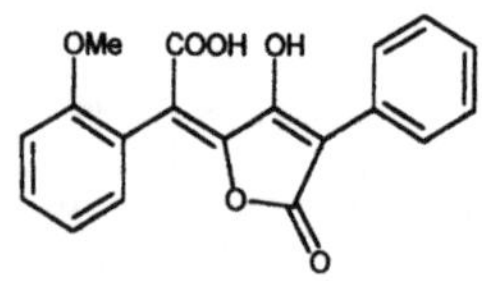

542 : Demethylleprapinic acid

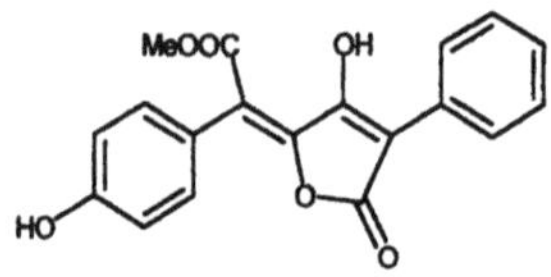

543 : 4-Hydroxyvulpinic acid

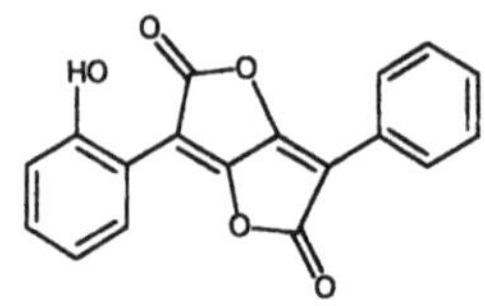

544 : 2-Hydroxypulvinic acid dilactone

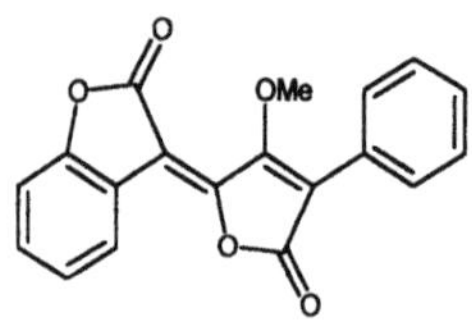

545 : Methylcalycin

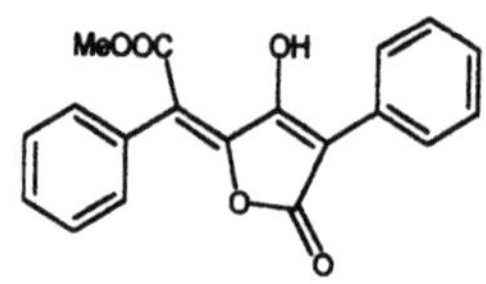

546 : Vulpinic acid

Fully methylated pulvinic acid derivatives are cleaved by alkali to 2-aryl-3-methoxymaleic anhydrides and arylacetic acids (see Scheme 169). The products formed indicate the structure of the particular pulvinic acid (*184*). By reaction of vulpinic acid with 2-hydroxyethoxyacetamide a mixture of pulvinic acid monolactam (**547**) and pulvinic acid dilactam (**548**) (see Scheme 170) has been obtained (*656*). Further new pulvinic acid derivatives are vulpinic acid isopropyl ether (**549**) (*660*) and the pyridazine **550** (*600*) (see Schemes 171 and 172). Chawla *et al.* (*67*) investigated the photochemical conversion of pulvinic acid to leprapinic acid.

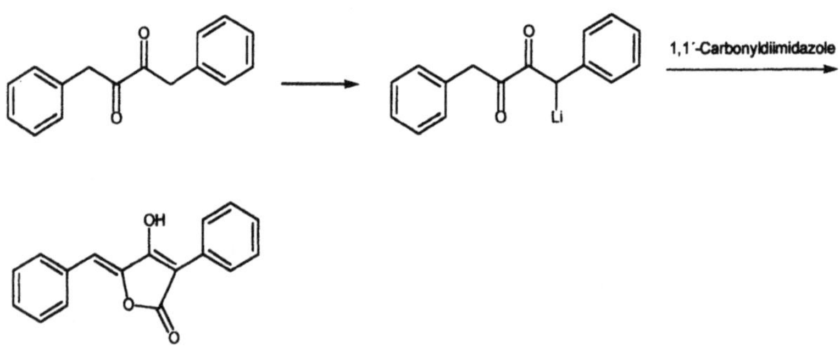

Scheme 165. Synthesis of pulvinones after Jerris *et al.* (*444*)

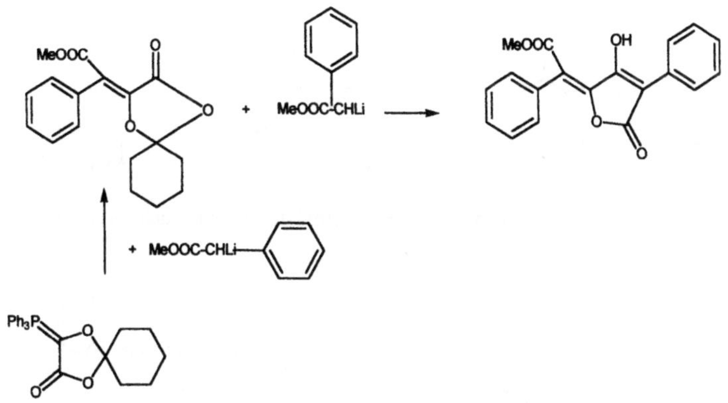

Scheme 166. Biomimetic synthesis of pulvinic acids after Ramage *et al.* (*606*)

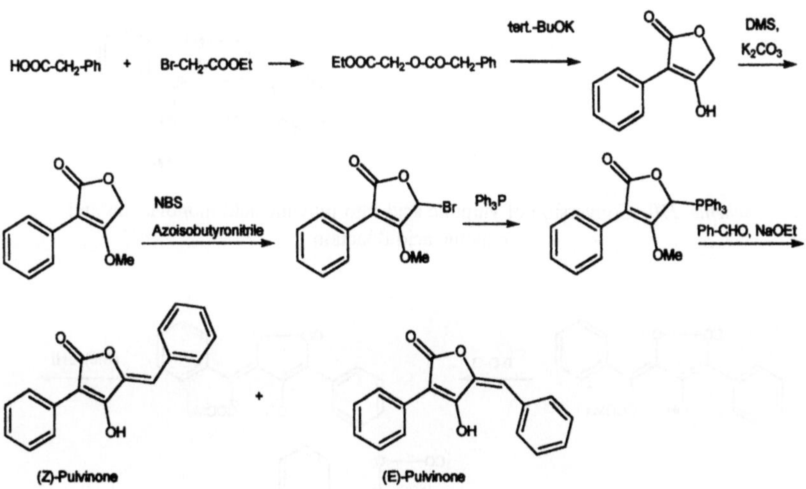

Scheme 167. Synthesis of pulvinones after Campbell *et al.* (*53*)

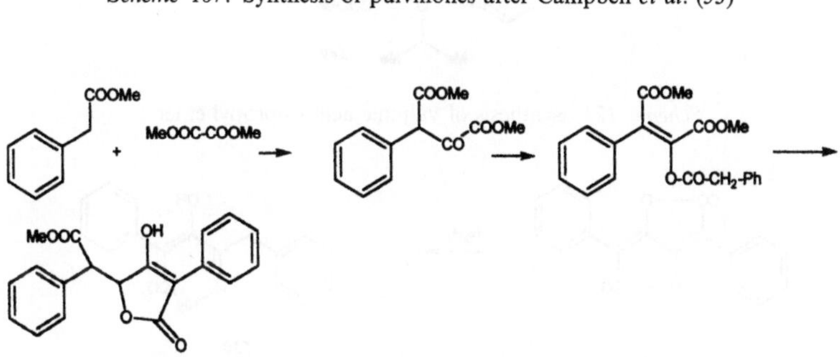

Scheme 168. Regiospecific synthesis of vulpinic acid after Weinstock (*717*)

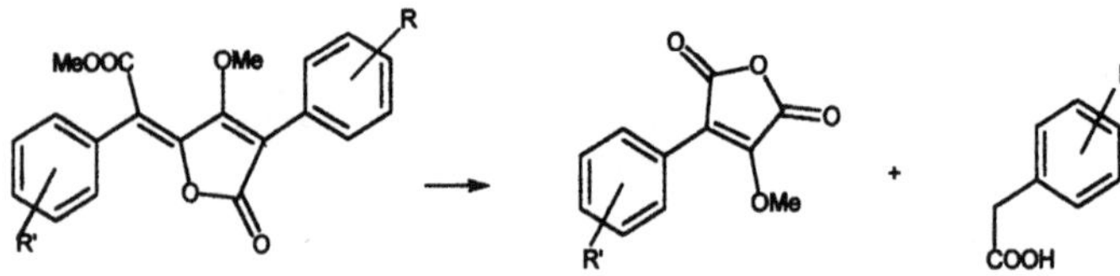

Scheme 169. Alkaline cleavage of fully methylated pulvinic acid derivatives

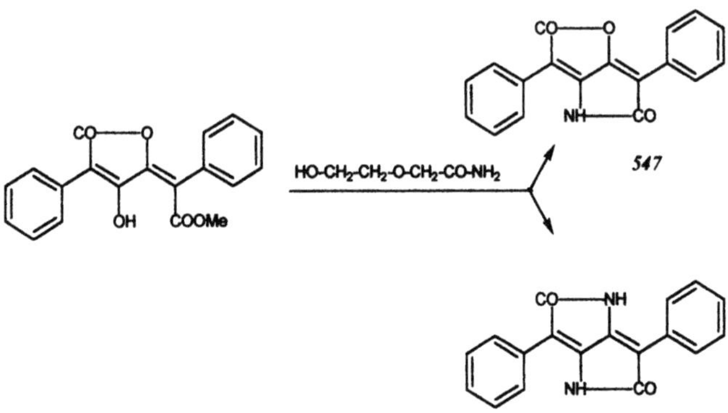

548

Scheme 170. Conversion of vulpinic acid into pulvinic acid monolactam and pulvinic acid dilactam

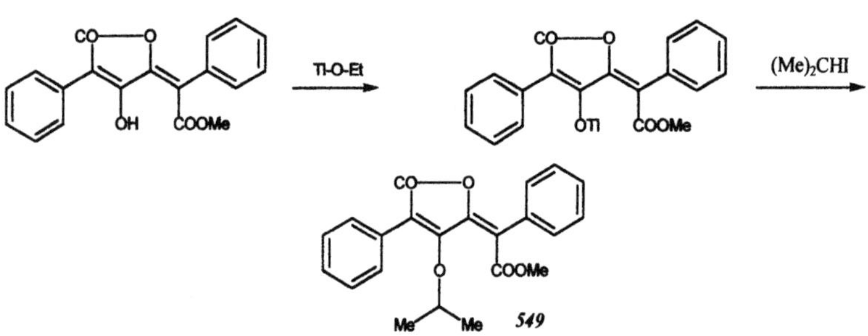

Scheme 171. Synthesis of vulpinic acid isopropyl ether

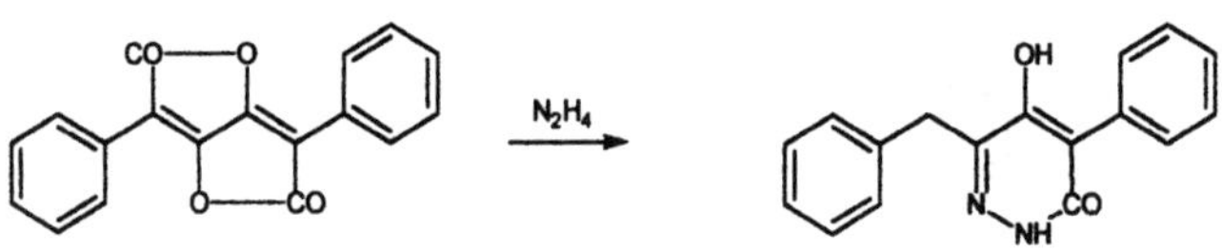

550

Scheme 172. Conversion of pulvinic acid dilactone to the pyridazine **550**

References, pp. 239–276

4.6. Terpenoids

4.6.1. Monoterpenoids

The following well-known monoterpenoids have been described from lichens, but without identification of their optical rotation: borneol, camphene, camphor, 1,8-cineol, *p*-cymene, fenchol, limonene, linalool, α-pinene, β-pinene, terpinen-4-ol, α-terpineol, α-thujone, β-thujone [all from *Evernia prunastri* (*319*)], and carvone [from *Cetraria islandica* (*658*)].

4.6.2. Sesquiterpenoids

Only a few sesquiterpenoids have been encountered in studies of lichens: β-caryophyllene, α-copaene, β-elemene, farnesane, β-gurjunene, longifolene, α-muurolene, β-selinene, and thujopsene from *Evernia prunastri* (*319*) and fukinanolide A (bakkanolide A) from *Cetraria islandica* (*658*). Abscisic acid (ABA) was detected in all 26 lichen species investigated (*175*).

4.6.3. Diterpenoids

Phytol and manool have been isolated from *Letharia vulpina* (*685*), rimuene was isolated from *Evernia prunastri* (*561*) and (−)-sandaracopimaric acid from *Ramalina hierrensis* (*330*).

4.6.4. Sesterterpenoids

Retigeranic acid, a sesterterpenoid from the lichen *Lobaria retigera*, was separated by HPLC into two stereoisomers, retigeranic acid A (**551**) and retigeranic acid B (**552**); the structures of both components were determined by X-ray analysis (*678*).

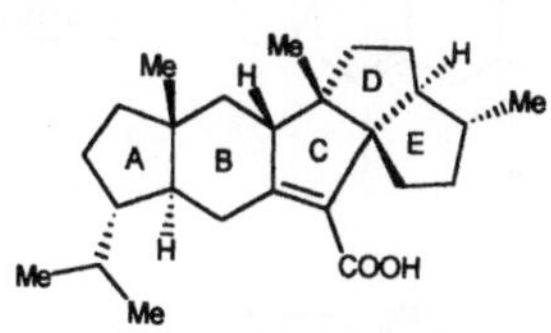

551 : (-)-Retigeranic acid A

mp 200-202°C

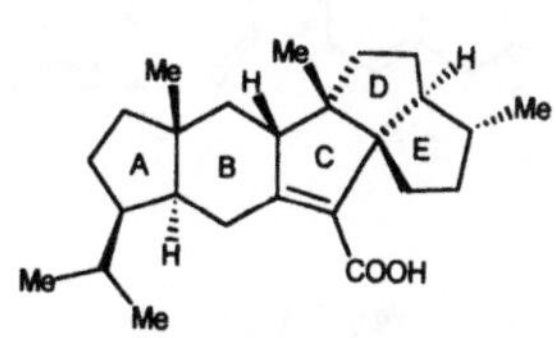

552 : (-)-Retigeranic acid B

mp 188-190°C

The structures of these rather complex sesterterpenoids were a challenge to organic chemists and between 1982 and 1990 no less than 5 syntheses of the retigeranic acids have been published. Corey *et al.*

(*83*) synthesized (±)-retigeranic acid *via* the ring annelation AB → ABC → ABCDE shown in Scheme 173. Paquette *et al.* (*570*) used (−)-S-limonene as a precursor of ring A for the synthesis of (−)-methyl retigeranate A (see Scheme 174). Hudlicky *et al.* (*377*, *378*) followed strategy A → AB + DE → ABCDE for their synthesis of (−)-retigeranic

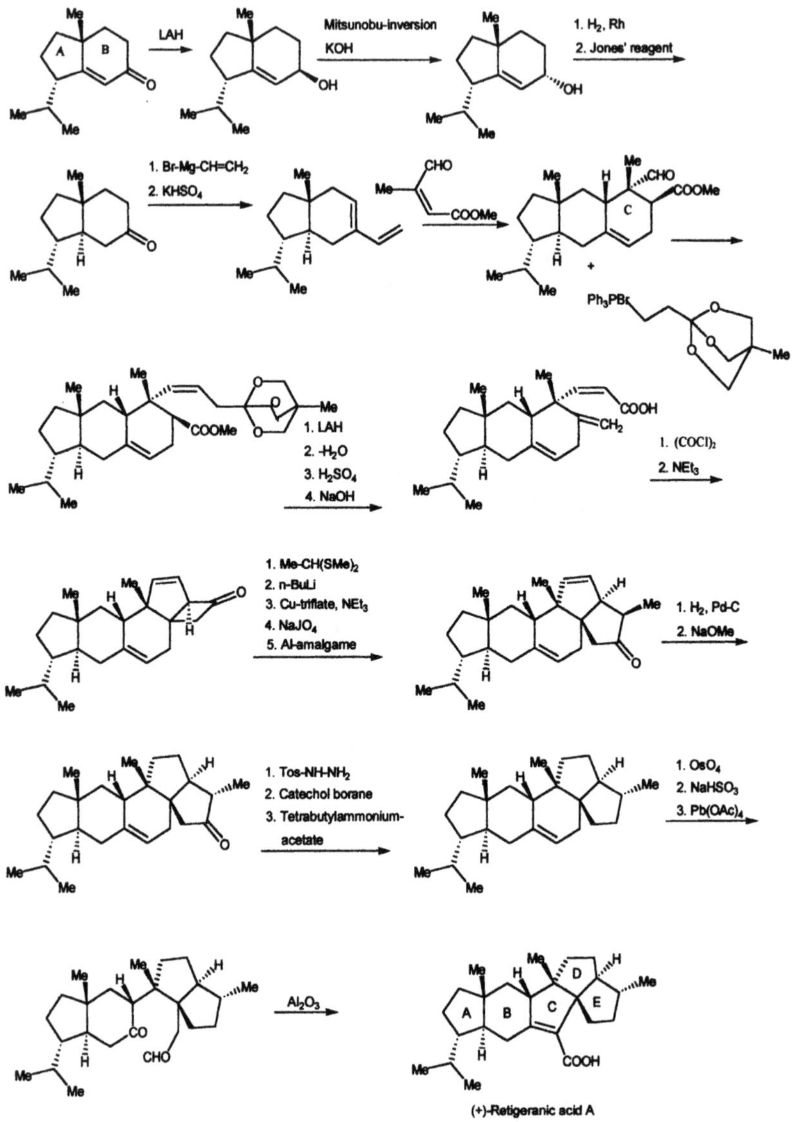

Scheme 173. Synthesis of (±)-retigeranic acid after Corey *et al.*

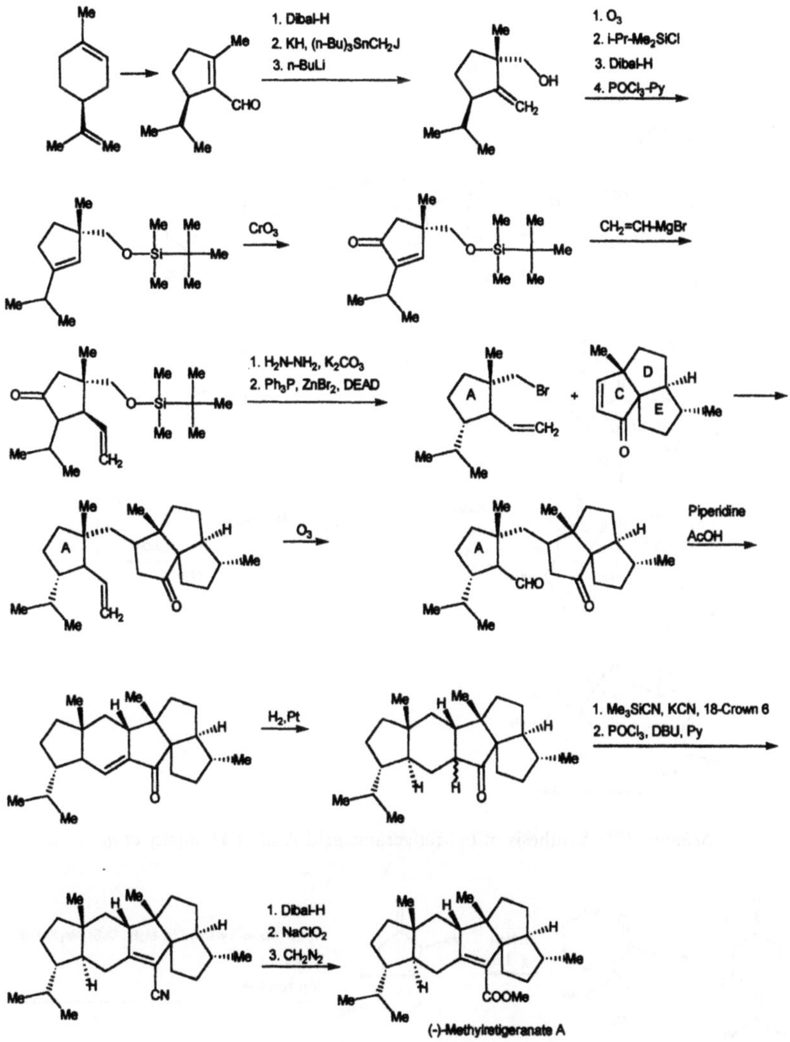

Scheme 174. Synthesis of (−)-methyl retigeranate A after Paquette *et al.*

acid A (see Scheme 175) and Wright *et al.* (*740*) used (+)-R-pulegone and (−)-S-limonene as starting materials for the synthesis of (−)-retigeranic acid (see Scheme 176), while a convergent synthesis of (−)-retigeranic acid A, based on a Diels-Alder and arene-alkene cycloaddition (CDE → ABCDE) strategy was described by Wender and Singh (*718*) (see Scheme 177).

Syntheses of potential retigeranic acid intermediates (*379*, *509*, *621*) are described in Schemes 178, 179 and 180.

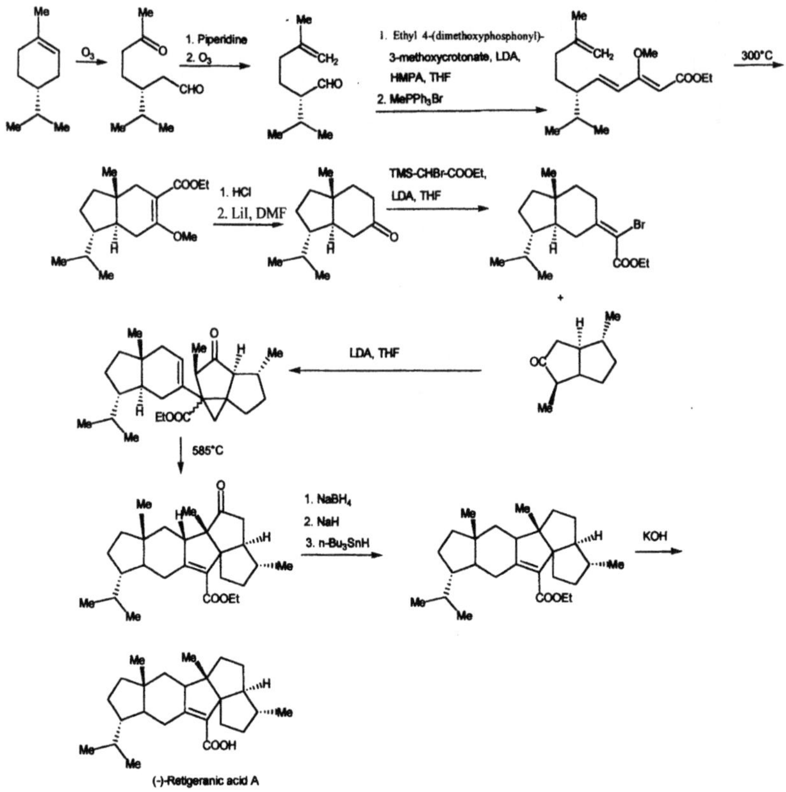

Scheme 175. Synthesis of (−)-retigeranic acid A after Hudlicky *et al.*

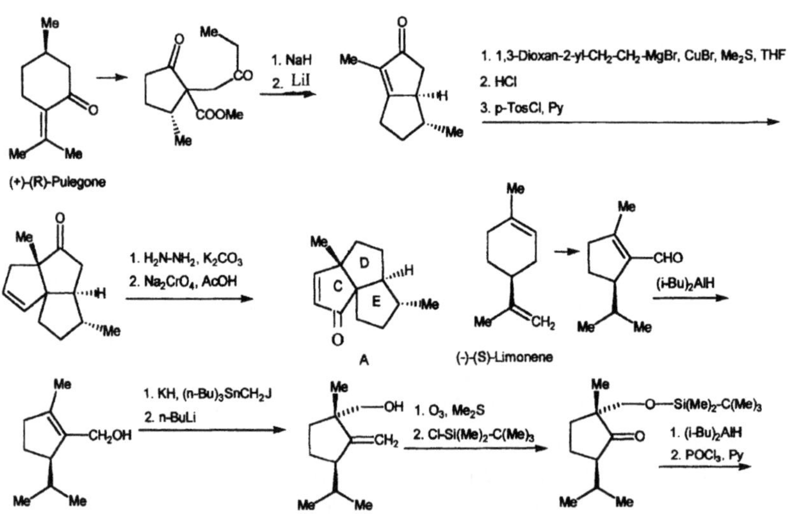

Scheme 176. Synthesis of (−)-retigeranic acid A after Wright *et al.*

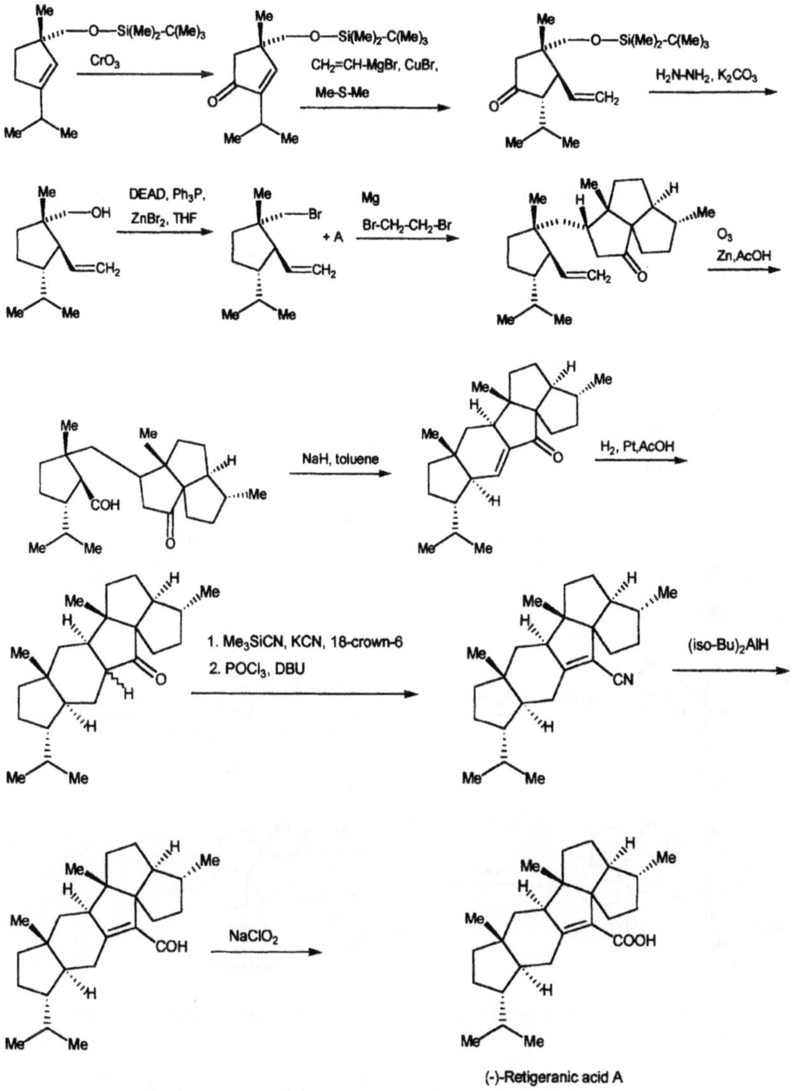

Scheme 176 (continued)

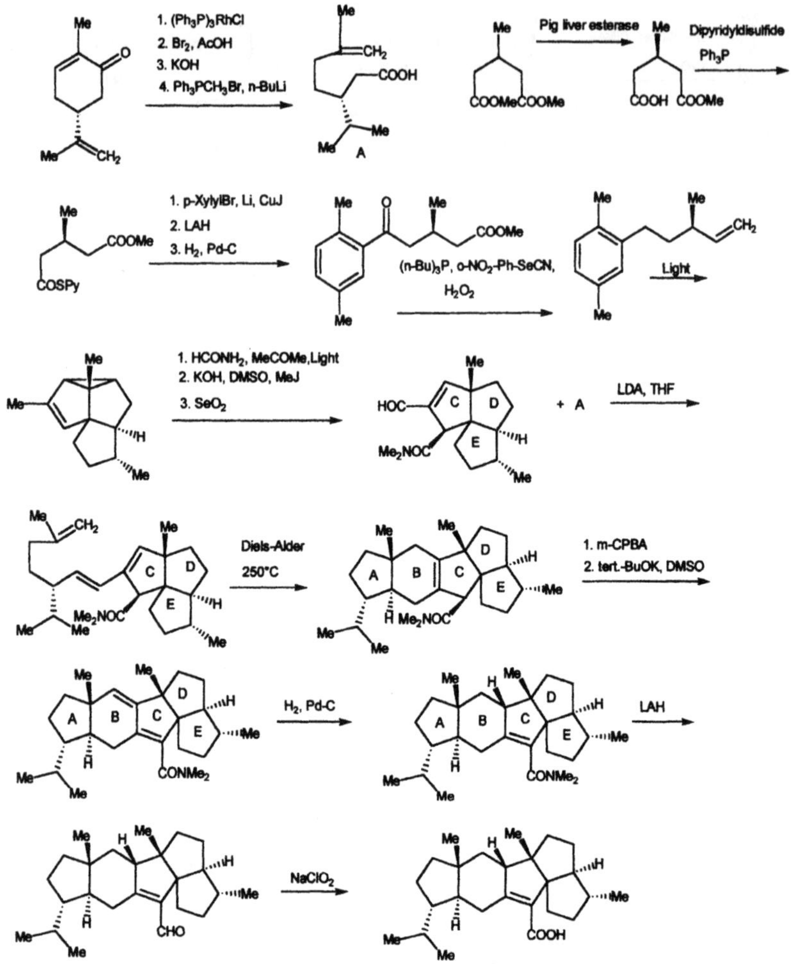

(-)-Retigeranic acid A

Scheme 177. Synthesis of (−)-retigeranic acid A after Wender and Singh

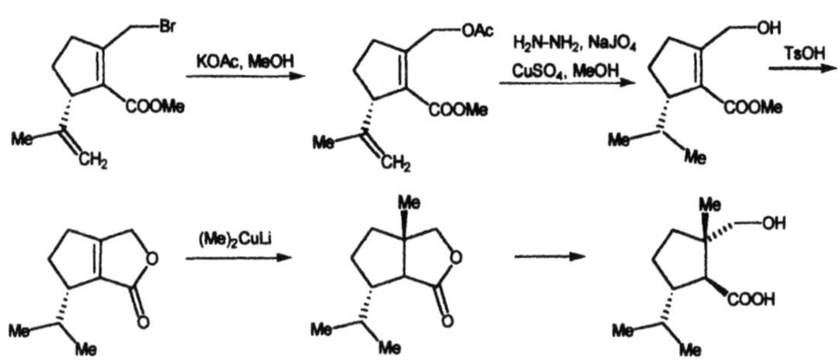

Scheme 178. Synthesis of a ring A intermediate of retigeranic acid A after Roberts

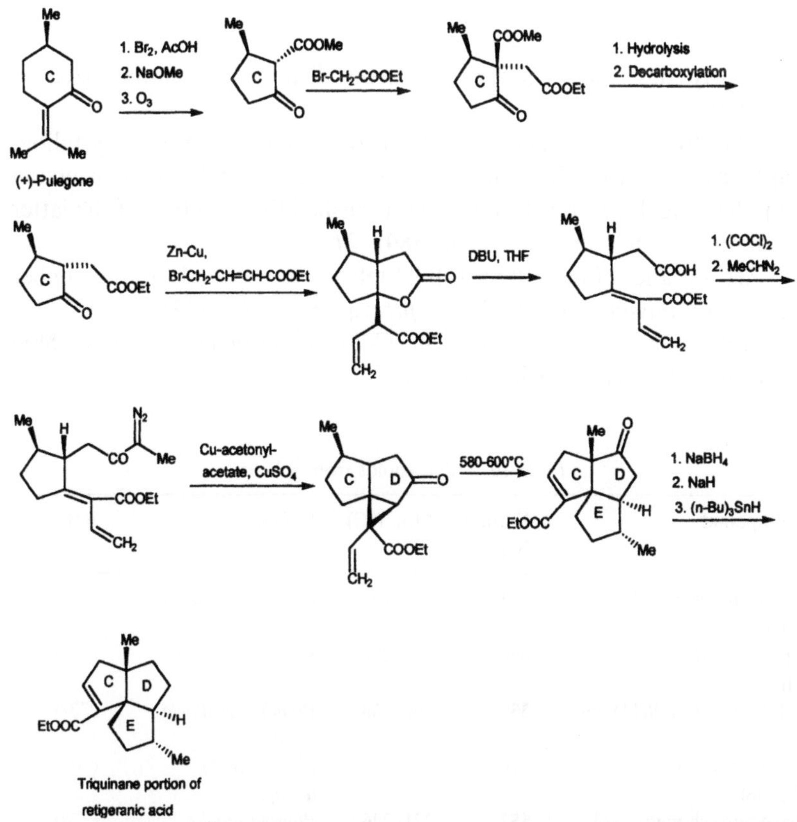

Scheme 179. Synthesis of the triquinane portion of retigeranic acid

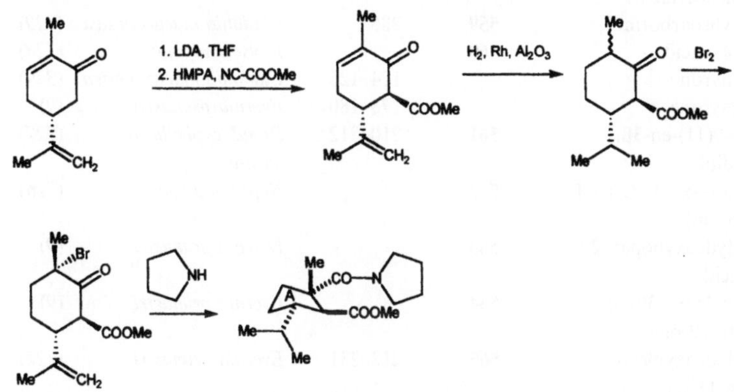

Scheme 180. Synthesis of ring A intermediate of retigeranic acid A after Llera and Fraser-Reid

4.6.5. Triterpenoids

Newly described triterpenoids from lichens are summarized in Table 10.

The stucture of phlebic acid D was determined by X-ray crystallography as 7β-acetoxy-22-hydroxyhopan-27-oic acid (**581**). Its conversion to phlebic acid C (see Scheme 181) established the structure of the latter as 22-hydroxyhopan-27-oic acid (**580**) (*24*).

Zeorin and hopan-3β,6α,15β,22-tetrol were found in the higher plants *Iris missouriensis* (*737*), *Tripterygium regelii* (*435*) and *Mollugo pentaphylla* (*663*) respectively. The C-70 polyterpene ficaprenol (**588**) was found in serveral lichens (*618*).

Table 10. Triterpenoids from lichens

Compound	Formula No.	M.p. (°C)	Lichen	Ref.
3β-Acetoxyfern-9(11)-en-12β-ol	**553**	230–232	*Pseudocyphellaria aurata*	(*727*)
3β-Acetoxyfern-9(11)-en-19β-ol	**554**	200–202	*Pseudocyphellaria aurata*	(*727*)
3β-Acetoxyfern-9(11)-en-12-one	**555**	266–268	*Pseudocyphellaria aurata*	(*727*)
3β-Acetoxyhopan-1β, 22-diol	**556**	273–275	*Canoparmelia texana*	(*403*)
6α-Acetoxyhopan-22-ol (Acetylzeorin, Lesdainin)	**557**	221–226	*Physcia aipolia*	(*728*)
			Lepraria lesdainii	(*484, 503*)
15α-Acetoxyhopan-22-ol (Dolichorrhizin)	**558**	200–205	*Nephroma parile*	(*617*)
Acetylisoarborinol	**559**	285	*Cladonia macaronesica*	(*329*)
Aipolic acid	**560**		*Physcia aipolia*	(*728*)
α-Amyrone		124–126	*Cladonia macaronesica*	(*329*)
β-Amyrone		178–180	*Evernia prunastri*	(*9*)
Fern-9(11)-en-3β, 12β-diol	**561**	210–212	*Pseudocyphellaria aurata*	(*727*)
Hopan-6α,7β, 22-triol (Nephrin)	**562**		*Nephroma* spec.	(*726*)
3β-Hydroxyhopan-29-oic-acid	**563**		*Evernia prunastri*	(*9*)
3β-Hydroxy-30-nor-hopan-22-one	**564**		*Evernia prunastri*	(*9*)
3β-Hydroxyolean-12-en-11-one	**565**	212–231	*Evernia prunastri*	(*727*)
22α-Hydroxy-3,4-secostictan-3-oic acid	**566**			(*82*)

Table 10 (continued)

Compound	Formula No.	M.p. (°C)	Lichen	Ref.
3β-Hydroxyserrat-14-en-21-one	**567**		*Evernia prunastri*	(*684a*)
21α-Hydroxyserrat-14-en-3-one	**568**		*Evernia prunastri*	(*684a*)
Isomultiflorenone	**569**	188–195	*Evernia prunastri*	(*9*)
Lanosterol		138–144	*Evernia prunastri, Usnea articulata*	(*353*)
Lupeone		160	*Evernia prunastri*	(*9*)
			Cladina macaronesica	(*329*)
21α-Methoxyserrat-14-en-3β,16β-diol	**570**		*Evernia prunastri*	(*684a*)
3β-Methoxyserrat-14-en-21α-ol	**571**		*Evernia prunastri*	(*684a*)
21α-Methoxyserrat-14-en-3β-ol	**572**		*Evernia prunastri*	(*684a*)
3β-Methoxyserrat-14-en-21-one	**573**		*Evernia prunastri*	(*684a*)
Methyl aipolate	**574**		*Physcia aipolia*	(*728*)
Moretenone	**575**		*Evernia prunastri*	(*562*)
30-Nor-hopan-3, 22-dione	**576**		*Rhizoplaca chrysoleuca*	(*687*)
			Usnea hesperina	(*353*)
30-Nor-21αH-hopan-3, 22-dione	**577**		*Evernia prunastri*	(*562*)
			Rhizoplaca chrysoleuca	(*687*)
30-Nor-21αH-hopan-22-one (Isoadiantone)	**578**	234–235	*Platismatia glauca*	(*428*)
30-Nor-21αH-7β-hydroxy-hopan-22-one	**579**		*Pseudocyphellaria sulphurea*	(*629*)
Phlebic acid C	**580**		*Peltigera aphthosa*	(*24*)
Phlebic acid D	**581**		*Peltigera aphthosa*	
			Conversion of phlebic acid D into phlebic acid C see Scheme 181	(*24*)
Serrat-14-en-3β,21α-diol	**582**		*Evernia prunastri*	(*684a*)
Serrat-14-en-3,21-dione	**583**		*Evernia prunastri*	(*684a*)
Stictan-3β,22α-diol (Retigeradiol)	**584**	283	*Pseudocyphellaria flavicans*	
			X-ray analysis	(*82*)
Taraxerone		240–242	*Cladina macaronesica*	(*329*)
22,29,30-Trinorhopan-3, 21-dione	**585**		*Rhizoplaca chrysoleuca*	(*687*)
Tylolupenylacetate	**586**		*Usnea articulata*	(*353, 684*)
Zeorinone	**587**	251–255	*Rinodina thiomela*	(*728*)

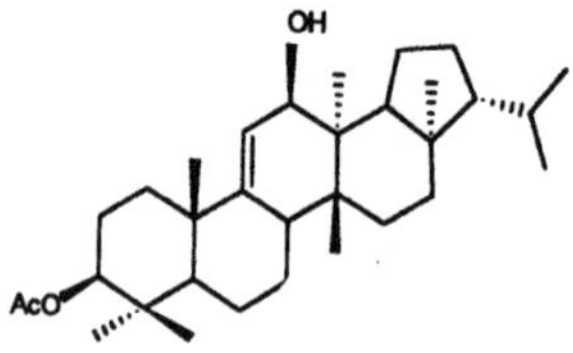

553

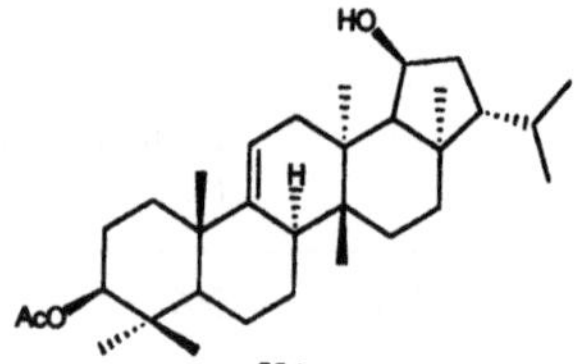

554

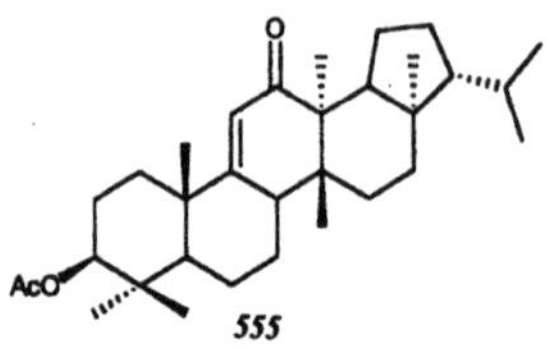

555

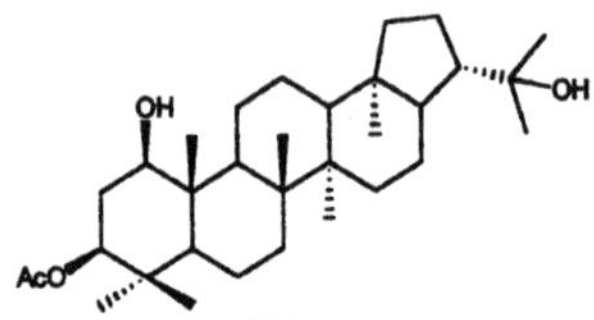

556

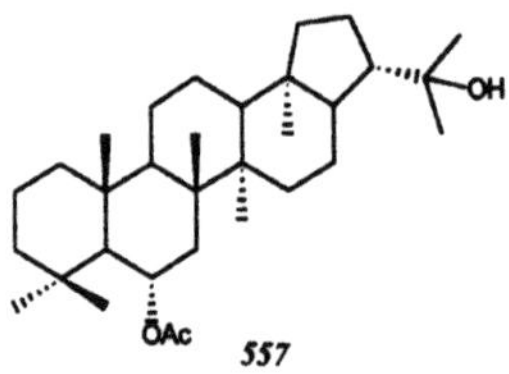

557

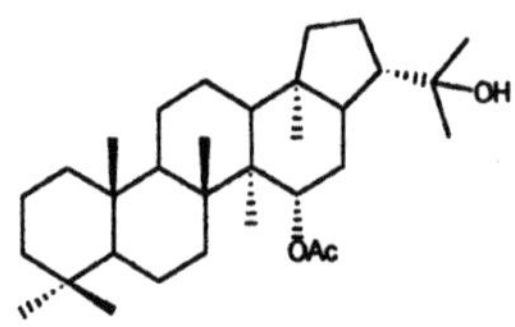

558

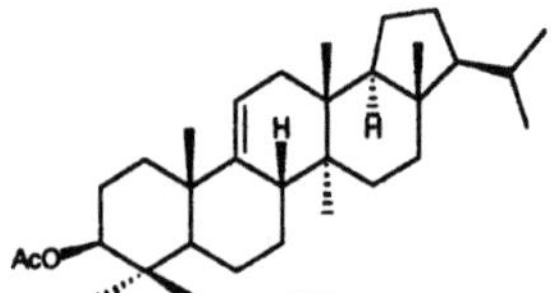

559

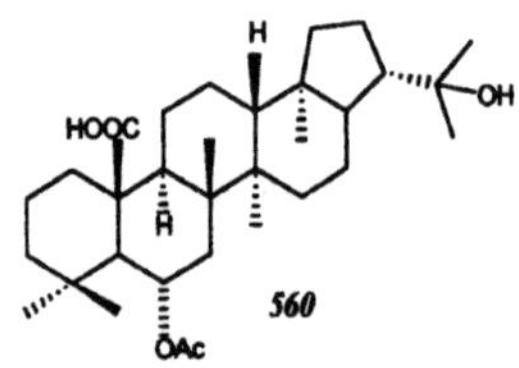

560

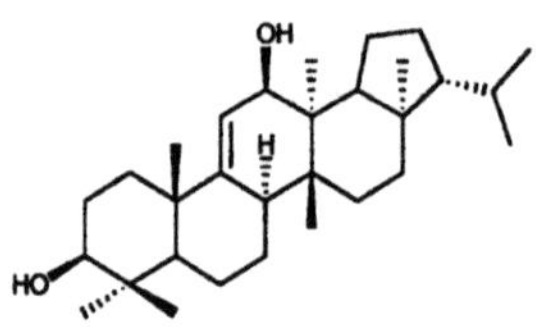

561

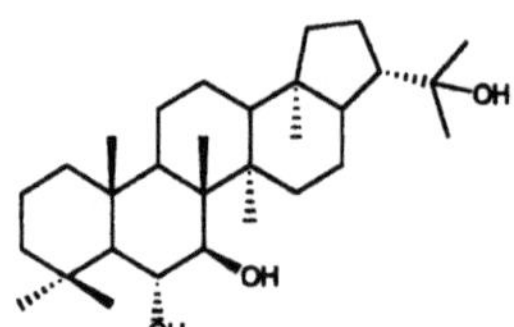

562

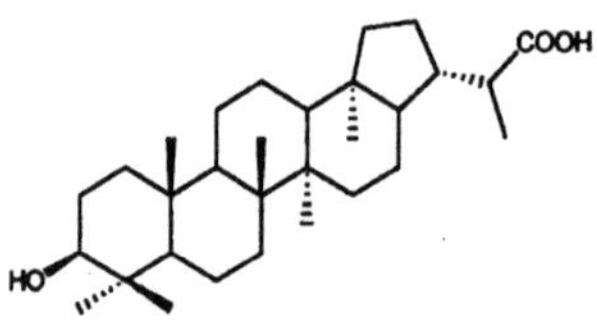

563

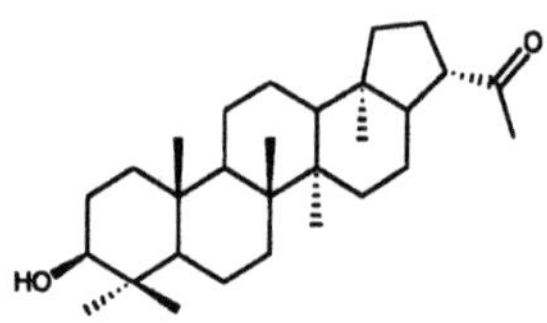

564

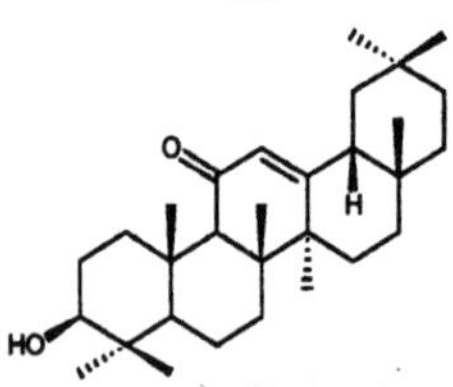

565

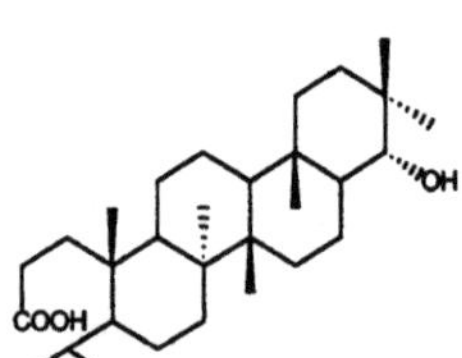

566

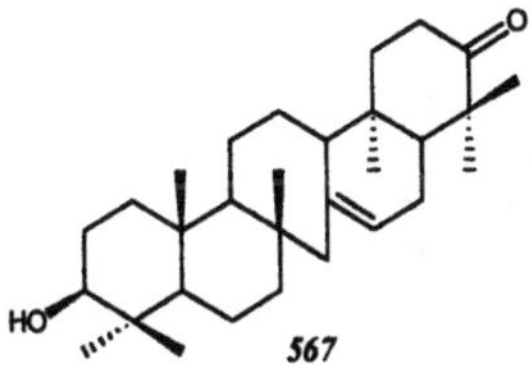

567

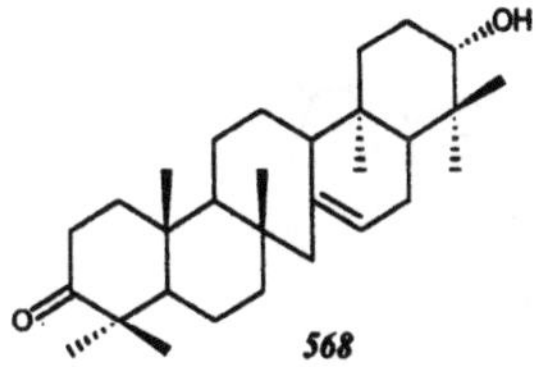

568

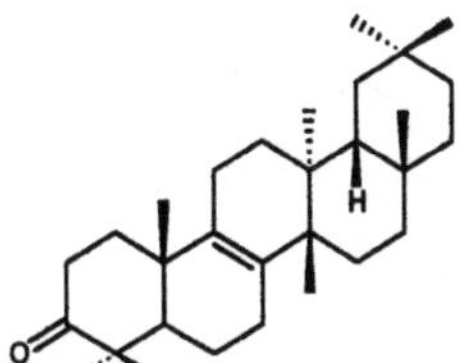

569

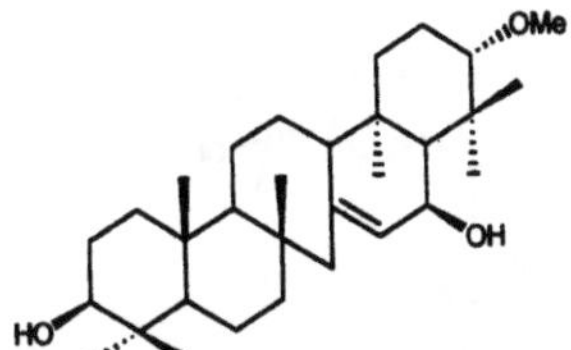

570

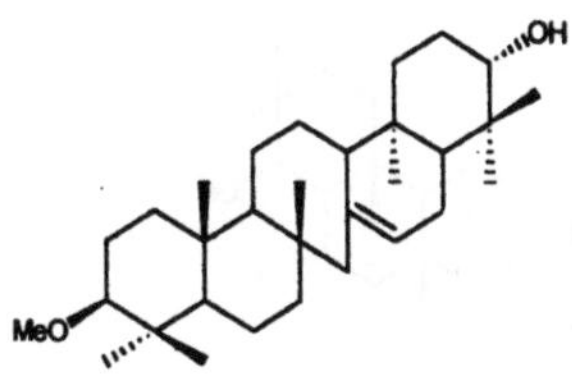

571

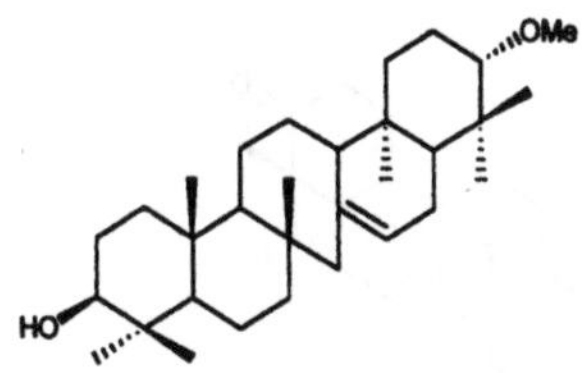

572

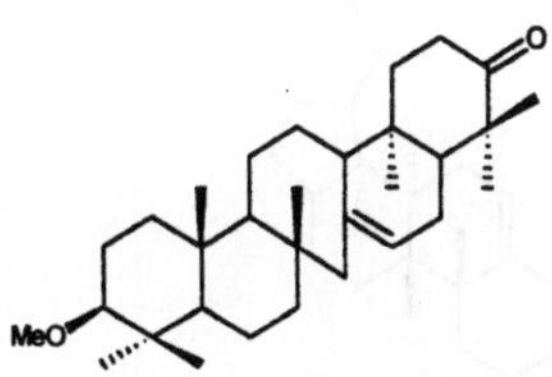

573

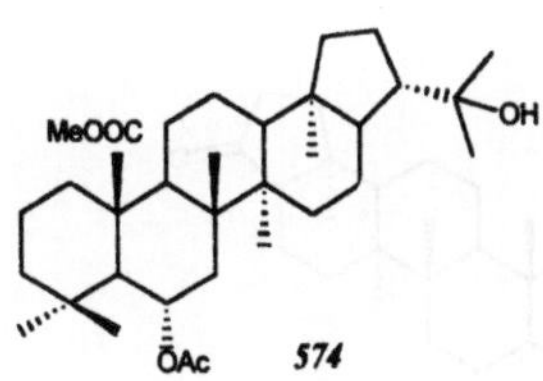

574

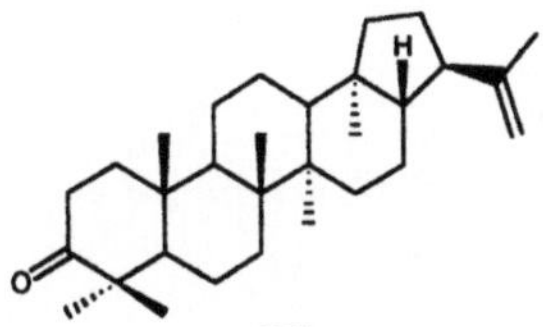

575

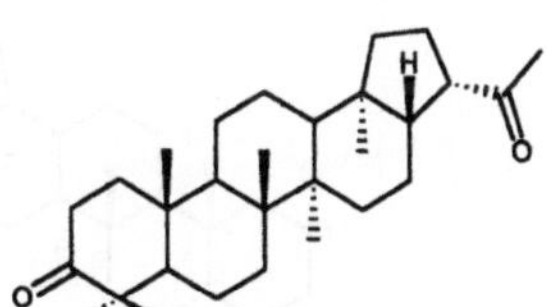

576

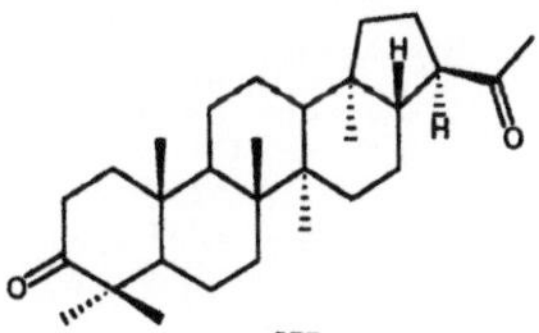

577

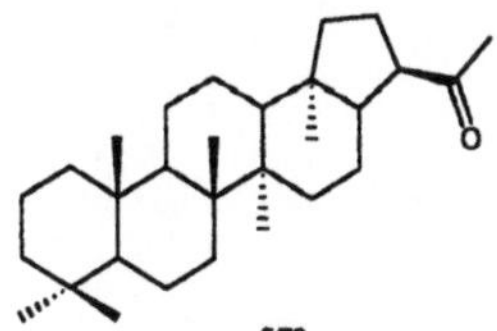

578

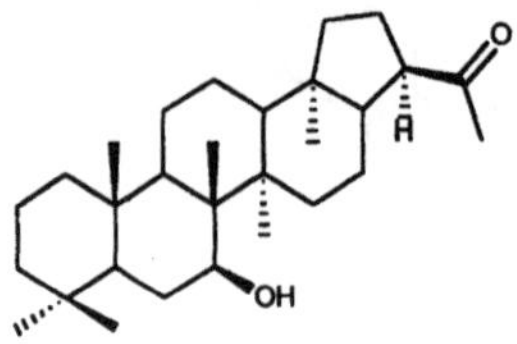

579

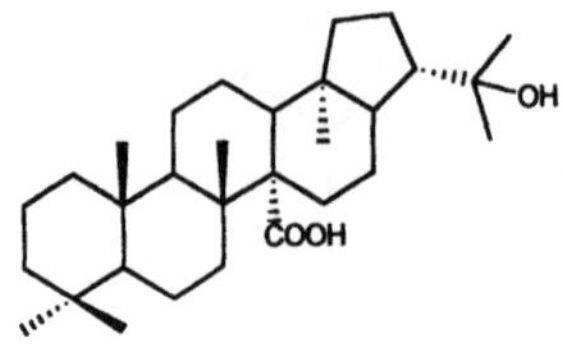

580

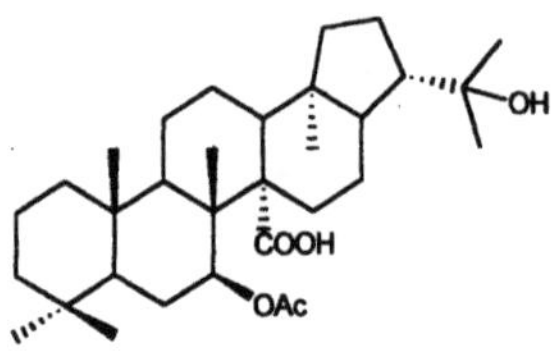

581

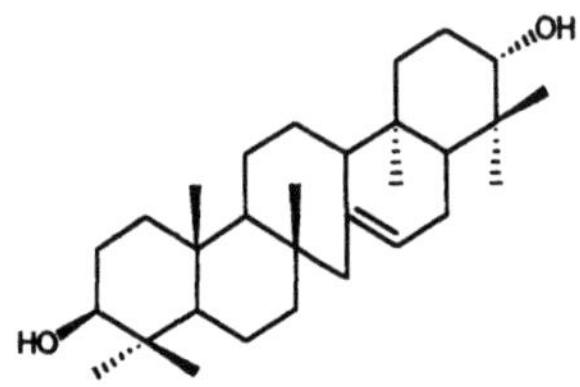

582

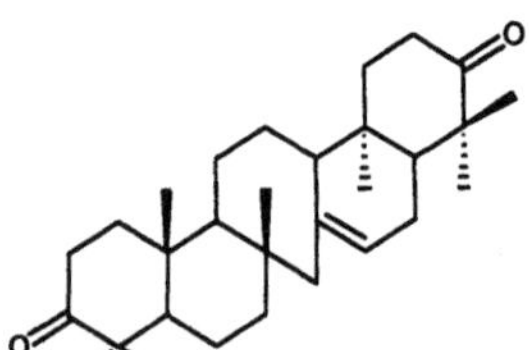

583

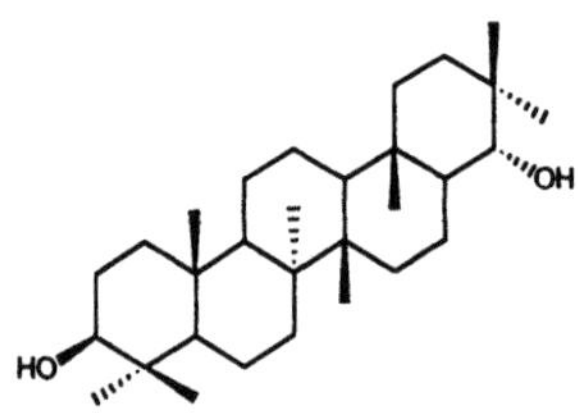

584

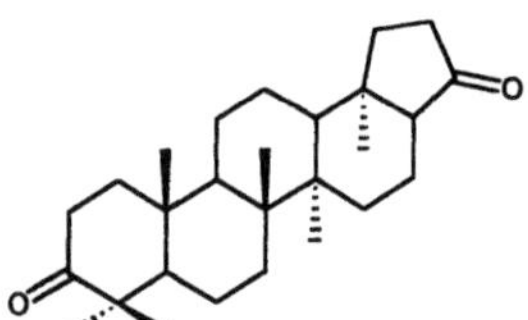

585

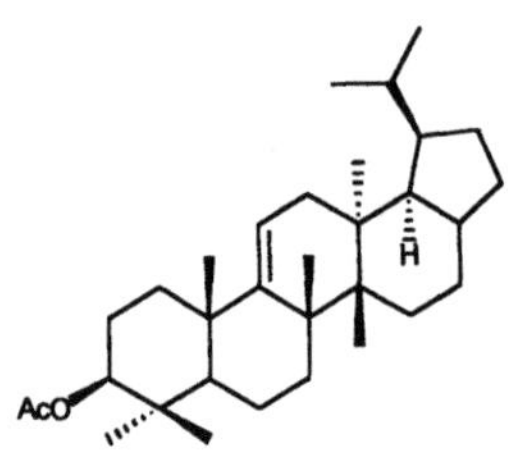

586

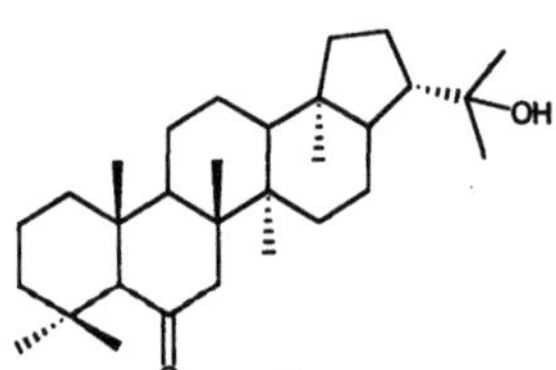

587

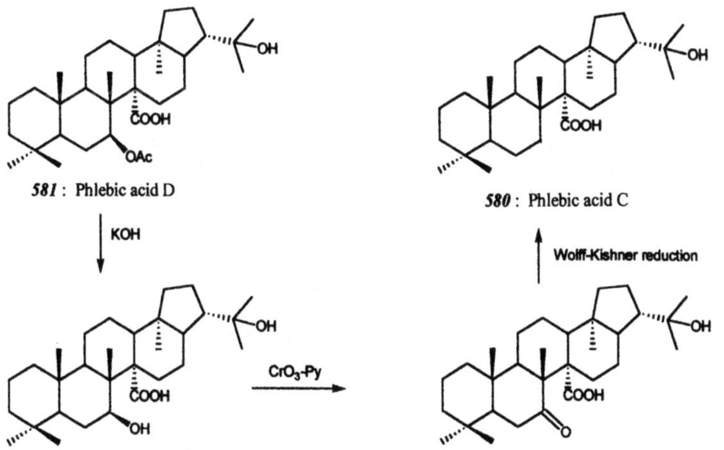

Scheme 181. Conversion of phlebic acid D to phlebic acid C

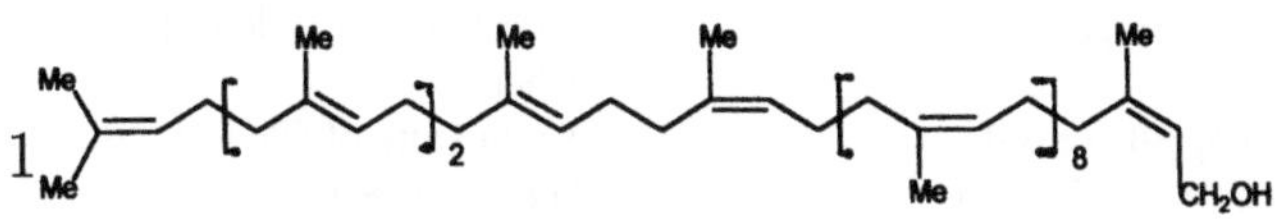

588 : Ficaprenol

4.6.6. *Steroids*

Newly described steroids are summarized in Table 11.

A synthesis of lichesterol (**599**) has been described (see Scheme 182) (*11*).

Table 11. Steroids from lichens

Compound	Formula No.	Lichen	Ref.
Cerevisterol	**589**	*Ramalina hierrensis*	(*330*)
Cholesta-3,5-dien-7-one		*Umbilicaria cylindrica*	(*406*)
Cholest-8-en-3β-ol		*Pseudevernia furfuracea*	(*339*)
Cholest-4-en-3-one		*Pseudevernia furfuracea*	(*339*)
Ergosta-5,22-dien-3β-ol		*Pseudevernia furfuracea*	(*339*)
		Lobaria pulmonaria, L. scrobiculata, Usnea longissima	(*626*)
Ergosta-7,22-dien-3β-ol		*Pseudevernia furfuracea*	(*339*)
		Lecanora stenotropa, Parmelia omphalodes, Usnea antartica	(*687*)

Table 11 (continued)

Compound	Formula No.	Lichen	Ref.
Ergosta-7,24(28)-dien-3β-ol		*Pseudevernia furfuracea*	(*339*)
Ergosta-8,22-dien-3β-ol		*Pseudevernia furfuracea*	(*339*)
Ergosta-3,5-dien-7-one		*Umbilicaria cylindrica*	(*406*)
Ergostan-3β-ol		*Pseudevernia furfuracea*	(*339*)
Ergosta-5,7,9(11),22-tetraen-3β-ol		*Pseudevernia furfuracea*	(*339*)
		Evernia mesomorpha, Lecanora stenotropa, Leptogium saturninum, Parmelia omphalodes, Ramalina terebrata, Umbilicaria decussata, Usnea antarctica	(*687*)
Ergosta-5,7,22-trien-3β-ol		*Pseudevernia furfuracea*	(*339*)
Ergost-7-en-3β-ol		*Pseudevernia furfuracea*	(*339*)
Ergosta-4-en-3-one		*Pseudevernia furfuracea*	(*339*)
Ergosterol-5α,8α-peroxide	**590**	Ergosterol peroxide from lichens	
Ergosterol-5β,8β-peroxide	**591**	is a mixture of **590** and **591**	(*18*)
24-Ethylcholesta-5,22-dien-3β-ol		*Lobaria pulmonaria, L. scrobiculata, Usnea longissima*	(*626*)
		Evernia mesomorpha, Lecanora stenotropa, Leptogium saturninum, Nephroma helveticum, Parmelia omphalodes, Ramalina terebrata, Umbilicaria decussata, Usnea antartica	(*687*)
Fremontol	**592**	*Bryoria fremontii*	(*655*)
24-Methyl-23-dehydrolophenol	**593**	*Umbilicaria cylindrica*	(*406*)
4α-Methylergosta-7,24(28)-dien-3β-ol	**594**	*Pseudevernia furfuracea*	(*339*)
4α-Methylergost-7-en-3β-ol	**595**	*Pseudevernia furfuracea*	(*339*)
Stigmasta-7,22-dien-3β-ol	**596**	*Pseudevernia furfuracea*	(*339*)
Stigmastan-3β-ol	**597**	*Pseudevernia furfuracea*	(*339*)
Stigmasta-3,5,22-trien-7-one	**598**	*Umbilicaria cylindrica*	(*406*)
Stigmast-5-en-3β-ol		*Pseudevernia furfuracea*	(*339*)
Stigmast-7-en-3β-ol		*Pseudevernia furfuracea*	(*339*)
		Collema polycarpon	(*405*)
Stigmast-4-en-3-one		*Pseudevernia furfuracea*	(*339*)
		Cladonia furcata	(*9*)
		Umbilicaria cylindrica	(*406*)
Stigmasterol		*Usnea articulata*	(*353*)
		Lobaria pulmonaria	(*659*)

References, pp. 239–276

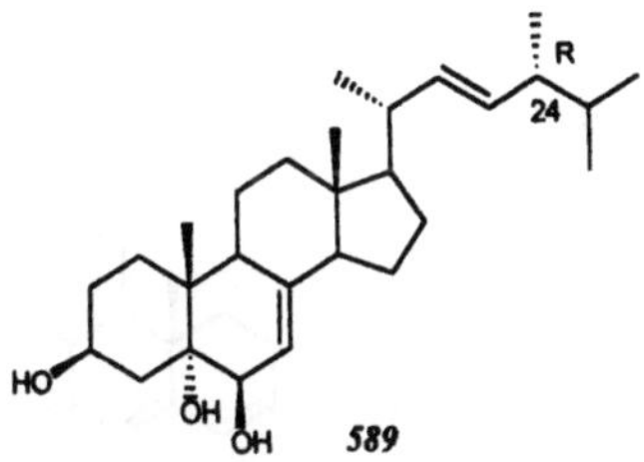

589

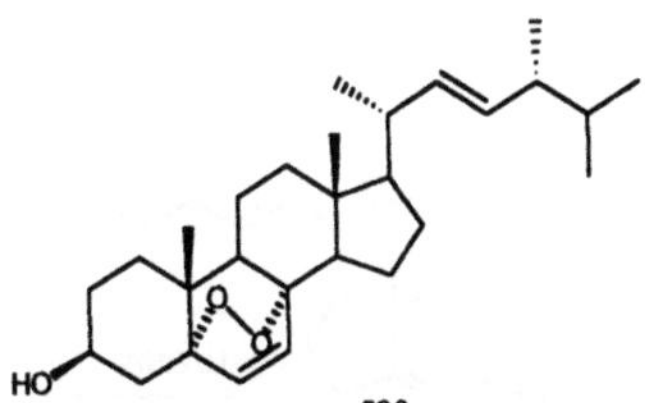

590

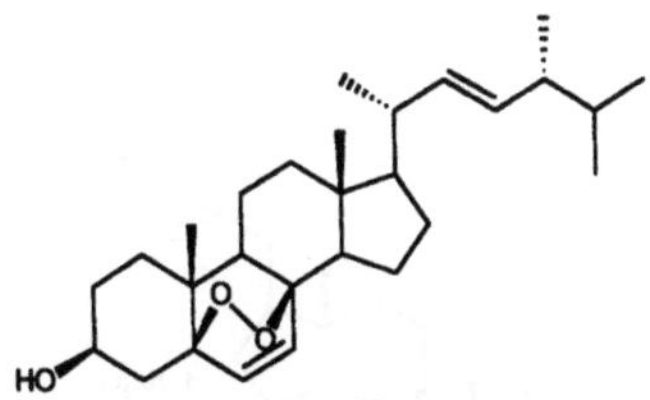

591

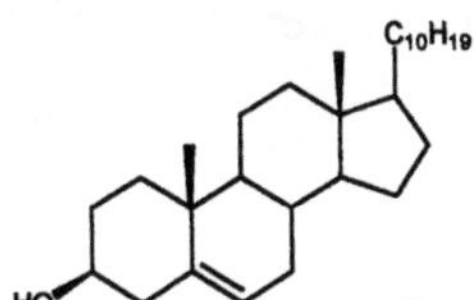

592

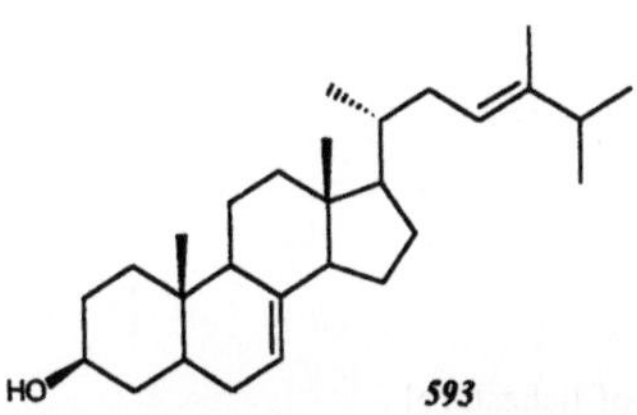

593

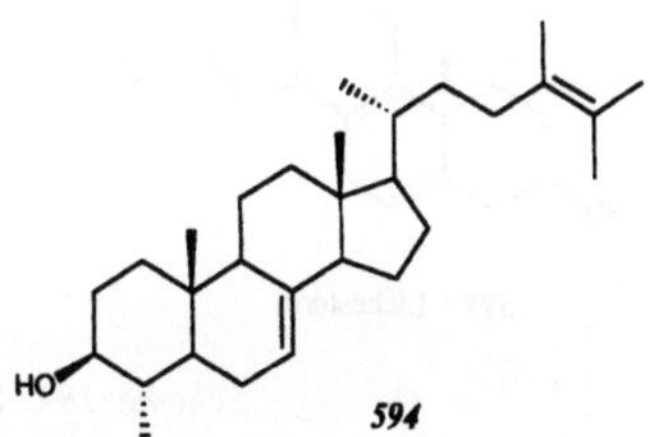

594

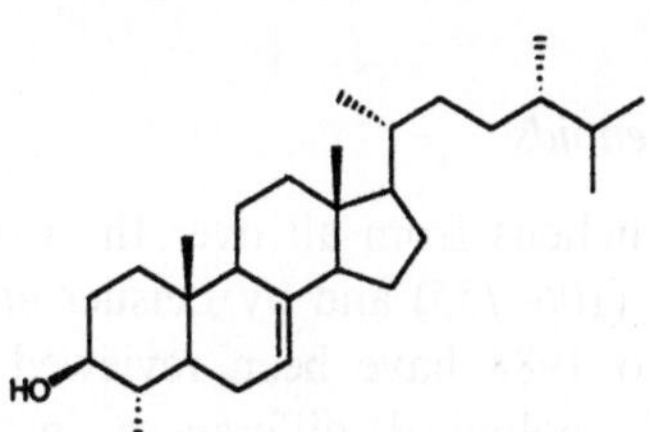

595

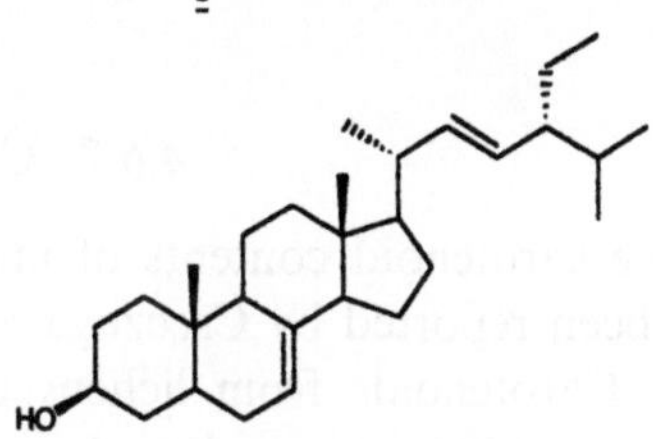

596

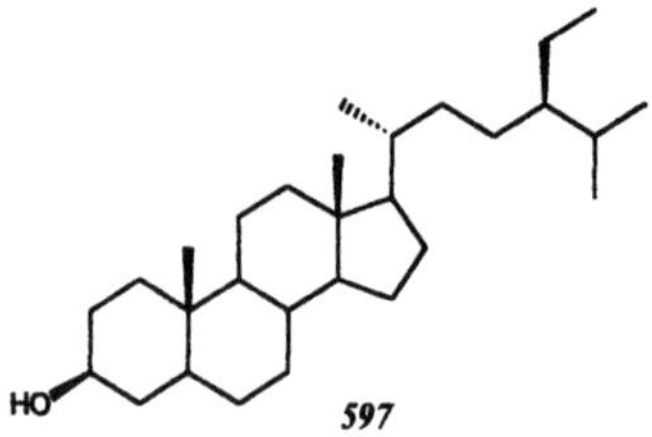

597

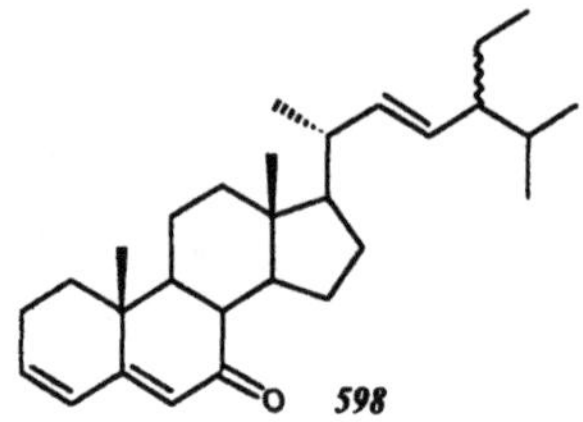

598

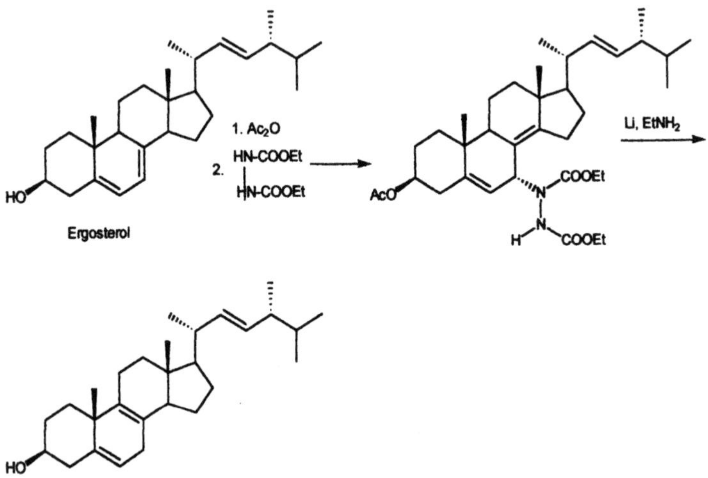

599 : Lichesterol

Scheme 182. Synthesis of lichesterol

4.6.7. Carotenoids

The carotenoid contents of many lichens from all over the world have been reported by Czeczuga *et al.* (*106–155*) and by Leisner *et al.* (*495*). Carotenoids from lichens up to 1988 have been reviewed by Czeczuga (*109*) who also discussed ecological differences in the carotenoid content (*111*).

Table 12. Metabolites from lichen mycobionts

Lichen mycobiont	Metabolites	Ref.
Aspergillus terreus	β-Orcinolcarboxylic acid	(*691*)
Baeomyces rufus	Ergosterol, triglyceride	(*616*)
Caloplaca ferruginea	7-Chloroemodin	(*615*)
Cetraria islandica	Usnic acid	(*757*)
Cetraria nivalis	Squamatic acid, barbatic acid	(*757*)
Cladonia grayi	4-*O*-Demethylsphaerophorin, grayanic acid, 4-*O*-demethylgrayanic acid	(*90*)
Cladonia polycarpia	Homoheveadride	(*105*)
Cladonia polycarpoides	Homoheveadride	(*105*)
Cladonia vulcani	Cytochromes	(*307*)
Evernia prunastri	Usnic acid	(*757*)
Graphis desquamescens	Graphenone, graphisquinone	(*549*)
Graphis scripta	Graphenone, graphisquinone, graphislactone A–D	(*549*, *693*)
Lecanora dispersa	Pannarin, dechloropannarin	(*497*)
Pseudephebe pubescens	Usnic acid	(*757*)
Ramalina duriaei	Usnic acid	(*757*)
Ramalina siliquosa	4-*O*-Demethylbarbatic acid, norbarbatic acid, norobtusatic acid, norisoobtusatic acid, hypoprotocetraric acid, 4-*O*-Demethylbarbatic acid, norisonotatic acid, convirensic acid, virensic acid, protocetraric acid, hyposalazinic acid, connorstictic acid, norstictic acid, cryptostictic acid, stictic acid, salazinic acid, hypostictic acid, hypoconstictic acid, consalazinic acid, constictic acid, menegazzianic acid, usnic acid	(*98*, *344*, *346*)
Ramalina subbreviuscula	4-*O*-Demethylbarbatic acid	(*343*)
	Usnic acid	(*757*)
Ramalina subcomplanata	4-*O*-Demethylbarbatic acid	(*343*)
Ramalina yasudae	Usnic acid	(*748*)
Stereocaulon curtatum	Lecanoric acid	(*347*)
Usnea filipendula	Usnic acid	(*455*)
Usnea flexilis	Usnic acid	(*455*)
	Protocetraric acid, salazinic acid	(*758*)
Usnea hirta	Usnic acid, atranorin	(*455*, *457*, *757*)
Usnea montis-fuji	Usnic acid	(*757*)
Usnea orientalis	Isostrepsilic acid, hypostrepsilic acid	(*469*)
Usena roseola	Usnic acid	(*455*)
Usnea rubescens	Usnic acid	(*748*)
Usnea rubicunda	Usnic acid	(*455*)
Xanthoria elegans	Parietin	(*757*)
Xanthoria fallax	Parietin	(*757*)
Xanthoria mandschurica	Parietin	(*757*)
Xanthoria mawsonii	Parietin	(*757*)

5. Culture and Chemistry of Lichen Symbionts and Cell Culture of Lichens

5.1. Culture of Mycobionts

Crittenden and Porter (*87*) have stressed the value of lichen-forming fungi as potential sources of novel metabolites; the same group (*86*) has studied preparation of pure cultures of 1183 species of lichen-forming and lichenicolous fungi covering 77 families and 280 genera. Cell growth and pigment production in suspension cultures of the mycobiont from the lichen *Cladonia cristatella* was investigated by Yamamoto *et al.* (*742*) while Hamada *et al.* (*345*) studied different media for growing the mycobionts of *Lecanora bicincta*, *Lecidea tessellata*, *Rhizocarpon cupreoatrum* and *Usnea mutabilis*. Metabolites found in lichen mycobionts are summarized in Table 12.

5.2. Lichen Tissue Cultures and their Metabolites

The use of lichen tissue cultures and the development of lichen thalli *in vitro* have been discussed by Japanese workers (*741*, *745*, *759*). The method for cultivating lichen tissue (Yamamoto method) is demonstrated in Fig. 11. The cultivated lichens and the corresponding references are summarized in Table 13.

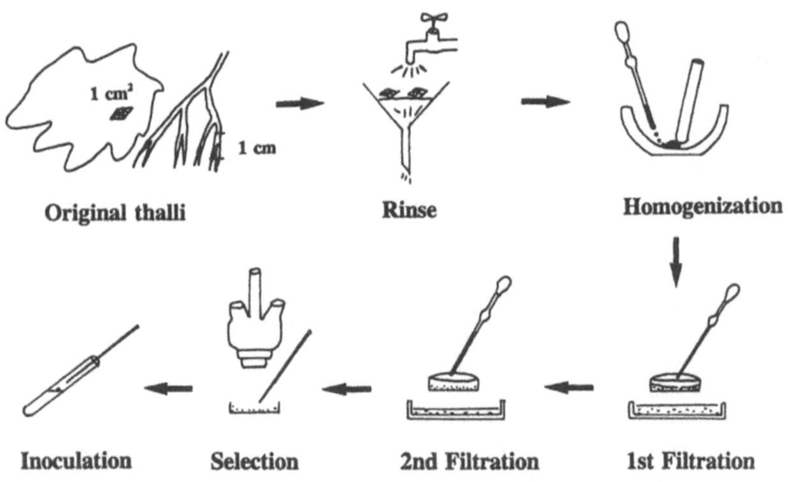

Fig. 11. Yamamoto method for cultivating lichen tissue

Table 13. Cultivated lichens

Lichen	Ref.
Alectoria lata	(*747*)
A. ochroleuca	(*747, 749*)
Evernia esorediosa	(*747, 749*)
E. prunastri	(*747, 749*)
Peltigera aphthosa	(*760*)
P. canina	(*760*)
P. malacea	(*760*)
P. praetextata	(*760, 761*)
P. pruinosa	(*760*)
Ramalina boninensis	(*747, 749*)
R. pacifica	(*747, 749*)
R. yasudae	(*747, 749*)
Rhizocarpon flavum	(*756*)
Umbilicaria aprina	(*749, 756*)
U. caroliniana	(*753*)
U. cylindrica	(*753*)
U. decussata	(*749, 756*)
U. esculenta	(*753*)
U. hyperborea	(*753*)
U. kisovana	(*753*)
U. muehlenbergii	(*749, 756*)
U. pensylvanica	(*756*)
U. polyrrhiza	(*753*)
U. proboscidea	(*753*)
U. vellea	(*753*)
Usnea bayleyi	(*749*)
U. bismolliuscula	(*747, 749*)
U. diffracta	(*747, 749*)
U. flexilis	(*749*)
U. longissima	(*747, 749*)
U. misaminensis	(*749*)
U. rubescens	(*747, 749*)
U. rubicunda	(*747, 749*)

5.3. Culture of Lichen Phycobionts

Yoshimura and Kurokawa (*753*) isolated the phycobionts (*Trebouxia*) from *Umbilicaria caroliniana*, *U. esculenta*, *U. hyperborea*, *U. muehlenbergii* and *U. pensylvanica*. Härmälä *et al.* (*354*) have cultivated the phycobionts of *Cetraria islandica*, *Cladina mitis*, *C. stellaris*, *Cladonia cornuta*, *C. gracilis* and *C. pleurota* but found no phenolic acids in the algae when these were analyzed by HPLC. The phycobionts of *Anzia colpota*, *A. hypoleucoides*, *A. japonica*, *A. opuntiella*, *A. ornata*, and *A. stenophylla* were studied taxonomically by Ihda *et al.* (*434*).

Kinoshita *et al.* (*454*) have described improved conditions for the isolation of protoplasts from the lichens *Cladia aggregata*, *Cladonia bellidiflora*, *C. boryi*, *C. pleurota* and *C. stellaris*.

5.4. Metabolites from Lichen Phycobionts

It is generally assumed that it is only the mycobiont which synthesize the typical lichen substances. Culberson *et al.* (*94*) examined the hypothesis that the algal partner in a lichen regulates the chemotype of the fungal partner and used *Cladonia chlorophaea* and *Trebouxia erici* from *Cladonia cristatella* as symbionts. They found that the production of cryptochlorophaeic acid or grayanic acid, which are typical metabolites of *Cladonia chlorophaea*, was not regulated by the phycobiont. Avalos and Vicente (*22*) reported the occurrence of usnic acid, atranorin and chloroatranorin in the phycobiont of *Evernia prunastri*. The group of Feige (personal communication of H.T. Lumbsch, Essen) repeated the procedure given by these workers and were unable to confirm their results. Because it is very important to know the eventual influence of lichen phycobionts of lichen metabolites further analyses of phycobionts should be performed.

Two chlorine-containing metabolites, nostoclide I (**600**) and II (**601**) have been isolated from the culture of a symbiontic blue-green alga (*Nostoc* spec.) in *Peltigera canina*, a common lichen. Their structures were determined by spectroscopy and X-ray diffraction (*750*).

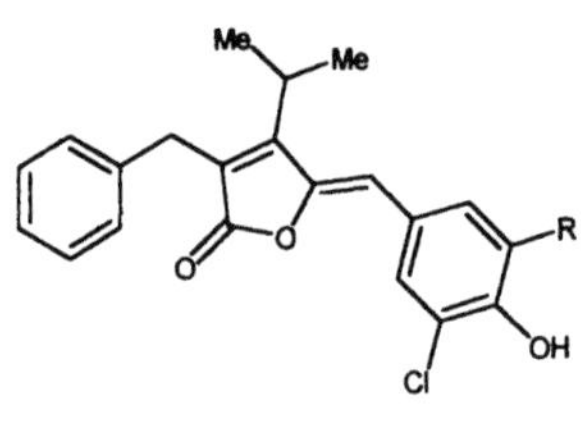

600 : Nostoclide I , R = Cl, mp 186-187°C

601 : Nostoclide II , R = H, mp 132-133°C

6. Biosynthesis of Lichen Substances

Only a few papers concerning the biosynthesis of lichen substances have been published between 1984 and 1998. Biogenesis of the diphenyl derivative contortin (see Scheme 183) very likely proceeds by phenolic coupling of two molecules of C-methylphloracetophenone; methylation could occur before or after diphenyl fomation (*222*). In lichens containing lecanoric and gyrophoric acids (*e.g. Lasallia pensylvanica, Parmelia tinctorum, Roccella canariensis, Umbilicaria caroliniana* and *U. muehlenbergii*) small amounts of orsellinic acid are always present, an observation which confirms that this compound is a biogenetic precursor of both depsides (*754*). According to Elix (*236, 211*) *para*-depsides can undergo acyl migration to the corresponding *meta*-depsides which can be converted into diaryl ether by an intramolecular Smiles rearrangement; subsequent cyclization leads to a depsidone. The

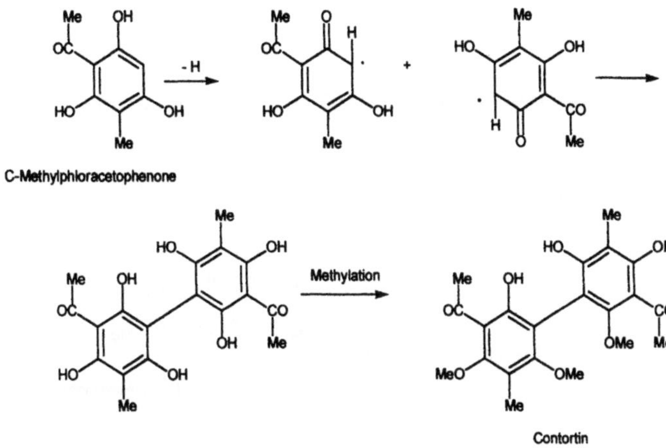

Scheme 183. Biosynthesis of contortin

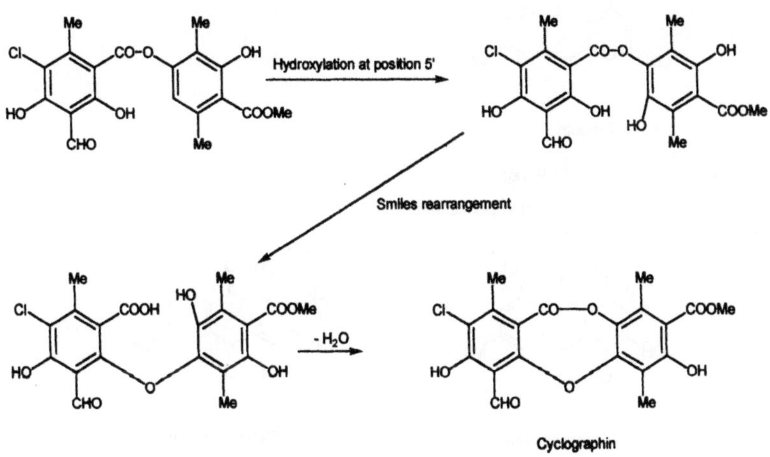

Scheme 184. Assumed biosynthesis of the depsidone cyclographin

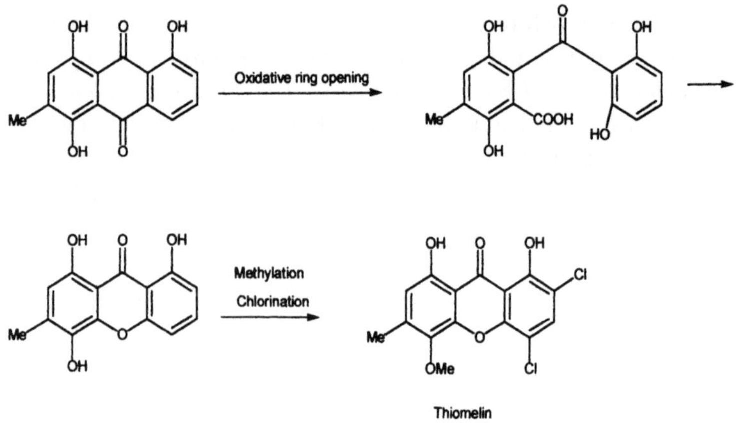

Scheme 185. Possible biosynthesis of the xanthone thiomelin

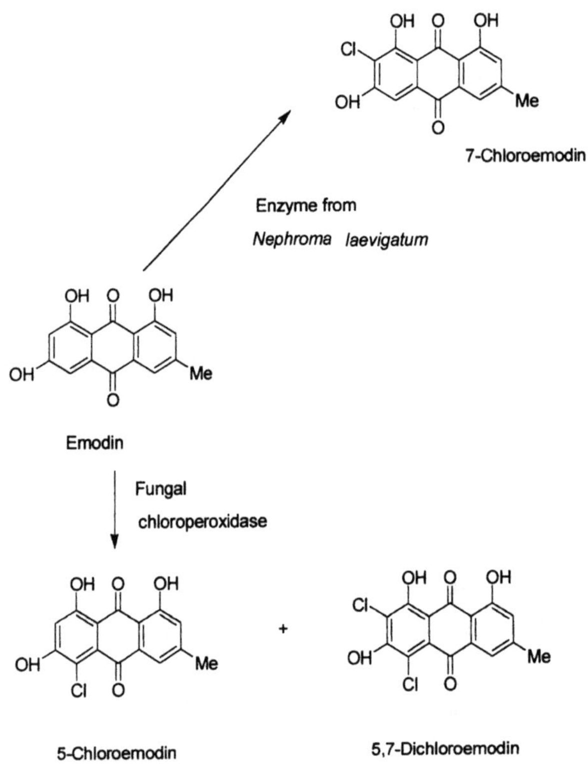

Scheme 186. Enzymatic chlorination of emodin

biosynthesis of cyclographin could follow such a route (see Scheme 184). Elix *et al.* (*218*) have also proposed the route shown in Scheme 185 for biosynthesis of the lichen xanthone thiomelin. By supplying acetate or glucose to *Pseudevernia furfuracea* thalli, it has been found that the accumulation of atranorin is independent of the photosynthetic activity but is repressed by glucose; lecanoric acid is produced with glucose (*316*). Orsellinate depside hydrolase has been isolated from *Evernia prunastri* (*331*).

Nephroma laevigatum from British Columbia contained the following anthraquinones: emodin, 7-chloroemodin, 7-chloro-1-*O*-methylemodin, 7-chloro-1-*O*-methylcitreorosein, 5-chloroemodin, 5-chloro-l-*O*-methylemodin, 5-chloro-l-*O*-methylcitreorosein and 5-chlorocitreorosein. Feeding experiments with Na[2-14C]-acetate and Na[1-13C]-acetate showed that these anthraquinones are biosynthesized *via* the polyketide pathway (*75*).

The chlorination of emodin and 7-chloroemodin was studied by Cohen and Towers (*76*) with an enzyme from *Nephroma laevigatum* and a fungal chloroperoxidase. The lichen enzyme chlorinated emodin only at C-7 while the fungal enzyme converted emodin to a mixture of 5-chloro-, 7-chloro- and 5,7-dichloroemodin (see Scheme 186).

7. Chemotaxonomy of Lichens

The use of chemotaxonomy in classification of lichens has been discussed frequently (*88*, *496*, *185*, *44*, *45*, *91*). Vainshtein (*704*) has published a guide to lichen chemotaxonomy. After a short introduction and a chapter on analytical methods for identification of lichen substances the main part of the booklet contains the names of lichens and their secondary products. Use of metabolites in taxonomy of lichens has been discussed by Lumbsch (*514*). According to Carlin (*57*) lichen chemistry and lichen biology need to be studied more thoroughly before use of lichen compounds in taxonomy can be properly evaluated. This author questions whether chemical diversity in lichens is caused by selection pressure and whether metabolic regulatory effects may be operating. It cannot be concluded with certainty whether or not the presence of different lichen compounds reflects genetic or metabolic differences. Rogers (*622*) has expressed the opinion that no taxonomic status should be accorded to entities which differ only in products from a single biosynthetic pathway, but varietal status should be given to those which have different biosynthetic pathways. Species status is justified if chemistry is correlated with morphological or physiological differences or if more than one major biosynthetic system is involved. So far, so good, but the problem is that the biosynthesis of many lichen substances is known only insufficiently. Feige and Lumbsch (*290*) distinguish three types of chemical variation in lichens: 1) in morphologically well defined species (*e.g. Hypogymnia bitteri, H. farinacea, H. laminisorediata, H. physodes, H. tubulosa, H. vittata*) the qualitative occurrence and relative concentrations of lichen substances are stable, 2) other taxa contain chemosyndromes (*e.g.* the *Lecanora subfusca* group), 3) variation of metabolites occurs within a thallus (*e.g. Lasallia* and *Roccella* species). A selection of papers on the chemosystematics of lichens arranged by taxa is listed in Table 14.

Table 14. Chemosystematic studies of lichen taxa

Family or genus	Contents	Ref.
Basidiolichenes	Eleven analysed Basidiolichenes were free of lichen substances	(*298*)
Buellia	Chemistry of xanthone-containing species of the genus *Buellia*	(*504*)
Canoparmelia	The genus *Canoparmelia* was segregated from *Pseudoparmelia* for chemical reasons	(*187*)
Cladonia	Analysis of 18 species of *Cladonia*, sect. Helopodium and 5 species of subsect. Foliosae	(*423*)
	Analysis of 32 species of the genus *Cladonia*, sect. Perviae	(*421*)
	Analysis of 40 species of the genus *Cladonia*, sect. Coccifera	(*422*)
	Analysis of 65 species of the genus *Cladonia*, sect. Cladonia and 14 species of the sect. Fuscata	(*424*)
Dimelaena	*Dimelaena australiensis* contained the depside sulphurellin	(*506*)
Diploicia	*Diploicia canescens* proved to be a complex of *D. canescens*, *D. subcanescens* and *D. canescens* ssp. *australasica*	(*229*)
Haematomma	Chemistry of 35 species of the genus *Haematomma*	(*665*)
Hypogymnia	Analysis of 6 species from Turkey	(*762*)
Lecanactis	Analysis of 51 specimens of *Lecanactis latebrarum* from Central Europe	(*487*)
Lecanora	Analysis of the *Lecanora rupicola* group in Europe	(*507*)
	Analysis of the *Lecanora subfusca* group in Australasia	(513)
	Analysis of 2 species of the *Lecanora subfusca* group	(516)
	Analysis of the *Lecanora subfusca* group in Japan	(*551*)
	Analysis of 1384 specimens of *Lecanora frustulosa* and *L. argòpholis*	(*706*)
	Chemistry of the *Lecanora dispersa* group from the Eastern alpine region	(*578*)
	Chemistry of *Lecanora radiosa*	(*359*)

Table 14 (continued)

Family or genus	Contents	Ref.
Lecidella	Analysis of 23 species	(*499*)
	Analysis of 25 specimens of 9 species	(*467*)
	Analysis of 19 saxicolous species from America	(*465*)
	Analysis of 170 specimens of *Lecidella carpathica* and 18 specimens of *L. viridans*	(*501*)
	Chemistry and distribution of the saxicolous species of the genus *Lecidella*	(*466*)
	Analysis of 200 specimens of the *Lecidella asema* group	(*500*)
	Analysis of European saxicolous *Lecidella* species	(*498*)
Lepraria	Analysis of 183 specimens of *Lepraria incana* from Central Europe	(*485*)
	Analysis of 362 specimens of *Lepraria lobificans*	(*486*)
	Chemistry and identification of the British species of *Lepraria* and *Leproloma*	(*568*)
	Analysis of 2500 specimens of the genera *Lepraria* and *Leproloma* from Central Europe	(*503*)
	Analysis of 150 specimens of *Lepraria rigidula* from Central Europe	(*489*)
	Analysis of 22 specimens of *Lepraria lesdainii*	(*484*)
Leproloma	Analysis of 5 *Leproloma* species	(*502*)
	Analysis of 2500 specimens of the genera *Lepraria* and *Leproloma* from Central Europe	(*503*)
Ochrolechia	Analysis of 600 specimens from Europe	(*350*)
Miriquidica	Creation of the genus *Miriquidica*; all known 14 species contain miriquidic acid	(*360*)
Parmelia s.l.	Classification of *Parmelia*	(*187*, *188*)
Pilocarpaceae	Chemistry of the genera *Fellhanera* and *Badimia*	(*517*)
Pseudevernia	Of 728 analysed specimens of *Pseudevernia furfuracea* from Norway 49% contained olivetoric acid, 50% physodic and hydroxyphysodic acids and 1% olivetoric and physodic acids	(*342*)
	Chemistry and ecology of *Pseudevernia furfuracea*	(*104*)
Ramalina	Chemistry of the *Ramalina americana* complex	(*96*)
	Chemistry of the *Ramalina siliquosa* complex	(*98*, *99*)
Rhizoplaca	Separation of *Rhizoplaca* from the collective genus *Lecanora*	(*508*)
	Chemotypes of *Rhizoplaca* in North America	(*538*)
Rinodina	Analysis of 26 species	(*505*, *537*)
Roccellaceae	Chemotaxonomy of the order Arthoniales	(*387*)
	Roccella hypomecha comprises 5 chemotypes	(*293*, *299*)
Squamarina	Analysis of 15 *Squamarina* species from Europe	(*292*)
Stereocaulon	Chemotaxonomy on the base of polysaccharides	(*688*)
Umbilicaria	Analysis of 11 *Lasallia* species	(*583*)
	Analysis of 56 species of the family Umbilicariaceae	(*556*)
	Analysis of 11 *Umbilicaria* species from Spain	(*644*)
Xanthoparmelia	Chemistry of the genus *Xanthoparmelia* in Australasia	(*235*)

8. Biological Activities of Lichen Substances and Lichens

8.1. Activity against Microorganisms

8.1.1. Antiviral Activity

A sulfate prepared by chlorosulfonation of GE-3-, a partially acetylated β–(1 → 6) glucan of the lichen *Umbilicaria esculenta*, inhibited the cytopathic effect of human immunodeficiency virus (HIV) and suppressed the HIV-antigen expression in Molt-4 (clone 8) cells (*371*). Emodin, 7-chloroemodin, 7-chloro-l-*O*-methylemodin and 5,7-dichloroemodin have exhibited antiviral activity (*738*, *72*). Stübler *et al.* (*675*, *676*) investigated the antiviral activity of the glucan lichenan. Seventeen depsides and depsidones have been examined for their inhibitory activity against HIV integrase (*559*). Yamamoto *et al.* (*743*) tested the extracts of numerous lichens on their inhibition of Epstein-Barr virus activation.

8.1.2. Activity against Bacteria and Fungi

The activity of lichen substances and lichens against bacteria and fungi is summarized in Table 15.

Whiton and Lawrey found that ascospore germination of the lichens *Cladonia cristatella* (*724*), *Graphis scripta* and *Caloplaca citrina* (*725*) was much reduced in the presence of lichen substances. Spore germination of *Cladonia cristatella* and *Caloplaca citrina* was inhibited by vulpinic acid, evernic acid and atranorin respectively; stictic acid showed no activity.

Table 15. Antifungal and antibacterial activities of lichen substances and lichens

Lichen substance	Activity against	Ref.
Alectosarmentin	*Staphylococcus aureus*, *Mycobacterium smegmatitis*	(*327*)
Atranorin	*Mycobacterium aurum*	(*436a*)
1′-Chloropannarin	*Leishmania*	(*302*)
Depside hydrolysis products	bacteria and fungi	(*436*)
Emodin	*Bacillus brevis*	(*13*)
Erythroglaucin	*Arthrobacter citreus*, *Bacillus brevis*, *B. subtilis*, *Streptomyces viridochromogenes*	(*12*)
Evernic acid	*Staphylococcus aureus*, *Bacillus subtilis*, *B. megaterium*	(*493*)
Leprapinic acid derivatives	Gram positive and Gram negative bacteria	(*601*)
Lobaric acid	*Mycobacterium aurum*	(*436a*)

Table 15 (continued)

Lichen substance	Activity against	Ref.
Methyl haematommate	*Epidermophyton floccosum, Microsporum canis, M. gypseum, Trichophyton rubrum, T. mentagrophytes, Verticillium achliae*	(*366*)
Methyl and ethyl orsellinate and methyl β-orsellinate	*Bacillus subtilis, Staphylococcus aureus, Pseudomonas aeruginosa, Escherichia coli, Candida albicans, Aspergillus niger*	(*436*)
Norlichexanthone	bacteria	(*43*)
Pannarin	*Leishmania*	(*302*)
Physcion	bacteria	(*13*)
(+)-Protolichesterinic acid	*Helicobacter pylori*	(*437*)
	Mycobacterium aurum	(*436a*)
Pulvinic acid lactone derivatives	*Drechslera rostrata, Alternaria alternata*	(*601–603, 613*)
	aerobic and anaerobic microorganisms	(*491*)
Salazinic acid	*Mycobacterium aurum*	(*436a*)
Usnic acid and derivatives	*Staphylococcus aureus, Bacillus subtilis, B. megaterium*	(*493, 80, 585, 302*)
	Mycobacterium aurum	(*436a*)
	aerobic and anaerobic microorganisms	(*491*)
Lichen		
Alectoria capillaris	bacteria, fungi	(*84*)
A. ochroleuca	*Staphylococcus aureus, Bacillus subtilis, Candida albicans*	(*436*)
Aspicilia gibbosa	*Bacillus subtilis, B. megaterium*	(*493*)
Cetraria islandica	*Staphylococcus aureus, Bacillus subtilis*	(*436*)
Cladia aggregata	Candida albicans	(*436*)
Cladia retipora	*Bacillus subtilis*	(*511*)
Cladonia arbuscula	*Staphylococcus aureus, Bacillus subtilis, Candida albicans*	(*436*)
Cladonia subulata	*Bacillus subtilis*	(*52*)
Cladonia uncialis		
Cornicularia aculeata	*Staphylococcus aureus, Bacillus subtilis, Candida albicans*	(*436*)
Evernia prunastri	bacteria and fungi	(*84*)
Flavoparmelia baltimorensis	*Staphylococcus aureus, Bacillus subtilis, B. megaterium*	(*493*)
Hypogymnia lugubris	*Bacillus subtilis*	(*511*)
Lasallia papulosa	*Bacillus subtilis, B. megaterium*	(*493*)
Nephroma arcticum (water extract)	various fungi	(*489a*)
Peltigera aphthosa	*Staphylococcus aureus, Bacillus subtilis, Escherichia coli, Pseudomonas aeruginosa, Candida albicans*	(*436*)
P. canina		
P. leucophlebia	*Bacillus subtilis, Escherichia coli, Pseudomonas aeruginosa*	(*436*)

Table 15 (continued)

Lichen substance	Activity against	Ref.
Pertusaria oculata	*Staphylococcus aureus, Escherichia coli, Pseudomonas aeruginosa*	(*436*)
Pseudevernia furfuracea	bacteria and fungi	(*84*)
Pseudocyphellaria colensoi		
P. degelii		
P. rufovirescens		
Ramalina celestri		
R. geniculata	*Bacillus subtilis*	(*52*)
Stereocaulon alpinum		
S. arcticum	*Staphylococcus aureus, Bacillus subtilis*	(*436*)
Stereocaulon colensoi		
S. corticulatum		
S. ramosum	*Bacillus colensoi*	(*52*)
Stereocaulon vanoyei		
Thamnolia subuliformis	*Staphylococcus aureus, Bacillus subtilis, Pseudomonas aeruginosa, Escherichia coli, Candida albicans*	(*436*)
Xanthoparmelia cumberlandii	*Staphylococcus aureus, Bacillus subtilis, Bacillus megaterium*	(*493*)

8.2. Activity against Algae

Sodium roccellate (10^{-7} M) increased the growth of *Ulva lactuca* (Chlorophyta) (*314*). Some lichen acids stimulated temporarily leakage of organic substances from *Trebouxia erici* cells (*703*). (−)-Usnic acid inhibited the growth of the green alga *Chlamydomonas reinhardii* (*636*).

8.3. Activity on Bryophytes

The lichen *Lecidea albocaerulescens* inhibited the growth of *Porella platyphylla* (liverwort), *Hedwigia ciliata* and *Anomodon attenuatus* (mosses) (*358*).

8.4. Activity on Higher Plants

Numerous lichen substances and lichen extracts are strong inhibitors of germination and growth of higher plants. The results of such studies are listed in Table 16.

Huneck and Schreiber (*409*) tested 33 lichen substances (aliphatic acids, depsides, depsidones, dibenzofurans, diphenylbutadienes and xanthones) for their growth inhibitory activity on the following higher plants: *Lepidium sativum*, *Pisum sativum*, *Avena sativa*, *Cucurbita pepo*

Table 16. Lichen substances and extracts as growth inhibitors of higher plants

Lichen substance and extract	Activity	Ref.
Barbatic acid 4-*O*-Demethylbarbatic acid Diffractaic acid Evernic acid Lecanoric acid 4-*O*-Methylorsellinic acid Orsellinic acid	Inhibition of growth of lettuce seedlings	(*565*)
Ergochrome AA (Secalonic acid A)	Phytotoxin	(*667*)
Evernic acid	Decreases the chlorophyll contents in spinach leaves	(*614*)
Lecanoric acid	Produces abnormalities in *Allium cepa* Root tips	(*677*)
Lichen phenols	Allelopathic activity	(*494*)
Quinone from *Pyxine* spec.	Inhibition of mitosis in *Allium cepa* roots	(*554*)
(–)-Usnic acid	Inhibition of germination and growth of *Lepidium sativum*	(*25*)
Usnic acid	Inhibition of mitosis in *Allium*	(*427*)
	Antitranspirant	(*56*)
Usnic acid (K-salt)	Inhibition of germination and growth of *Phaseolus mungo* and *Triticum aestivum*	(*157*)
Cladonia gracilis *C. impexa* *Cornicularia muricata* (water extracts)	Inhibition of germination of *Pinus sylvestris*	(*608*)
Evernia prunastri	Inhibition of germination of *Nigella damascena* seeds	(*609*)
	Inhibition of germination of seeds of *Agropyrum rigidum, Agrostis stolonifera, Festuca rubra, Lepidium sativum, Poa trivialis, Polygonum lapathifolium, Silene vulgaris, Sinapis alba*	(*610, 611*)
Letharia vulpina	Inhibition of germination and growth of *Triticum sativum* and *Allium sativum*	(*161*)
Parmelia conspersa (water extract)	Stimulation of the formation of new radicular cells of *Peperomia magnoliaefolia*	(*50*)

and *Phaseolus coccineus*. Roccellic acid, atranorin, evernic acid, dichloroorsellinic acid, fumarprotocetraric acid, psoromic acid, virensic acid, α-collatolic acid, (–)-usnic acid, (–)-isousnic acid, vulpinic acid, leprapinic acid, rhizocarpic acid, and thiophanic acid were strong growth inhibitors at concentrations between 10^{-3} and 10^{-5} M. The influence of usnic acid on sunflower and maize plantlets was investigated by Lascéve *et al.* (*490*).

8.5. Acitivity on Animals

Caperatic acid defended the lichen *Cetraria oaksiana* against attack by the slug *Pallifera varia* (*492*). *Flavoparmelia baltimorensis* and *Xanthoparmelia cumberlandia* also contained compounds acting as antiherbivores (*493*). Methyl β-orcinolcarboxylate, ethyl haematommate and ethyl 5-chlorohaematommate showed strong nematocidal activity against the second-stage larvae of the dog roundworm (*Toxocara canis*) (*4*). Giez *et al.* (*324*) and Emmerich *et al.* (*278*) studied the effect of numerous lichen substances on growth and development of the polyphageous herbivore insect *Spodoptera littoralis* (Noctuidae). The following compounds were studied: evernic acid, atranorin, psoromic acid, physodic acid, 3-hydroxyphysodic acid, fumarprotocetraric acid, stictic acid, norstictic acid, salazinic acid, vulpinic acid, pulvinic acid dilactone, calycin, rhizocarpic acid, parietin, and usnic acid. Whereas larval survival was usually not affected most of the compounds studied caused pronounced reduction of larval growth in chronic feeding experiments with neonate larvae. Usnic acid, vulpinic acid and 3-hydroxyphysodic acid were the most active compounds.

Hesbacher *et al.* (*363*) analysed 103 imagines caught in the wild and representing 8 different genera and 16 different species of Arctiidae (insects) for the presence of lichen compounds. Parietin, atranorin and methyl β-orcinolcarboxylate were the most frequently detected compounds. The ecological role of the sequestered lichen compounds is unknown. The linear homoglucan PC-2 from *Parmelia caperata* enhanced the hippocampal long-term potentiation in rats (*654*). Some lichens such as *Nephroma arcticum*, *Peltigera canina*, *P. malacea*, *Placynthium asperellum*, *Psoroma hypnorum*, *Stereocaulon vesuvianum* and *Xanthoria candelaria* contain lectins (*352*).

8.6. Acitivity on Man

Schindler (*638*) has published an article on the history of the medical use of lichens and Hanssen and Schädler (*351*) have described the use of lichens and lichen substances in Chinese medicine. (+)-Usnic acid effected proliferation of polychromatic erythrocytes in mice (*8*) and was identified together with diffractaic acid as the analgesic and antipyretic component of the lichen *Usnea diffracta* (*567*). Numerous lichen substances are allergenes and cause contact dermatitis; among them are atranorin, barbatic acid, diffractaic acid, evernic acid, fumarprotocetraric acid, lobaric acid, perlatolic acid, physodic acid, physodalic acid, protolichesterinic acid, salazinic acid, stictic acid, and usnic acid (*39*, *286*, *696*).

8.7. Tumor and Anti-Tumor Activity

Physodalic acid exhibited mutagenicity on *Salmonella typhimurum* (*646*). The following lichen substances have shown anti-tumor activity: acetylated pustulanes (*564*, *716*), lichenan derivatives (*171*, *172*), barbatic acid, 4-*O*-demethylbarbatic acid, diffractaic acid, evernic acid, lichesterinic acid, methyl gyrophorate, solorinic acid (*746*), nephrosterinic acid, (+)-protolichesterinic acid (*372*), chrysophanol and emodin derivatives (*471*), usnic acid and derivatives (*575*, *689*, *746*) and an extract from *Usnea fasciata* (*575*). Orcinol, olivetol, 1-*n*-heptyl-3,5-dihydroxybenzene, 1-*n*-tridecyl-3,5-dihydroxybenzene (*19*) and scabrosin esters (*283*, *732*) were found to exhibit cytotoxic activity.

8.8. Inhibition of Enzymes

A number of lichen substances listed in Table 17 are enzyme inhibitors.

Table 17. Lichen substances and lichens as inhibitors of enzymes

Lichen substance and lichen	Inhibited enzyme	Ref.
4- and 5-Alkylresorcinols	tyrosinase	(*535*)
Atranorin	trypsin and pankreaselastase	(*584*)
	phosphorylase	(*1*)
Baeomycesic acid	5-lipoxygenase	(*439*)
Bis-(2,4-dihydroxy-6-*n*-propyl-phenyl) methane	tyrosinase	(*535*)
9-*cis*-Octadecenamide	enzymes involved in arachidonate metabolism	(*436b*)
Chrysophanol	glutathione reductase	(*33*)
Confluentic acid	monoamin oxidase B	(*280*)
Divarinol	tyrosinase	(*453*)
Lichen acids	HIV-1 integrase	(*559*)
Lichen cultures of *Cetraria juniperina*, *Hypogymnia physodes* and *Letharia vulpina*	tyrosinase	(*368*)
Lichens	superoxide dismutase	(*743*)
4-*O*-Mcthylcryptochlorophaeic acid	prostaglandin synthetasc	(*648*)
2′-*O*-Methylperlatolic acid	monoamin oxidase B	(*280*)
(+)-Protolichesterinic acid	HIV-1 reverse transcriptase	(*574*)
Usnic acid	phosphorylase	(*1*)
Vulpinic acid	phosphorylase	(*1*)

8.9. Antioxidant Activity

Atranorin, divaricatic acid, 1′-chloropannarin, pannarin and physodic acid show antioxidant activity (*367*, *433*).

9. Harmful Effects of Lichens

Lichens can deteriorate or destroy ancient frescoes, old buildings and colored windows of cathedrals. The effect is due to the production of oxalic acid which converts the calcium carbonate of the substrate to calcium oxalate, thus changing the structure of the surface. The topic (biodeterioration) has been thoroughly discussed by Nimis *et al.* (*563*).

10. Commercial Uses of Lichens and Lichen Substances

10.1. Uses in Perfume Production and in Cosmetics

Large amounts of two lichens are processed for production of essential ingredients in the perfume industry: thus in 1997 1900 tons of *Pseudevernia furfuracea* and 700 tons of *Evernia prunastri* were processed, mainly at Grasse, the centre of the essential oil industry in France (*447*). Usnic acid is used as a preservative in cosmetic creams (*643*) and atranorin, divaric acid, gyrophoric acid, pannarin, and usnic acid are potential sunscreen materials (*294*).

10.2. Lichens as Biomonitors

Lichens are useful indicators for monitoring air pollution (*312*, *361*, *362*, *446*, *458*). The literature concerning air pollution and lichens is summarized continously in the journal "Lichenologist".

10.3. Lichens for Dyeing

Dyeing with lichens, although an old industry, is nowadays practiced only by amateurs. Booklets by Feddersen-Fieler (*289*), Casselman (*62*, *63*) and Fraser (*303*) contain detailed instructions.

Acknowledgement

I am grateful to Prof. W. Herz (Tallahassee) for grammatical help, to Ing. H.G. König (Halle/Saale) for help with the computer, to Dr. D. Joulain (Grasse) for information concerning the consumption of lichens in the perfume industry and the "Fonds der Chemischen Industrie e. V." (Frankfurt/Main) for a research grant.

Notes Added in Proof

Chapter 2.7. X-Ray Analysis. Hybocarpone (Ernst-Russell MA, Elix JA, Chai CLL, Willis AC, Hamada N, Nash III TH (1999) Hybocarpone, a Novel Cytotoxic Naphthazarin Derivative from Mycobiont Cultures of the Lichen *Lecanora hybocarpa*. Tetrahedron Letters **40**: 6321).

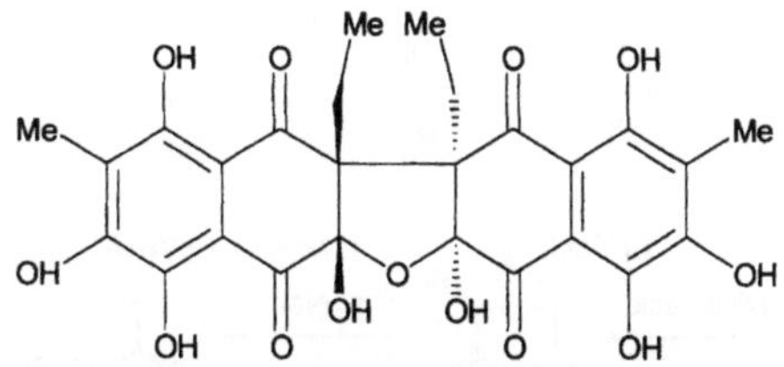

Hybocarpone, orange prisms, mp 167-168°C

$[\alpha]_D$ - 19.2 ($CHCl_3$)

Chapter 4.1. Nitrogen Containing Compounds. Diacylglyceryl-*N*,*N*,*N*-trimethylhomoserine was found in *Cladonia fimbriata*, *Parmelia saxatilis*, *Peltigera canina*, *P. collina*, and *P. horizontalis* (Künzler K, Eichenberger W (1997) Betaine Lipids and Zwitterionic Phospholipids in Plants and Fungi. Phytochemistry **46**: 883).

Chapter 4.2. Phosphorus Containing Compounds. Phosphatidylcholine and phosphatidylethanolamine (PE) have been detected in *Cladonia fimbriata*, *Parmelia saxatilis* (except PE), *Peltigera canina*, *P. collina*, and *P. horizontalis* (Künzler K, Eichenberger W, loc. cit.).

Chapter 4.4.5. γ-Lactonic Acids. Maier et al. isolated from the lichen *Punctelia microsticta* gyrophoric acid, (–)-isomuronic acid and

(−)-dihydropertusaric acid and revised the stereostructure of the latter compound on the basis of spectroscopic evidence and molecular modeling to

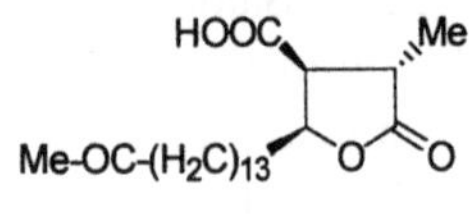

(-)-Dihydropertusaric acid

(Maier MS, González Marimon DI, Stortz CA, Adler MT (1999) A Revised Structure for (−)-Dihydropertusaric Acid, a γ-Butyrolactone Acid. J Nat Prod **62**: 1565).

Total syntheses of racemic protolichesterinic and roccellaric acids were achieved with the use of a tungsten-π-ally complex in the key step (see Scheme) (Chen M-J, Liu R-S (1998) Efficient Total Syntheses of (±)-Protolichesterinic Acid and (±)-Roccellaric Acid *via* Tungsten-π-allyl Complexes. Tetrahedron Letters **39**: 9465).

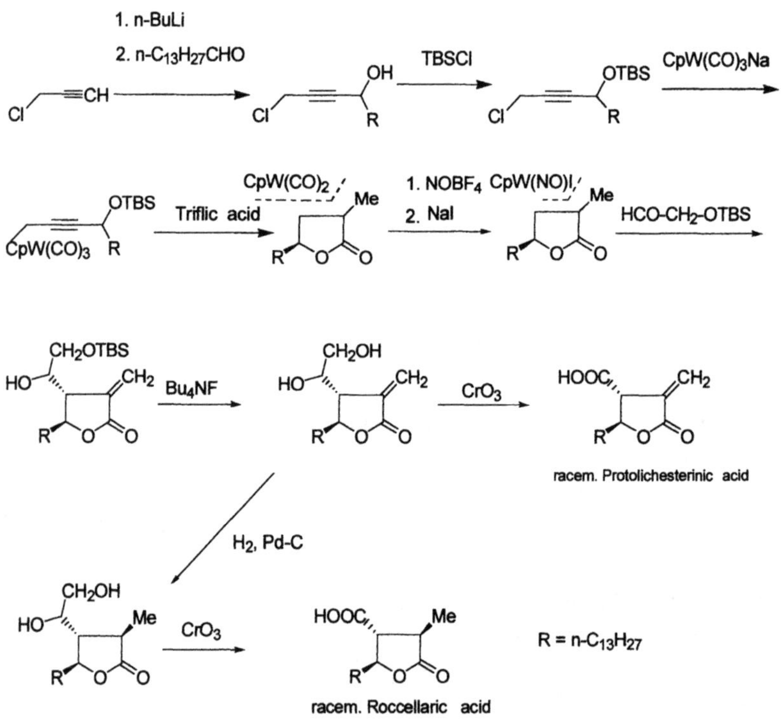

Scheme: Synthesis of (±)-protolichesterinic and (±)-roccellaric acids *via* a tungsten-π-allyl complex

Chapter 4.5.6. Depsides. The tridepsides methyl ovoate, methyl 2″-*O*-methylgyrophorate and methyl 2′,2″-di-*O*-methylgyrophorate have been detected in *Pseudocyphellaria billardierei* (Elix JA, Mc Caffery LF (1999) Three New Tridepsides in the Lichen *Pseudocyphellaria billardierei*. Australasian Lichenology **45**: 12).

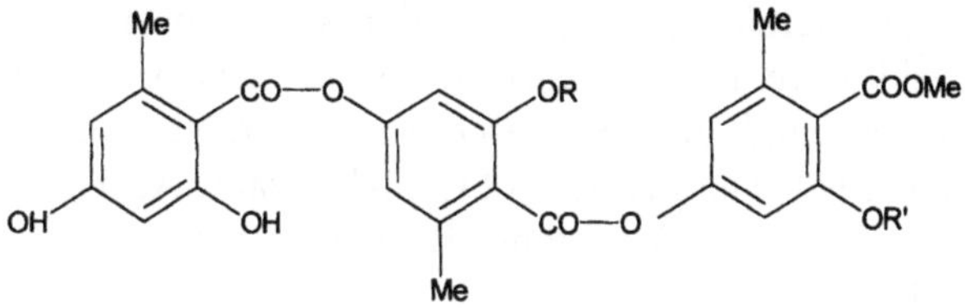

Methyl ovoate: R = Me, R' = H

Methyl 2"-O-methylgyrophorate: R = H, R' = Me

Methyl 2',2"-di-O-methylgyrophorate: R = R' = Me

Norbaeomycesic acid is a new depside from *Hypotrachyna orientalis* (Elix JA, Griffin FK, Louwhoff SHJJ (2000) Norbaeomycesic Acid, a New Depside from the Lichen *Hypotrachyna orientalis*. Australasian Lichenology **46**: 25).

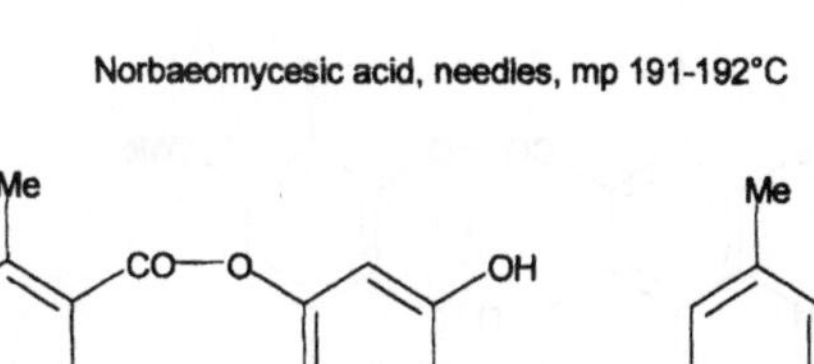

Norbaeomycesic acid, needles, mp 191-192°C

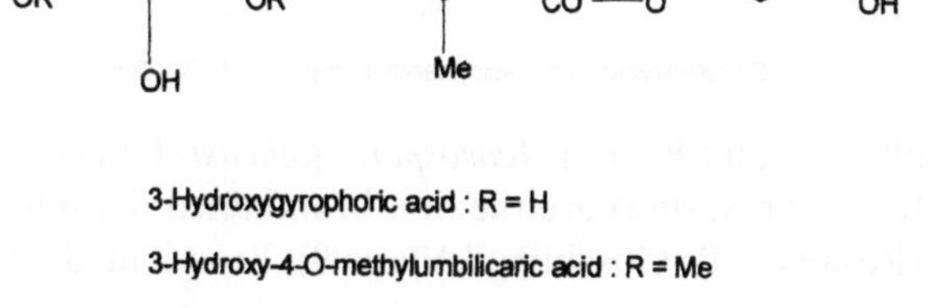

3-Hydroxygyrophoric acid : R = H

3-Hydroxy-4-O-methylumbilicaric acid : R = Me

In *Parmelinopsis* species the tridepsides 2,4-di-*O*-methylhiascic acid (*P. spumosa*), 2,5-di-*O*-methylhiascic acid (*P. neodamaziana*), 3-hydroxygyrophoric acid (*P. subfatiscens*) and 3-hydroxy-4-*O*-methylumbilicaric acid (*P. subfatiscens*) have been detected (Elix JA, Wardlaw JH (2000) Four New Tridepsides from *Parmelinopsis* species. Australasian Lichenology **47**: 8).

Chapter 4.5.7. Depsidones. The depsidones chalybaeizanic acid and quaesitic acid have been isolated from the lichens *Xanthoparmelia amphixanthoides* and *Hypotrachyna quaesita* respectively (Elix JA, Wardlaw JH (1999) The structure of Chalybaeizanic Acid and Quaesitic Acid, Two New Lichen Depsidones Related to Salazinic Acid. Aust J Chem **52**: 713).

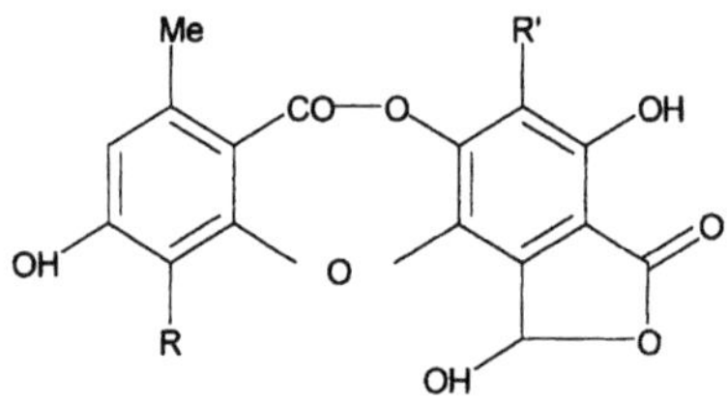

Chalybaeizanic acid : R = R' = CHO, mp> 300°C, dec.

Quaesitic acid : R = CHO, R' = CH_2-O-CO-CH=CH-COOH, mp > 350°C, dec.

The depsidone 2-methoxypsoromic acid has been identified in *Sulcaria sulcata* and a *Pertusaria* spec. from Papua New Guinea (Elix JA, Wardlaw JH, Archer AW, Obermayer W (1999) 2-Methoxypsoromic Acid, a New Lichen Depsidone from *Pertusaria* and *Sulcaria* Species. Aust J Chem **52**: 717).

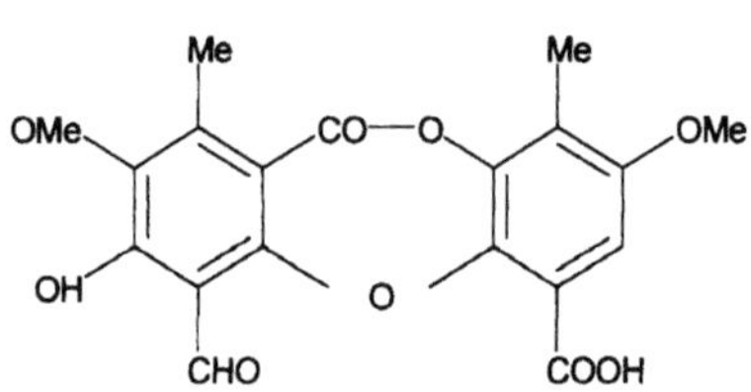

2-Methoxypsoromic acid, needles, mp > 350°C, dec.

Acarogobien A and B from *Acarospora gobiensis* (collected near the lake Ysyk Köl, Kyrgyzstan) are the first brominated depsidones found in lichens (Řezanka T, Guschina IA (1999) Brominated Depsidones from *Acarospora gobiensis*, a Lichen from Central Asia. J Nat Prod **62**: 1675). Interestingly both compounds belong to a group of depsidones which were known only from fungi, e.g. aurantiacin A and B, emeguisin A, B and C, nidulin and unguinol from *Preussia aurantiaca*, *Emericella unguis* and *Aspergillus nidulans* respectively.

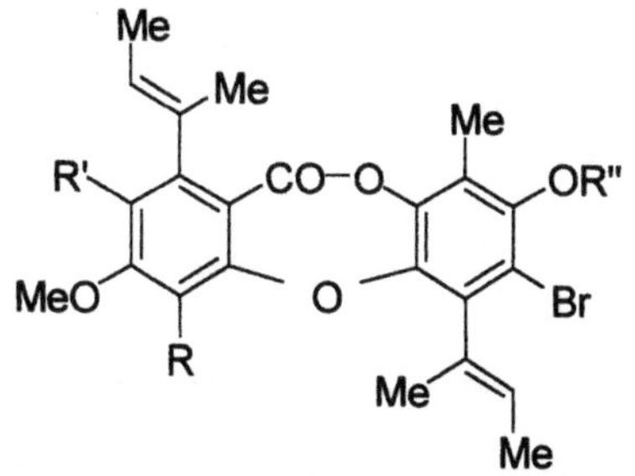

Acarogobien A: R = Me, R' = R" = H; mp 172 °C

Acarogobien B: R = CHO, R' = Br, R" = CH_2-CH_2-CH=C(Me)$_2$; mp 197°C

Chapter 4.5.9. Chromanes and Chromones. Takenaka et al. isolated from mycobiont cultures of the lichen *Graphis scripta* (collected in Japan) the chromones 5-hydroxy-2,3-dimethyl-7-methoxychromone, 5-hydroxy-3-hydroxymethyl-2-methyl-7-methoxychromone and 5-hydroxy-2-hydroxymethyl-3-methyl-7-methoxychromone (Takenaka Y, Tanahashi T, Nagakura N, Hamada N (2000) 2,3-Dialkylchromones from Mycobiont Cultures of the Lichen *Graphis scripta*. Heterocycles *53*: 1589).

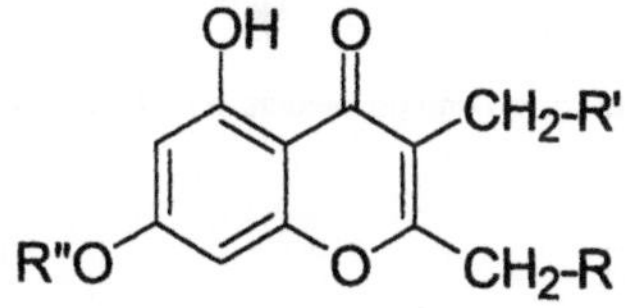

5-Hydroxy-2,3-dimethyl-7-methoxychromone: R = R' = H, R" = Me; mp 120-121°C

5-Hydroxy-3-hydroxymethyl-2-methyl-7-methoxychromone: R = H, R' = OH, R" = Me; mp 147°C

5-Hydroxy-2-hydroxymethyl-3-methyl-7-methoxychromone: R = OH, R`-H, R" = Me; mp 205°C

Chapter 4.5.10. Xanthones. Cultures of the spore-derived mycobionts of *Pyrenula japonica* and *P. pseudobufonia* synthesized the new xanthones 1,5,8-trihydroxy-3-methylxanthone and 1,8-dihydroxy-5-methoxy-3-methylxanthone beside the known 1,7-dihydroxy-3-methylxanthone (Tanahashi T, Takenaka Y, Ikuta Y, Tani K, Nagakura N, Hamada N (1999) Xanthones from the Cultured Lichen Mycobionts of *Pyrenula japonica* and *Pyrenula pseudobufonia*. Phytochemistry **52**: 401).

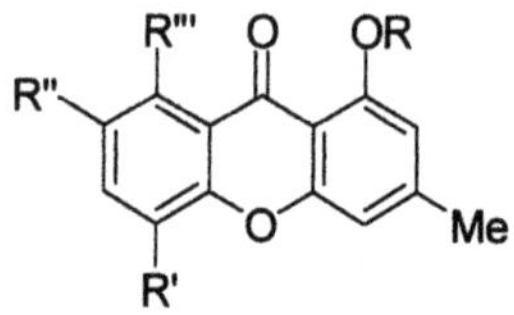

1,5,8-Trihydroxy-3-methylxanthone: R = R'' = H, R' = R''' = OH; yellow needles, mp 276.5-278°C

1,8-Dihydroxy-5-methoxy-3-methylxanthone: R = R'' = H, R' = OMe, R''' = OH; yellow needles, mp 214-215°C

Chapter 4.5.14. Naphthopyranes. Euplectin and coneuplectin are new naphthopyrones from *Flavoparmelia euplecta*. (Ernst-Russell MA, Chai CLL, Wardlaw JH, Elix J A (2000) Euplectin and Coneuplectin, New Napthopyrones from the lichen *Flavoparmelia euplecta*. J Nat Prod **63**: 129).

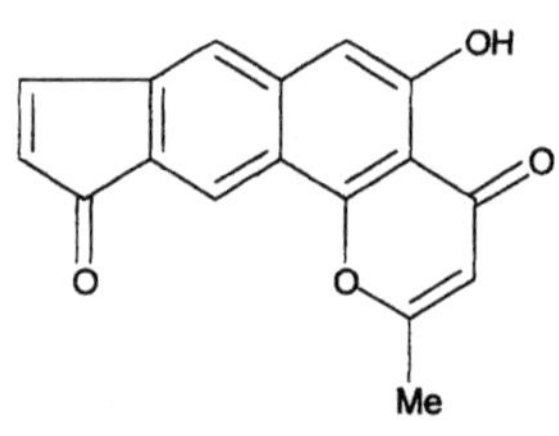

Euplectin, red-orange needles, mp 320°C, dec.

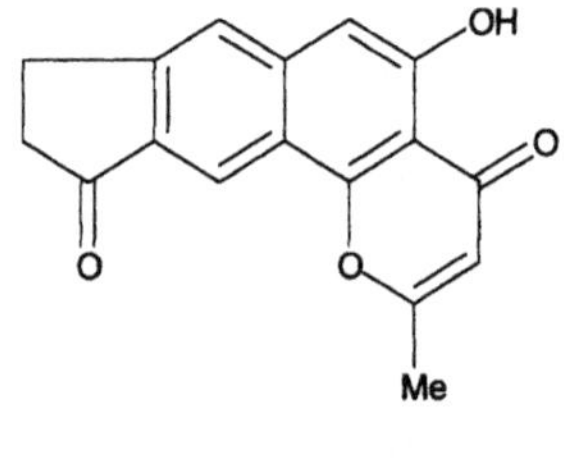

Coneuplectin, yellow needles, mp 295°C

Chapter 4.5.17. Isocoumarins. From cultures of the spore-derived mycobionts of the lichen *Graphis scripta* var. *pulverulenta* the isocoumarins graphislactone A–C and the 6H-dibenzo/b,d/pyran-6-one graphislactone D were isolated (*693*).

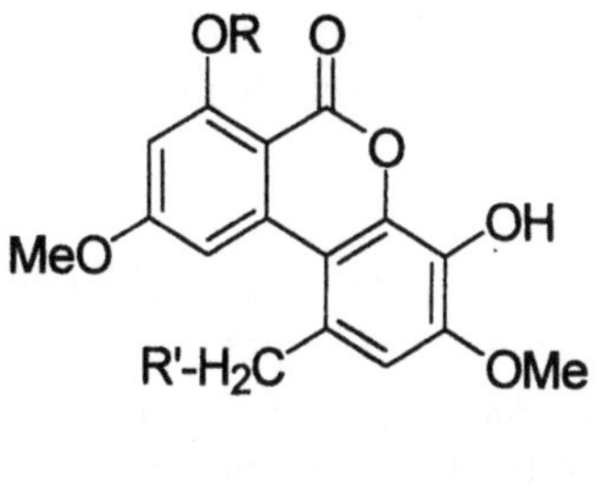

Graphislactone A: R = R' = H; mp 236-237°C

Graphislactone B: R = Me, R' = H; mp 65-67°C

Graphislactone C: R = H, R' = OH; mp 214-216°C

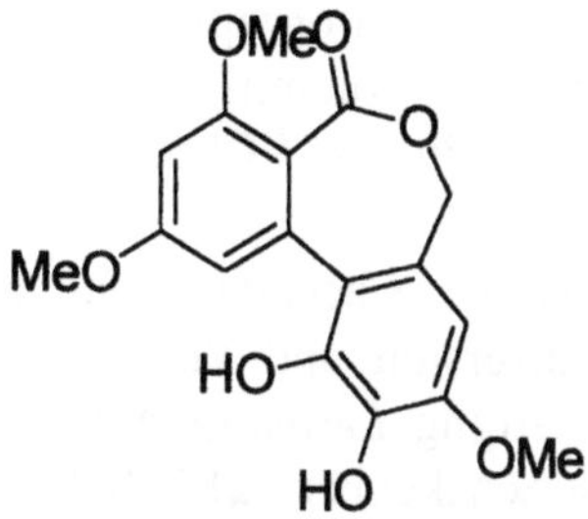

Graphislactone D; mp 110-112°C

Tanahashi et al. found in the cultured lichen mycobiont of a *Graphis* species from the Philippines two further isocoumarins, 8-methyldichlorodiaportin and 6,8-dimethylcitreoisocoumarin (Tanahashi T, Takenaka Y, Nagakura N, Hamada N, Miyawaki H (2000) Two Isocoumarins from the Cultured Lichen Mycobiont of *Graphis* sp. Heterocycles **53**: 723)

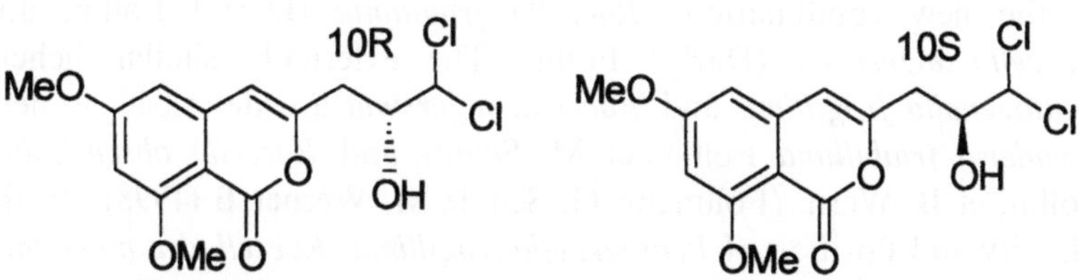

8-O-Methyldichlorodiaportin (mixture of 10R : 10S = 19 : 1); mp 155°C, $[\alpha]_D^{25}$ + 36 ($CHCl_3$)

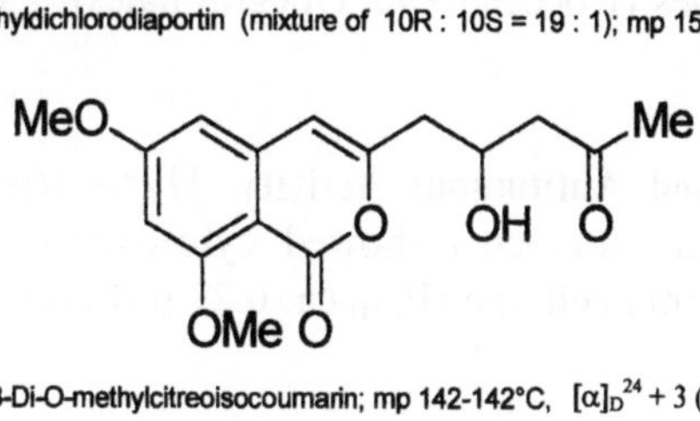

6,8-Di-O-methylcitreoisocoumarin; mp 142-142°C, $[\alpha]_D^{24}$ + 3 ($CHCl_3$)

Chapter 5.4. Metabolites from Lichen Mycobionts. The volatile compounds (ca. 130) produced by a *Nostoc* species from a *Collema* lichen were separated and identified by GC-MS; most abundant were cyclohexane and cyclopentane derivatives (Dembitsky VM, Shkrob I, Dor I (1999) Separation and Identification of Hydrocarbons and other Volatile Compounds from Cultured Blue-Green Alga *Nostoc* spec. by Gas Chromatography – Mass Spectrometry Using serially coupled Capillary Columns with consecutive Nonpolar and Semipolar Stationary Phases. J Chromatogr A **682**: 221).

Chapter 7. Chemotaxonomy of Lichens. Lumbsch (Lumbsch HT (1999) Biodiversität und Verbreitungsmuster tropischer Gruppen der Flechtengattung *Lecanora* Ach. (Lecanorales, Ascomycota). Cour Forsch-Inst Senckenberg **215**: 143) described the biodiversity and distribution patterns of tropical groups in the lichen genus *Lecanora* with special reference to Australasian species.

Rambold and Triebel (Rambold G, Triebel D (1999) Generic Concepts in Lichenized and Lichenicolous Ascomycetes since 1950 – a Historical Approach. Symb Bot Ups **32**: 2) discussed the generic concepts in lichenized and lichenicolous ascomycetes, including chemical characters.

Crespo *et al.* (Crespo A, Gavilán R, Elix JA, Gutiérrez G (1999) A Comparison of Morphological, Chemical and Molecular Characters in some Parmelioid Genera. Lichenologist **31**: 451) analysed the morphological and chemical data (cell-wall chemistry and medullary chemistry) from 64 genera of Parmeliaceae.

Chemical and morphological analyses of the critical lichen taxa *Pentagenella fragillima* Darb. and *Roccellodea nigerrima* Darb. resulted in the new combinations *Roccella fragillima* (Darb.) Follm. and *Roccella nigerrima* (Darb.) Follm. The externally similar lichens *Pentagenella fragillima* and *Roccella nigerrima* are described as new: *Ingaderia friabillima* Follm. et M. Schulz and *Roccella obscurissima* Follm. et B. Wern. (Follmann G, Schulz M, Werner B (1998) On the Identity and Position of *Pentagenella fragillima, Roccellodea nigerrima*, and some Related Species (Roccellaceae, Opegraphales). J Hattori Bot Lab **85**: 245).

Chapter 8.7. Tumour and Antitumour Activity. Hybocarpone (Ernst-Russell MA, CLL *et al.*, loc. cit.) showed cytotoxicity against the murine P-815 mastocytoma cell line (IC_{50} ca. 0.27 μM and 1.67 μg/mL respectively).

References

1. Abo-Khatwa AN, Al-Robai AA, Al-Jawhari DA (1996) Lichen Acids as Uncoupler of Oxidative Phosphorylation of Mouse-liver Mitochondria. Natural Toxins **4**: 96
2. Addison JB (1985) Application of Pyrolysis-Chemical Ionisation Mass Spectrometry to Lichenology. Analyst **110**: 933
3. Aghoramurthy K, Seshadri TR (1952) Nuclear Oxidation in Flavones and Related Compounds. Part XXXIX. Occurrence of Nuclear Oxidation in the Biogenesis of Lichen Acids. Proc Ind Acad Sci **35A**: 327
4. Ahad AM, Goto Y, Kiuchi F, Tsuda Y, Kondo K, Sato T (1991) Nematocidal Principles in "Oakmoss Absolute" and Nematocidal Activity of 2,4-Dihydroxybenzoates. Chem Pharm Bull **39**: 1043
5. Ahmadjian V (1982) The Nature of Lichens. Natural History **91**: 31
6. Ahmadjian V (1993) The Lichen Symbiosis. Wiley & Sons, New York
7. Alagna L, Prosperi T, Tomlinson AAG, Kjøsen H, Mo F (1990) An EXAFAS, and Preliminary X-ray Crystallographic Investigation of an Iron-containing Product from the Lichen *Cladonia deformis*. Biochim Biophys Acta **1036**: 71
8. Al-Bekairi AM, Quereshi S, Chaudhry MA, Krishna DR, Shah AH (1991) Mitodepressive, Clastogenic and Biochemical Effects of (+)-Usnic acid in Mice. J Ethnopharmacol **33**: 217
9. Allemand P (1988) Contribution à l' Étude de la Composition de Lichens: Application de la Spectrométrie de Masse en Tandem (MS-MS) à l' Identification de Triterpènes et Steroides. Thesis. Fac des Sciences, Univ Neuchatel
10. Ammann W, Richarz R, Wirthlin T, Wendisch D (1982) ^{1}H and ^{13}C Chemical Shifts and Coupling Constants of Lupane. Org Magn Res **20**: 260
11. Anastasia M, Fiecchi A (1981) Synthesis of Lichesterol [Erogosta-5,8(9),22-trien-3β-ol]. J Chem Soc Perkin I: 2125
12. Anke H, Kolthoum I, Laatsch H (1980) Metabolic Products of Microorganisms. 192. The Anthraquinones of the *Aspergillus glaucus* Group. II. Biological Activity. Arch Microbiol **126**: 231
13. Anke H, Kolthoum I, Zähner H, Laatsch H (1980). The Anthraquinones of the *Aspergillus glaucus* Group. I. Occurrence, Isolation, Identification and Antimicrobial Activity. Arch Microbiol **126**: 223
14. Appa Rao AVN, Rao PS, Huneck S (1984) Isolation of Demethylleprapinic Acid from *Lepraria* sp. Fitoterapia **55**: 241
15. Aptroot A, Berendsen W (1989) The Nature of Pruina on Pyxinaceae (Lichenes). Proc Kon Nederl Akad v Wetenschapen **92**: 409
16. Archer AW (1993) Identification of Orcinol *para*-Depsides in the Lichen Genus *Pertusaria* by Thin Layer Chromatography. Mycotaxon **46**: 1
17. Archer AW, Taylor WC (1987) Homoheveadride, a Cyclononadiene *Bis*-anhydride from *Cladonia polycarpoides*. Phytochemistry **26**: 2117
18. Arditti J, Fisch REMH, Flick BH (1972) Ergosterol Peroxide from *Rhizoctonia repens:* Composition, Conformation and Origin. J Chem Soc Chem Comm 1217
19. Arisawa M, Ohmura K, Kobayashi A, Morita N (1989) A Cytotoxic Constituent of *Lysimachia japonica* Thunb. (Primulaceae) and the Structure-Activity Relationship of Related Compounds. Chem Pharm Bull **37**: 2431
20. Arnone A, Nasini G, De Pava OV (1991) A Reinvestigation of the Structure of Biruloquinone, a 9,10-Phenanthrenequinone Isolated from *Mycosphaerella rubella*. Phytochemistry **30**: 2729

21. Arup U, Ekman S, Lindblom L, Mattson J-E (1993) High Performance Thin Layer Chromatography (HPTLC), an Improved Technique for Screening Lichen Substances. Lichenologist **25**: 61
22. Avalos A, Vicente C (1987) The Occurrence of Lichen Phenolics in the Photobiont Cells of *Evernia prunastri*. **6**: 74
23. Bachelor FW, Cheriyan UO, Wong JD (1979) Cleavage of Depsides by *tert*-Butyl Alcohol. Phytochemistry **18**: 487
24. Bachelor FW, King GG, Richardson J (1990) Phlebic Acids C and D, Lichen Acids from *Peltigera aphthosa*. Phytochemistry **29**: 601
25. Barbalić L (1963) Beitrag zur Kenntnis der Einwirkung von L-Usninsaure auf höhere Pflanzen. Qual Plant Mater Veg **9**: 286
26. Baron M, Gorin PAJ, Iacomini M (1988) Isolation and Identification of a Linear (1→3)-Linked β-D-Glucan and Other Carbohydrate Components of the Lichen *Stereocaulon ramulosum* (Sw.) Räusch. Carbohydrate Res **177**: 235
27. Baron M, Gorin PAJ, Iacomini M (1989) Structural Studies on a Galactomannan Isolated from the Lichen *Stereocaulon ramulosum*. Agric Biol Chem **53**: 1751
28. Baron M, Iacomini M, Fanta ES, Gorin AJ (1991) Galactomannan, Lichenan and Isolichenan from the Polysaccharide-rich Lichen *Neuropogon aurantiaco-ater*. Phytochemistry **30**: 3125
29. Bartlett AJ, Holker JH, O'Brien E, Simpson TJ (1983) Synthesis and Regiospecific Deoxygenation of β-Resorcylic Ester Derivatives to 4-Hydroxybenzoates. J Chem Soc Perkin I 667
30. Behrens U, Hencken G, Kopf J (1976) Usnic Acid, $C_{18}H_{16}O_7$. Cryst Struct Comm **5**: 51
31. Benn MH, Lorimer SD, Perry NB (1998) Revision of Structure of Rangiformic Acid. Phytochemistry **47**: 1649
32. Birkbeck AA, Sargent MV, Elix JA (1990) The Structure of the Lichen Depsidones Fulgidin and Isofulgidin. Aust J Chem **43**: 419
33. Bironaite DA, Cenas NK, Medentsev AG, Akimenko VK (1991) The Inhibition of Glutathione Reductase by Quinones. Z Naturforsch **46c**: 966
34. Blaser D, Gunzinger J, Stöckli-Evans H, Tabacchi R (1992) Structure of 2-Methoxy Methyl β-Orcinol Carboxylate (Methyl 4-Hydroxy-2-methoxy-3,6-dimethylbenzoate). Acta Cryst **C 48**: 1126
35. Blaser D, Stöckli-Evans H (1992) Structure of the Methyl Ester of Everninic Acid (Methyl 2-Hydroxy-4-methoxy-6-methylbenzoate). Acta Cryst **C 48**: 1124
36. Bol L, Zhongwan L, Handong S (1990) Chemical Constituents from *Lobaria isidiophora*. Acta Bot Yunnan **12**: 447
37. Bodo B, Hebrard P, Molho L, Molho D (1973) Un Nouvel Acide Aliphatique des Lichens, *Desmaziera evernioides* et *Ramalina bourgeana*. Tetrahedron Letters 1631
38. Brandänge S, Josephson S, Mörch L, Vallen S (1977) Absolute Configurations of Alkylcitric Acids. Acta Chem Scand **B 31**: 307
39. Brasch J, Jacobsen P (1991) Flechten and ihre Allergene. Allergologie **14**: 99
40. Brassy C, Bachet B, Bodo B, Molho D (1982) Structure de l' Atranorine. Acta Cryst **B 38**: 3126
41. Brassy C, Bachet B, Molho L, Molho D (1985) Methyl α-(3-Hydroxy-5-oxo-4-phenyl-2-(2H)-furylidene phenylacetate, Vulpinic Acid, $C_{19}H_{14}O_5$. Acta Cryst **C 41**: 781
42. Braun M (1981) Regioselektive Synthese der natürlich vorkommenden Anthrachinone Digitopurpon, Islandicin, Catenarin und Erythroglaucin. Liebigs Ann Chem 2247

43. Broadbent D, Mabelis RP, Spencer H (1975) 3,6,8-Trihydroxy-l-methylxanthone – an Antibacterial Metabolite from *Penicillium patulum*. Phytochemistry **14**: 2082
44. Brodo IM (1978) Changing Concepts Regarding Chemical Diversity in Lichens. Lichenologist **10**: 1
45. Brodo IM (1986) Interpreting Chemical Variation in Lichens for Systematic Purposes. Bryologist **89**: 132
46. Brodo IM, Elix JA (1993) *Lecanora jamesii* and the Relationship between *Lecanora* s.str. and *Straminella*. Bibl Lichenol **53**: 19
47. Bruun T, Lamvik A (1971) Haemoventosin. Acta Chem Scand **25**: 483
48. Buckingham J (1994) Dictionary of Natural Products, vol 1. Chapman and Hall, London, p 570
49. Buemi G, Zuccarello F (1992) Molecular Conformations, Hydrogen-bond Strength's and Electronic Structure of Usnic Acid: an AM1 and CNDO/S Study. Theochem **68**: 89
50. Buffard Y, Rondon Y (1977) Effects de l` Extrait du Lichen *Parmelia conspersa* (Ehrht.) Ach. sur la Rhizogenèse de Bautures de *Peperomia magnoliaefolia* Dietr. Bull Soc bot Fr **124**: 161
51. Caccamese S, Compagnini A, Toscano RM, Cascio O (1986) Methyl β-Orcinolcarboxylate and Atranol from the Lichen *Stereocaulon vesuvianum*. J Nat Prod **46**: 1159
52. Calder VL, Cole ALJ, Walker JRL (1986) Antibiotic Compounds from New Zealand Plants. III.: A Survey of some New Zealand Plants for Antibiotic Substances. J Royal Soc New Zealand **16**: 169
53. Campbell AC, Maidment MS, Pick JH, Stevenson DFM (1985) Synthesis of (E)- and (Z)-Pulvinones. J Chem Soc Perkin I 1567
54. Cannon JR, Cresp TM, Metcalf BW, Sargent MV, Vinliguerra G, Elix JA (1971) The Bromination of some Derivatives of Resorcinol. J Chem Soc (C) 3495
55. Canonica L, Rindone B, Santaniello E, Scolastico C (1972) A Total Synthesis of Mycophenolic Acid, some Analogues and some Biogenetic Intermediates. Tetrahedron **28**: 4395
56. Carbonnier J, Batcho E, Shaker-Bazarnov H, Molho D (1993) Proprietes de l`Acide Usnique, Antitranspirant d` Origine Naturelle. Rev Rés Amélior Prod Agr Milieu Aride **5**: 19
57. Carlin G (1987) On the Use of Chemical Characters in Lichen Taxonomy. Graphis Scripta **1**: 72
58. Carlson RM, Oyler AR (1976) Direct Methods for α-Methylene Lactone Synthesis Using Itaconic Acid Derivatives. J Org Chem **41**: 4065
59. Carpenter RC, Sotheeswaran S, Sultanbawa MUS, Ternai B (1980) ^{13}C-NMR Studies of some Lupane and Taraxerane Triterpenes. Org Magn **14**: 462
60. Carvalho CF, Sargent MV (1984) Naturally Occurring Dibenzofurans. Part 5. Synthesis of Melacarpic Acid. J Chem Soc Perkin I 1613
61. Carvalho CF, Sargent MV (1984) Naturally Occurring Dibenzofurans. Part 6. Synthesis of Didymic Acid. J Chem Soc Perkin I 1621
62. Casselman KL (1993) Craft of the Dyer. Colour from Plants and Lichens, 2nd edn. Dover Publ Inc, New York
63. Casselman KL (1994) Lichen Dyes: Preparation and Dyeing. Maine Naturalist **2**: 105
64. Chamy MC, Gambaro V, Garbarino JA, Quilhot W (1985) Studies on Chilean Lichens. VII. The Phenolic Constituents of *Protusnea malacea*. J Nat Prod **48**: 307

65. Chandrasenan K, Neelakantan S, Seshadri TR (1965) Naturally Occurring Dibenzoquinones. Bull Natl Inst Sci India **28**: 92

66. Chapman BE, Roser DJ, Seppelt RD (1994) ^{13}C-NMR Analysis of Antarctic Cryptogam Extracts. Antarctic Science **6**: 295

67. Chawla HM, Chibber SS, Niwas S (1980) Novel Photochemical Conversion of Pulvinic Acid to Leprapinic Acid. Tetrahedron Letters **21**: 2089

68. Chawla HM, Gambhir I, Kathuria L (1979) Separation and Identification of Lichen Acids of the Pulvinic Acid Series. J High Resol Chromatogr & Chromatogr Comm **2**: 673

68a. Chênevert R, Lavoie M, Dasser M (1997) Use of Aldolases in the Synthesis of Non-carbohydrate Natural Products. Stereoselective Synthesis of Aspicilin C-3–C-9 Fragment. Can J Chem **75**: 68

69. Chester DO, Elix JA, Kennedy JM (1986) Isodidymic Acid, a New Dibenzofuran from the Lichen *Cladonia didyma.* Aust J Chem **39**: 1759

70. Chiarello J, Joullie MM (1988) Synthetic Routes to Cristatic Acid and Derivatives. Tetrahedron **44**: 41

71. Cleaver L, Croft JA, Ritchie E, Taylor WC (1976) Chemical Studies of the Protaceae. IX. Synthesis of 5-Alkylresorcinols from Aliphatic Precursors. Aust J Chem **29**: 1989

72. Cohen PA, Hudson JB, Towers GHN (1996) Antiviral Activities of Anthraquinones, Bianthrones and Hypericin Derivatives from Lichens. Experientia **52**: 180

73. Cohen PA, Towers GHN (1995) Anthraquinones and Phenanthroperylenequinones from *Nephroma laevigatum.* J Nat Prod **58**: 520

74. Cohen PA, Towers GHN (1995) The Anthraquinones of *Heterodermia obscurata.* Phytochemistry **40**: 911

75. Cohen PA, Towers GHN (1996) Biosynthetic Studies on Chlorinated Anthraquinones in the Lichen *Nephroma laevigatum.* Phytochemistry **42**: 1325

76. Cohen PA, Towers GHN (1997) Chlorination of Anthraquinones by Lichen and Fungal Enzymes. Phytochemistry **44**: 271

77. Comber MF, Sargent MV, Skelton BW, White AH (1989) Depsidone Synthesis. Part 24. The Synthesis of Epiphorellic Acid 2. Pseudodepsidone and X-ray Crystal Structure of a Grisadienedione Epoxide. J Chem Soc Perkin I 441

78. Connolly JD, Freer AA, Huneck S (1984) Crystal Structure of (−)-Placodiolic Acid, a Dibenzofuran Derivative from the Lichen *Rhizoplaca chrysoleuca.* Phytochemistry **23**: 702

79. Connolly JD, Freer AA, Kalb K, Huneck S (1984) Eriodermin, a Dichlorodepsidone from the Lichen *Erioderma physcioides* – Crystal Structure Analysis. Phytochemistry **23**: 857

80. Conover MA, Mierzawa R, King A, Loebenberg D, Bishop WR, Puar M, Patel M, Coval SJ, Hershenhorn J, Strobel GA (1992) Usnic Acid Amide, a Phytotoxin and Antifungal Agent from *Cercosporidium henningsii.* Phytochemistry **31**: 2999

81. Cooper AB, Wang J, Saksena AK, Girijallabhan V, Ganguly AK, Chan T-M, McPhail AT (1992) Synthesis of (+)-8-Methyl Cercosporamide: Stereochemical Correlation of Natural (−)-Cercosporamide with (+)-Usnic Acid. Tetrahedron **48**: 4757

82. Corbett RE, Simpson J, Groh EM, Nicholson BK, Wilkins AL, Robinson WT (1982) Lichens and Fungi. Part 16. The Crystal and Molecular Structures of Stictane-3β, 22α-diol and 22α-Hydroxy-3,4-secostictan-3-oic Acid. J Chem Soc Perkin I 1339

83. Corey EJ, Desai MC, Engler TA (1985) Total Synthesis of (±)-Retigeranic Acid. J Am Chem Soc **107**: 4339
84. Cosar G, Tumbay E, Zeybek N, Ozer A (1988) The Antibacterial and Antifungal Effect of some Lichens Growing in Turkey. Part I. *Evernia prunastri, Pseudevernia furfuracea* and *Alectoria capillaris.* Fitoterapia **59**: 505
85. Cresp TM, Sargent MV, Elix JA, Murphy DPH (1973) Formylation and Bromination *ortho* to the Hydroxy Group of 2-Carbonyl-substituted Phenols in the Presence of Titanium (IV) chloride. J Chem Soc Perkin I 340
86. Crittenden PD, David JC, Hawksworth DL, Campbell FS (1995) Attempted Isolation and Success in the Culturing of a Broad Spectrum of Lichen-forming and Lichenicolous Fungi. New Phytol **130**: 267
87. Crittenden PD, Porter N (1991) Lichen-forming Fungi: Potential Sources of Novel Metabolites. Tibtech **9**: 409
88. Culberson CF (1986) Biogenetic Relationships of the Lichen Substances in the Framework of Systematics. Bryologist **89**: 91
89. Culberson CF, Ammann K (1979) Standardmethode zur Dünnschichtchromatographie von Flechtensubstanzen. Herzogia **5**: 1
90. Culberson CF, Armaleo D (1992) Induction of a Complete Secondary Product Pathway in a Cultured Lichen Fungus. Exper Mycol **16**: 52
91. Culberson CF, Culberson WL (1976) Chemosyndromic Variation in Lichens. Syst Bot **1**: 325
92. Culberson CF, Culberson WL, Gawn S, Johnson A (1987) New Depsides from Lichens: Microchemical Methodologies Applied to the Study of New Natural Products Discovered in Herbarium Specimens. Am J Bot **74**: 403
93. Culberson CF, Culberson WL, Johnson A (1985) Orcinol-type Depsides and Depsidones in the Lichens of the *Cladonia chlorophaea* Group (Ascomycotina, Cladoniaceae). Bryologist **88**: 380
94. Culberson CF, Culberson WL, Johnson A (1985) Does the Symbiont Alga Determine Chemotype in Lichens? Mycologia **77**: 657
95. Culberson CF, Culberson WL, Johnson A (1986) Two New Lichen Products, Elatinic Acid and Methyl Barbatate, from the Genus *Haematomma* (Ascomycotina, Haematommaceae). Mycologia **78**: 888
96. Culberson CF, Culberson WL, Johnson A (1990) The *Ramalina americana* Complex (Ascomycotina, Ramalinaceae): Chemical and Geographic Correlations. Bryologist **93**: 167
97. Culberson CF, Culberson WL, Johnson A (1990) A Novel *meta*-Depside from the *Ramalina subfraxinea* Complex (Ascomycotina, Ramalinaceae). Bryologist **93**: 193
98. Culberson CF, Culberson WL, Johnson A (1992) Characteristic Lichen Products in Cultures of Chemotypes of the *Ramalina siliquosa* Complex. Mycologia **84**: 705
99. Culberson CF, Culberson WL, Johnson A (1993) Occurrence and Histological Distribution of Usnic Acid in the *Ramalina siliquosa* Species Complex. Bryologist **96**: 181
100. Culberson CF, Elix JA (1989) Lichen Substances. In: Dey PM, Harborne JB (eds) Methods in Plant Biochemistry, vol 1. Plant Phenolics. Academic Press, London, p 509
101. Culberson CF, Hale ME, Tønsberg T, Johnson A (1984) New Depsides from the Lichen *Dimelaena oreina* and *Fuscidea viridis.* Mycologia **76**: 148
102. Culberson CF, Johnson A (1982) Substitution of Methyl *Tert.*-butyl Ether for Diethyl Ether in the Standardized Thin-layer Chromatographic Method for Lichen Products. J Chromatogr **238**: 483

103. Culberson CF, Johnson A, Patwardhan PG, Makhija U (1990) New Depsides in *Stirtonia ramosa* (Ascomycotina, Arthoniaceae). Bryologist **93**: 279
104. Culberson WL, Culberson CF, Johnson A (1977) *Pseudevernia furfuracea-olivetorina* Relationships: Chemistry and Ecology. Mycologia **69**: 604
105. Culberson WL, Culberson CF, Johnson A, Park YS (1993) New Chemistries in the *Cladonia cariosa* Complex and Homoheveadride in Natural Thalli and Single-spore Cultures of *C. polycarpoides* and *C. polycarpia.* Bibl Lichenol **53**: 43
106. Czeczuga B (1983) Mutatoxanthin, the Dominant Carotenoid in Lichens of the *Xanthoria* Group. Biochem Syst Ecol **11**: 329
107. Czeczuga B (1985) Carotenoids in Representatives of the Cladoniaceae. Biochem Syst Ecol **13**: 83
108. Czeczuga B (1986) Investigations on Carotenoids in Lichens. XI. Lichens from Lapland. Ann Bot Fenn **23**: 251
109. Czeczuga B (1988) Carotenoids. In: Galun M (ed) CRC Handbook of Lichenology, vol III. CRC Press, Boca Raton, p 25
110. Czeczuga B (1988) Investigations on Carotenoids in Lichens XXII. Lichens from the Upper Tracja Valley (Bulgaria). Acta Soc Bot Polon **57**: 447
111. Czeczuga B (1993) Carotenoids in Lichens. Bibl Lichenol **53**: 53
112. Czeczuga B (1984) Carotenoids in Certain Lichens of Bialowieza Forest. Acta Soc Bot Polon **63**: 21
113. Czeczuga B, Alstrup V (1987) The Carotenoid Content of Lichens from Greenland. Biochem Syst Ecol **15**: 297
114. Czeczuga B, Beckett RP, Pienaar RN (1988) Investigations on Carotenoids in Lichens. XVIII. Carotenoid Content in Certain Lichens from Southern Africa. South Afr J Bot **54**: 571
115. Czeczuga B, Bubrick P (1986) Investigations of Carotenoids in Lichens IX. Cartotenoids in Lichens from Israel. Israel J Bot **35**: 41
116. Czeczuga B, Caccamese S, Passadore MV (1989) Investigations on Carotenoids in Lichens XX. Carotenoids in Lichens from Various Italian Environment. Phyton **29**: 15
117. Czeczuga B, Calvelo S (1994) Investigations on Carotenoids in Lichens from *Nothofagus* Forests Along the Precipitation Gradient (North-Western Patagonia, Argentina). Feddes Rep **105**: 201
118. Czeczuga B, Calvelo S (1995) Investigations on Carotenoids of Fruticose Lichens from *Nothofagus* Forests, North-Western Patagonia (Argentina). Bol Soc Argent Bot **31**: 19
119. Czeczuga B, Christensen SN (1994) The Carotenoids of *Pseudevernia furfuracea* Along a North–South Gradient in Europe. Feddes Rep **105**: 473
120. Czeczuga B, Cifuentes B (1986) Investigations on Carotenoids in Lichens. XII. Some Species from the Pyrenaen Peninsula. Acta Soc Bot Polon **55**: 239
121. Czeczuga B, Cifuentes B, Reynaud PA (1988) Carotenoids in Lichens from the Canary Islands. Biochem Syst Ecol **16**: 117
122. Czeczuga B, Doll R (1991) The Carotenoid Content of Lichens of the *Cladonia* Genus from Mecklenburg. Feddes Rep **102**: 431
123. Czeczuga B, Egea JM (1990) Investigations on Carotenoids in Lichens. XXVII. Carotenoids in Lichens from North Africa and SE Spain. An Biol **16**: 131
124. Czeczuga B, Etayo J, Giralt M, Casares M, Lumbsch HT, Salema R (1996) Carotenoids in the Thalli of Lichen Species of the Iberian Peninsula. Feddes Rep **107**: 89

125. Czeczuga B, Eversman S (1993) Carotenoids in Lichens from Yellowstone National Park and Adjacent Forests in Montana and Wyoming. Bryologist **96**: 102

126. Czeczuga B, Ferraro De Corona L (1987) Carotenoids in Lichens from Argentina. Biochem Syst Ecol **15**: 519

127. Czeczuga B, Ferraro I, Baron M (1992) Carotenoids in some Lichen Species from South Brazil and Paraguay. Bol Bot **13**: 23

128. Czeczuga B, Gutkowski R, Czerpak R (1986) Investigations on Carotenoids in Lichens. V. Lichens from Antarctica. Polish Polar Res **7**: 295

129. Czeczuga B, Hammer S (1994) Carotenoids in Representatives of the Lichen Genus *Cladonia* in North America. Bryologist **97**: 67

130. Czeczuga B, Harada H, Smith CW (1997) Carotenoids in some Lichens from Hawaiian, Ogasawara (Bonin) and Mariana Islands (Pacific). J Hattori Bot Lab **80**: 263

131. Czeczuga B, Huneck S, Cogt U (1994) Carotenoids in Lichens from Mongolia. J Hattori Bot Lab **75**: 371

132. Czeczuga B, Jacobsen P (1993) Carotenoids in West Greenland Lichens. Arctic and Alpine Res **25**: 413

133. Czeczuga B, John EA (1993) Carotenoids from Nothern Canada. Feddes Rep **104**: 551

134. Czeczuga B, Kashiwadani H (1993) Carotenoids of *Parmotrema tinctorum* and *Rimelia clavulifera* (Parmeliaceae, Lichenes) from Various Localities in Japan. Bull Natl Sci Mus (Tokyo) **B 19**: 113

135. Czeczuga B, Kristinsson H (1992) Investigations on Carotenoids in Lichens XXXIII. Carotenoids in Lichens from Heathland in Iceland. Acta Bot Islandica **11**: 3

136. Czeczuga B, Kudratov I, Bai Bulatova NE (1993) Carotenoids in certain Lichens from Pamir-Alai and Tien-Shan Mountains. Feddes Rep **104**: 59

137. Czeczuga B, Lallement R (1993) Investigations on Carotenoids in Lichens. 44. Carotenoids of some Lichens from France. Acta Bot Gallica **140**: 35

138. Czeczuga B, Lim G, Koon Tan T, Streimann H, Jenie UA, Gruezo WS (1994) Carotenoids in Certain Species of Lichens from Tropical Region. Feddes Rep **105**: 99

139. Czeczuga B, Moberg R, Alstrup V (1992) Investigations on Carotenoids in Lichens XXXII. Carotenoids Occurring in the Thalli of Lichens from Kenya (Equatorial Africa). Acta Soc Bot Polon **61**: 231

140. Czeczuga B, Obermayer W, Upreti DK, Sharma LR (1996) Carotenoids in Lichens in Various Regions of the Himalayas. J Hattori Bot Lab **80**: 323

141. Czeczuga B, Olech M (1989) Investigations of Carotenoids in Lichens XXIV. Further Studies in Lichens of the Antarctica. Ser Cient Inst Antarct Chileno **39**: 91

142. Czeczuga B, Osorio HS (1989) Investigations on Carotenoids in Lichens XXI. Astaxanthin, the Dominant Carotenoid in some Lichens from Uruguay. Israel J Bot **38**: 115

143. Czeczuga B, Richardson DHS (1989) Carotenoids in some Lichen Species from Ireland. Lichenologist **21**: 363

144. Czeczuga B, Shchelkunova RP (1986) Carotenoids from the Taimyr Region of North Siberia. Biochem Syst Ecol **14**: 463

145. Czeczuga B, Skirina IF, Dzae RJ, Maximov OB, Stepanenko LS (1992) Carotenoids in Lichens from Primorski Territory (Vladivostok Region) and North Korea. Feddes Rep **103**: 579

146. Czeczuga B, Skirina IF, Maximov OB, Stepanenko LS (1989) Investigations on Carotenoids in Lichens. XIX. Carotenoids in Lichens of the Tundra of Kamchatka Region (Far East). Phyton **29**: 7

147. Czeczuga B, Skult H (1988) Carotenoids in Lichens of Southern Finland. Ann Bot Fenn **25**: 229

148. Czeczuga B, Stenross S, Christensen SN, Ahti T (1991) Variability of Carotenoid Composition in some Species of the Lichen Genera *Cladonia* and *Cladina*. Ann Bot Fenn **28**: 123

149. Czeczuga B, Taylor FJ (1991) Investigations on Carotenoids in Lichens. XXVIII. Carotenoids of some New Zealand Lichens. New Zealand J Bot **29**: 1

150. Czeczuga B, Upreti DK (1990) Carotenoids in Lichens of the *Lobaria* and *Peltigera* Genera from India. Bull Bot Survey India **32**: 80

151. Czeczuga B, Von Arb C, Lumbsch HT (1995) Carotenoids in Lichens from the Swiss Alps. Feddes Rep **106**: 173

152. Czeczuga B, Wilhelm G (1992) Carotenoids in certain Lichens from the State of Illinois. Transact Illin State Acad Sci **85**: 1

153. Czeczuga B, Xavier-Filho L (1987) Investigations on Carotenoids in Lichens. VII. Some Lichens from Brazil. Rev Brasil Biol **47**: 243

154. Czeczuga B, Xavier-Filho L (1987) Investigations on Carotenoids in Lichens. X. Luteoxanthin and Apo-12′-violaxanthal in Lichens from the Antarctica. Ser Cient Inst Antarct Chileno **36**: 151

155. Czeczuga B, Yoshida K (1991) Carotenoids in certain Lichens from the Chichibu Mountains, Central Japan. Feddes Rep **102**: 661

156. Dallacker F, Ditgens K (1975) Derivate des Methylendioxybenzols, 41. Synthese des Phlebiorubrons und der Polyporsäure. Z Naturforsch **30c**: 1

157. Dalvi RR, Singh B, Salunkhe DK (1972) Physiological and Biochemical Investigations on the Phytotoxicity of Usnic Acid. Phyton **29**: 63

158. Damon RE, Schlessinger RH (1976) An Efficient and Stereospecific Total Synthesis of DL-Protolichesterinic Acid. Tetrahedron Letters 1561

159. Da Silva MLC, Iacomini M, Gorin PAJ (1990) Identification of Lactose and D-Volemitol in the Lichen *Sticta damaecornis*. An Acad Brasil Ciencias **62**: 9

160. Da Silva MDLC, Iacomini M, Jablonski E, Gorin PAJ (1993) Carbohydrates, Glycopeptide and Protein Components of the Lichen *Sticta* sp. and Effect of Storage. Phytochemistry **33**: 547

161. Dauriac H, Rondon Y (1976) Quelques Aspects des Propriétés d' un Lichen, le *Letharia vulpina* (L.) Hue, en Biologie Végétale Experimentale. I. Action sur la Germination du Blé et de Développemente de l' Ail. Bull Soc bot Fr **123**: 235

162. David F, Elix JA, Samsudin MWB (1990) Two New Aliphatic Acids from the Lichen *Parmotrema praesorediosum*. Aust J Chem **43**: 1297

163. Dembitsky VM (1992) Lipids of Lichens. Progr Lipid Res **31**: 373

164. Dembitsky VM, Bychek IA, Kashin AG (1992) Chemical Constituents of some Lichen Species. J Hattori Bot Lab **71**: 255

165. Dembitsky VM, Bychek IA, Rozentsvet OA (1993) Diacylglyceryltrimethylhomoserines and Phospholipides of some Lichens. Phytochemistry **34**: 6

166. Dembitsky VM, Bychek IA, Shustov MV, Rozentsvet OA (1991) Phospholipid and Fatty Acid Composition of some Lichen Species. Phytochemistry **30**: 837

167. Dembitsky VM, Rezanka T, Bychek IA (1992) Fatty Acids and Phospholipids from Lichens of the Order Lecanorales. Phytochemistry **31**: 851

168. Dembitsky VM, Rezanka T, Bychek IA (1992) Lipid Composition of some Lichens. Phytochemistry **31**: 1617

169. Dembitsky VM, Rezanka T, Bychek IA, Shustov MV (1991) Identification of Fatty Acids from *Cladonia* Lichens. Phytochemistry **30**: 4015

170. Dembitsky VM, Rezanka T, Bychek IA, Shustov MV (1992) Fatty Acid Composition of *Parmelia* Lichens. Phytochemistry **31**: 841

171. Demleitner S, Kraus J, Franz G (1992) Synthesis and Antitumour Activity of Curdlan and Lichenan Branched at C-6. Carbohydr Res **226**: 239

172. Demleitner S, Kraus J, Franz G (1992) Synthesis and Antitumour Activity of Sulfoalkyl Derivatives of Curdlan and Lichenan. Carbohydr Res **226**: 247

173. Dertien BK, De Kok JL, Kuiper PJC (1977) Lipid and Fatty Acid Composition of Tree-growing Terrestrial Lichens. Physiol Plant **40**: 175

174. Devlin JP, Falshaw CP, Ollis WD, Wheeler R (1971) Phytochemical Examination of the Lichen *Lecanora rupicola* (L.) Zahlbr. J Chem Soc 1318

175. Dietz S, Hartung W (1998) Abscisic Acid in Lichens: Variation, Water Relations and Metabolism. New Phytol **138**: 99

176. Djura P, Sargent MV (1983) The Structure of Isopannarin. Aust J Chem **36**:1057

177. Dmitrenok PS, Elkin JN, Stepanenko PS, Krivoshekova OE (1987) Massenspektroskopische Ionisation von einigen Flechtenmetaboliten (in Russian). Chim Prirodn Soedin 295

178. Domon B, Hostettmann M, Hostettmann K (1982) Droplet Counter-current Chromatography with Non-aqueous Solvent Systems. J Chromatogr **246**: 133

179. Dornhagen J, Scharf H-D (1985) Synthesis of Dichloroisoeverninic Acid. Tetrahedron **41**: 173

180. Dornhagen J, Scharf H-D (1985) Verbesserte Synthese von Dichlorisoeverninsäure. Z Naturforsch **40b**: 1541

181. Duffley RP, Handrick GR, Uliss DB, Lambert G, Dalzell HC, Razdan RK (1980) Synthesis of 2′- and 3′-Acetoxyolivetols [5-(2- and 3-Acetoxypentyl)-1,3-benzene Diols]: Key Intermediates in the Synthesis of Tetrahydrocannabinol Derivatives. Synthesis 733

182. Durrani AA, Tyman HP (1980) Long-chain Phenols. Part 16. A Novel Synthesis of Homologous Orsellinic Acids and their Methyl Esters. J Chem Soc Perkin I 1658

183. Dyke HJ, Elix JA, Marcuccio SM, Whitton AA (1987) Oxidation of Alkyl 1,6-Dihydroorsellinates. A New Method for the Synthesis of Methyl Orsellinate and Homologues. Aust J Chem **40**: 431

184. Edwards RL, Gill M (1973) Constituents of Higher Fungi. Part XIII. 2-Aryl-3-methoxymaleic Anhydrides from Pulvinic Acid Derivatives. A Convenient Method for Determination of Structure of Fungal and Lichen Pulvinic Acid Derivatives. J Chem Soc Perkin I 1538

185. Egan RS (1986) Correlations and Non-correlations of Chemical Patterns with Lichen Morphology and Geography. Bryologist **89**: 99

186. Ejiri H, Sankawa U, Shibata S (1975) Graciliformin and its Acetates in *Cladonia graciliformis*. Phytochemistry **14**: 277

187. Elix JA (1993) Progress in the Generic Delimitation of *Parmelia* sensu lato Lichens (Ascomycotina: Parmeliaceae) and a Synoptic Key to the Parmeliaceae. Bryologist **96**: 359

188. Elix JA (1994) *Parmelia.* Flora Australia **55**: 1

189. Elix JA, Adler MT, Wardlaw JH (1996) A Further Three New Lichen Depsidones. Aust J Chem **49**: 1175

190. Elix JA, Barbero M, Giralt M, Lumbsch HT, McCaffery LF (1995) 2″-*O*-Methylgyrophoric Acid, a New Lichen Tridepside. Aust J Chem **48**: 1761

191. Elix JA, Barclay CE, Archer AW (1995) New Depsides in *Pertusaria* Lichens (Pertusariaceae, Lichenized Ascomycotina) and a New Species from Australia. Flechten Follmann, Contrib Lichenol in Honour of G Follmann. Geobot Phytotaxonom Study Group, Cologne, p 15
192. Elix JA, Barclay CE, David F, Criffin FK, Hill AM, McConnell DB, Wardlaw JH (1993) Synthesis of Further Lichen Depsides. Aust J Chem **46**: 301
193. Elix JA, Barclay CE, Lumbsch HT (1994) New Depsides from the Lichen *Lecanora planaica*. Aust J Chem **47**: 1199
194. Elix JA, Barclay CE, Lumbsch HT, Wardlaw JH (1997) New Chloro Depsides from the Lichen *Lecanora lividocinerea*. Aust J Chem **50**: 971
195. Elix JA, Benetts SA (1990) 6-*O*-Methylarthothelin and 1,3,6-Tri-*O*-methylarthothelin, Two New Xanthones from a *Dimelaena* Lichen. Aust J Chem **43**: 1587
196. Elix JA, Bennett SA, Jiang H (1991) 2,5-Dichloro-6-*O*-methylnorlichexanthone and 4,5-Dichloro-6-*O*-methylnorlichexanthone, Two New Xanthones from Australian *Dimelaena* Lichen. Aust J Chem **44**: 1157
197. Elix JA, Calanasan CA, Archer AW (1991) Subpicrolichenic Acid and Superpicrolichenic Acid, Two New Depsones from *Pertusaria* Lichens. Aust J Chem **44**: 1487
198. Elix JA, Chappell H-M, Jiang H (1991) Four New Lichen Xanthones. Bryologist **94**: 304
199. Elix JA, Chester DO, Gaul KL, Parker JL, Wardlaw JH (1989) The Identification and Synthesis of Further Lichen β-Orcinol *para*-Depsides. Aust J Chem **42**: 1191
200. Elix JA, Chester DO, Wardlaw JH, Wilkins AL (1990) The Identification of Two New β-Orcinol *para*-Depsides in the Lichen *Pseudocyphellaria norvegica*. Aust J Chem **43**: 191
201. Elix JA, Crook CE (1992) The Joint Occurrence of Chloroxanthones in Lichens, and a Further Thirteen New Lichen Xanthones. Bryologist **95**: 52
202. Elix JA, Crook CE, Jiang H, Zhi-Ning Z (1992) Synthesis of New Lichen Xanthones Aust J Chem **45**: 845
203. Elix JA, Crook CE, Lumbsch HT (1992) The Chemistry of Foliicolous Lichens. 1. Constituents of *Sporopodium vezdeanum* and *S. xantholeucum*. Mycotaxon **44**: 409
204. Elix JA, David F (1991) New Depsides from the Lichen *Biatora sorediosa*. Aust J Chem **44**: 1643
205. Elix JA, Engkanian U, Jones AL, Raston CL, Sargent MV, White AH (1978) Chemistry and Crystal Structure of Leprolomin, a Novel Diphenyl Ether from the Lichen *Psora leprolomum*. Aust J Chem **31**: 2057
206. Elix JA, Ernst-Russell KD (1993) A Catalogue of Standardized Thin Layer Chromatographic Data and Biosynthetic Relationships for Lichen Substances, 2nd Edn. Australian National University
207. Elix JA, Ernst-Russel MA (1996) Butlerins D, E and F – Three New Hexasubstituted Lichen *p*-Terphenyls. Aust J Chem **49**: 1247
208. Elix JA, Evans JE, Nash III TH (1988) New Depsides from *Dimelaena* Lichens. Aust J Chem **41**: 1789
209. Elix JA, Evans JE, Parker JL (1987) A One Step Synthesis of the Depsidone Furfuric Acid: Evidence for an Artefact in the Isolation from a Lichen. Aust J Chem **40**: 2129
210. Elix JA, Feige GB, Lumbsch HT, Mies B, Wardlaw JH (1995) The Structure Determination of Simonyellin – a New Naphthopyran. Aust J Chem **48**: 2035
211. Elix JA, Gaul KL (1986) The Interconversion of the Lichen Depsides *para*- and *meta*-Scrobiculin, and the Biosynthetic Implications. Aust J Chem **39**: 613

212. Elix JA, Gaul KL, Hockless DC, Wardlaw JH (1995) Structure Determination of Butlerins A, B and C-Three New Lichen *p*-Terphenyls. Aust J Chem **48**: 1049
213. Elix JA, Gaul KL, James PW (1985) α-Acetylhypoconstictic Acid, a New Depsidone from the Lichen *Menegazzia dispora*. Aust J Chem **38**: 1735
214. Elix JA, Gaul KL, James PW, Purvis OW (1987) Three New Lichen Depsidones. Aust J Chem **40**: 417
215. Elix JA, Gaul KL, Jiang H (1993) The Structure and Synthesis of some Minor Xanthones from the Lichen *Rinodina thiomela*. Aust J Chem **46**: 95
216. Elix JA, Gaul KL, Kantivilas G, James PW (1993) Siphulellic Acid, a New Depsidone from the Lichen *Siphulella coralloidea*. Bibl Lichenol **53**: 67
217. Elix JA, Gaul KL, Lumbsch HT (1987) Isolation of a Novel Lichen Xanthone from the Genus *Diploschistes* s.lat. Aust J Chem **40**: 1031
218. Elix JA, Gaul KL, Sterns M, Samsudin MW (1987) The Structure of the Novel Lichen Xanthone Thiomelin and its Cogenors. Aust J Chem **40**: 1169
219. Elix JA, Jayanthi VK (1986) Synthesis of the Lichen Depsides Miriquidic Acid and Normiriquidic Acid. Aust J Chem **39**: 791
220. Elix JA, Jayanthi VK (1987) Synthetic Confirmation of the Structure of the Lichen Benzyl Esters Alectorialic and Barbatolic Acids. Aust J Chem **40**: 1841
221. Elix JA, Jayanthi VK (1987) The Isolation of the Lichen Depside 4-*O*-Demethylmicrophyllinic Acid. Aust J Chem **40**: 1851
222. Elix JA, Jayanthi VK, Jones AJ, Lennard CJ (1984) A Novel Biphenyl from the Lichen *Psoroma contortum*. Aust J Chem **37**: 1531
223. Elix JA, Jayanthi VK, McCaffery LF, Yu J (1998) Synthesis of New Lichen Tridepsides. Aust J Chem **51**: 1045
224. Elix JA, Jayanthi VK, Wardlaw JH (1989) 2-*O*-Methylhiascic Acid, a New Tridepside in the Lichen *Parmelinopsis neodamaziana*. Aust J Chem **42**: 1423
225. Elix JA, Jenie UA (1989) A Synthesis of the Lichen Diphenyl Ether Epiphorellic Acid 1. Aust J Chem **42**: 987
226. Elix JA, Jenie UA, Arvidsson L, Jørgensen PM, James PW (1986) New Depsidones from the Lichen Genus *Erioderma*. Aust J Chem **39**: 719
227. Elix JA, Jenie UA, Jenkins GA (1987) Three New Depsidones from the Lichen *Neofuscelia subincerta*. Aust J Chem **40**: 2031
228. Elix JA, Jenie UA, Parker JL (1987) A Novel Synthesis of the Lichen Depsidones Divaronic Acid and Stenosporonic Acid, and the Biosynthetic Implications. Aust J Chem **40**: 1451
229. Elix JA, Jenkins GA, Lumbsch HT (1988) Chemical Variation in the Lichen Genus *Diploicia* (Ascomycotina). Mycotaxon **33**: 457
230. Elix JA, Jenkins GA, Venables DA (1990) New Chlorine-containing Depsidones from Lichens. Aust J Chem **43**: 197
231. Elix JA, Jiang H (1990) 5,7-Dichloro-3-*O*-methylnorlichexanthone, a New Xanthone from the Lichen *Lecanora broccha*. Aust J Chem **43**: 1591
232. Elix JA, Jiang H, Portelli VJ (1990) Structure and Synthesis of the Lichen Xanthone Isoarthothelin (2,5,7-Trichloronorlichexanthone). Aust J Chem **43**: 1291
233. Elix JA, Jiang H, Wardlaw JH (1990) A New Synthesis of Xanthones. 2,4,7-Trichloronorlichexanthone and 4,5,7-Trichloronorlichexanthone, Two New Lichen Xanthones. Aust J Chem **43**: 1745
234. Elix JA, Jin Y, Adler MT (1989) 3-Hydroxyumbilicaric Acid and 3-Methoxyumbilicaric Acid from the Lichen *Parmelinopsis bonariensis*. Aust J Chem **42**: 765
235. Elix JA, Johnston J, Armstrong PA (1986) A Revision of the Lichen Genus *Xanthoparmelia* in Australasia. Bull Brit Museum **15**: 163

236. Elix JA, Jones AL, Lajide L, Coppins BJ, James PW (1984) Two New Diphenyl Ethers and a New Depside from the Lichen *Micarea prasina* Fr. Aust J Chem **37**: 2349

237. Elix JA, Kennedy JM (1985) Synthesis of the Lichen Dibenzofuran Subdidymic Acid. Aust J Chem **38**: 1857

238. Elix JA, Lajide L (1984) The Identification of Further New Depsidones in the Lichen *Parmelia notata*. Aust J Chem **37**: 857

239. Elix JA, Lajide L (1984) Synthesis of the Lichen Depsides Pseudocyphellarins A and B. Aust J Chem **37**: 2153

240. Elix JA, Lajide L, Galloway DJ (1982) Metabolites from the Genus *Psoroma*. Aust J Chem **35**: 2325

241. Elix JA, Lajide L, Wardlaw JH (1995) The structure of Echinocarpic Acid. A Benzyl Ester from the Lichen *Parmelia norcrambidiocarpa*. Aust J Chem **48**: 1213

242. Elix JA, Lumbsch HT, Lücking R (1995) The Chemistry of Foliicolous Lichens. 2. Constituents of some *Byssoloma* and *Sporopodium* Species. Bibl Lichenol **58**: 81

243. Elix JA, Lumbsch HT, Wardlaw JH (1995) Conhypoprotocetraric Acid, a New Lichen β-Orcinol Depsidone. Aust J Chem **48**: 1479

244. Elix JA, Mahadevan I, Wardlaw I, Wardlaw JH, Arvidsson L, Jørgensen PM (1987) New Depsides from *Erioderma* Lichens. Aust J Chem **40**: 1581

245. Elix JA, Mayrhofer H, Wippel A (1995) 5-Chlorodivaricatic Acid, a New Depside from the Lichen Genus *Dimelaena*. Australasian Lichenol Newsletter **36**: 25

246. Elix JA, McCaffery LF (1997) Epiphorellic Acid 3, a New Lichen Diphenyl Ether. Aust J Chem **50**: 1101

247. Elix JA, Naidu R (1995) Identification of some Minor Dibenzofurans in the Lichen *Combea californica*. Bibl Lichenol **57**: 117

248. Elix JA, Naidu R, Laundon JR (1992) A Synthesis of the Lichen Dibenzofuran Pannaric Acid 2-Methyl Ester and its Isomer 3-*O*-Methylpannaric Acid. Aust J Chem **45**: 785

249. Elix JA, Naidu R, Laundon JR (1994) The Structure and Synthesis of 4-Oxypannaric Acid 2-Methyl Ester, a Dibenzofuran from the Lichen *Leproloma diffusum*. Aust J Chem **47**: 703

250. Elix JA, Naidu R, Thor G (1995) Cyclographin, a New Depsidone from the Lichen *Catarraphia dictyoplaca*. Aust J Chem **48**: 635

251. Elix JA, Parker JL, Tearne PD, Wardlaw JH (1985) Synthesis of Further Divarinol and Olivetol Lichen *para*-Depsides. Aust J Chem **38**: 1863

252. Elix JA, Portelli VJ (1990) A Synthesis of the Lichen Xanthone Thiomelin. Aust J Chem **43**: 1773

253. Elix JA, Pratt SJ, Laundon JR (1990) A Synthesis of the Lichen Dibenzofuran Pannaric Acid 6-Methyl Ester. Aust J Chem **43**: 2107

254. Elix JA, Robertson F, Wardlaw JH, Willis AC (1994) Isolation and Structure Determination of Demethylchodatin – a New Lichen Xanthone. Aust J Chem **47**: 2291

255. Elix JA, Senanayke BS, Kalb K (1998) The Structure of Isoplacodiolic Acid and Isopseudoplacodiolic Acid, Two New Dibenzofuran Derivatives from the Lichen Genus *Haematomma*. Herzogia **13**: 145

256. Elix JA, Venables DE (1993) 4-*O*-Methyllividic Acid, a New Lichen Depsidone. Mycotaxon **47**: 275

257. Elix JA, Venables DA, Archer AW (1994) Further New Depsones from the Lichen *Pertusaria truncata*. Aust J Chem **47**: 1345

258. Elix JA, Venables DA, Brako L (1990) New Chlorine-containing Depsidones from the Lichen *Phyllopsora corallina* var. *ochroxantha.* Aust J Chem **43**: 1953

259. Elix JA, Venables DA, Lumbsch HT, Brako L (1994) Further New Metabolites from Lichens. Aust J Chem **47**: 1619

260. Elix JA, Venables DA, Wedin M (1994) New Dibenzofurans and Depsides from the Lichen *Bunodophoron patagonicum.* Aust J Chem **47**: 1335

261. Elix JA, Wardlaw JH (1986) The Synthesis of New *meta*-Depsides from *Ramalina* Lichens. Aust J Chem **39**: 227

262. Elix JA, Wardlaw JH (1987) Synthesis of Lichen *para*-Sphaerophorol Depsides. Aust J Chem **40**: 425

263. Elix JA, Wardlaw JH (1996) The Structure of Dissectic Acid, a β-Orcinol *meta*-Depside from the Lichen *Heterodermia dissecta.* Aust J Chem **49**: 539

264. Elix JA, Wardlaw JH (1996) Hypoalectorialic Acid and Conechinocarpic Acid, Two New Benzyl Esters from Lichens. Aust J Chem **49**: 727

265. Elix JA, Wardlaw JH (1996) Synthesis of Depsides Present in the Lichen *Porpidia glaucophaea.* Aust J Chem **49**: 917

266. Elix JA, Wardlaw JH (1997) Synthesis of Further β-Ketoalkyl Lichen Depsides. Aust J Chem **50**: 479

267. Elix JA, Wardlaw JH (1997) New Depsides from the Lichen *Neofuscelia depsidella.* Aust J Chem **50**: 1145

268. Elix JA, Wardlaw JH (1998) Methyl Pseudoalectoronate, a New Depsidone from the Lichen *Parmotrema poolii.* Australasian Lichenol **43**: 12

269. Elix JA, Wardlaw JH, Archer AW (1999) New Depsides from *Pertusaria* Lichens. Australasian Lichenol **44**: 9

270. Elix JA, Wardlaw JH, Archer AW, Lumbsch HT, Plümper M (1997) Four New Depsidones from *Pertusaria* and *Lecanora* Lichens. Australasian Lichenol **44**: 22

271. Elix JA, Wardlaw JH, David F (1996) The Structure of Subconfluentic Acid, a Depside from the Lichen *Lecidella* cf. *cyanosarca.* Aust J Chem **49**: 159

272. Elix JA, Wardlaw JH, Yoshimura I (1997) Sublobaric Acid and Oxolobaric Acid, Two New Depsidones from the Lichen *Anzia hypoleucoides.* Aust J Chem **50**: 763

273. Elix JA, Whitton AA (1989) Synthesis of the Lichen Depsides Glomelliferic Acid, Loxodellic Acid, Glomellic Acid, their 4-*O*-Demethyl Analogues and Oxostenosporic Acid. Aust J Chem **42**: 1969

274. Elix JA, Whitton AA, Sargent MV (1984) Recent Progress in the Chemistry of Lichen Substances. Progr Chem Org Nat Prod **45**: 104

275. Elix JA, Wilkins AL, Wardlaw JH (1987) Five Fully Substituted Depsides from the Lichen *Pseudocyphellaria pickeringii.* Aust J Chem **40**: 2023

276. Elix JA, Yu J (1993) New Lichen β-Orcinol Depsidones and their Congeners. J Hattori Bot Lab **74**: 317

277. Elix JA, Yu J, Tønsberg T (1991) 4-*O*-Methylhiascic Acid and 5-*O*-Acetyl-4-*O*-methylhiascic Acid, Two New Lichen Tridepsides. Aust J Chem **44**: 157

278. Emmerich R, Giez I, Lange OL, Proksch P (1993) Toxicity and Antifeedant Activity of Lichen Compounds Against the Polyphagous Herbivorous Insect *Spodoptera littoralis.* Phytochemistry **33**: 1389

279. Enders D, Prokopenko OF (1995) Diastereo- and Enantioselective Synthesis of the 18-Membered Lichen Macrolide (+)-Aspicilin. Liebigs Ann Chem 1185

280. Endo Y, Hayashi H, Sato T, Maruno M, Ohta T, Nozoe S (1994) Confluentic Acid and 2′-*O*-Methylperlatolic Acid, Monoamine Oxidase B Inhibitors in a Brazilian Plant, *Himatanthus sucuuba.* Chem Pharm Bull **42**: 1198

281. Epstein E, Sagee O, Cohen JD, Garty J (1986) Endogenous Auxin and Ethylene in the Lichen *Ramalina duriaei*. Plant Physiol **82**: 1122

282. Epstein E, Sagee O, Cohen JD, Garty J (1991) Endogenous Auxin and Ethylene in the Lichen *Ramalina duriaei*. Ann Bot **68**: 317

283. Ernst-Russell MA, Chai CLL, Hurne AM, Waring P, Hockless DCR, Elix JA (1999) Structure Revision and Cytotoxic Activity of the Scabrosin Esters, Epidithiopiperazinediones from the Lichen *Xanthoparmelia scabrosa*. Aust J Chem **52**: 279

284. Ernst-Russell MA, Chai CLL, Hockless DCR, Elix JA (1998) Synthesis of the Pentasubstituted Lichen *p*-Terphenyls Butlerins A, B and C. Aust J Chem **51**: 1037

285. Estévez MP, Legaz E, Olmeda L, Pérez FJ, Vicente C (1981) Purification and Properties of a New Enzyme from *Evernia prunastri*, which Reduces L-Usnic Acid. Z Naturforsch **36c**: 35

286. Evans FJ, Schmidt RJ (1980) Plants and Plant Products that Induce Contact Dermatitis. Planta Med **38**: 289

287. Evans JE (1987) Identification and Synthesis of Secondary Lichen Metabolites. Thesis, Austral Natl Univ, Canberra

288. Fahselt D (1994) Secondary Biochemistry of Lichens. Symbiosis **16**: 117

289. Feddersen-Fieler G (1982) Farben aus Flechten. Schaper, Hannover

290. Feige GB, Lumbsch HT (1995) Some Types of Chemical Variation in Lichens. Crypt Bot **5**: 31

291. Feige GB, Lumbsch HT, Huneck S, Elix JA (1993) Identification of Lichen Substances by a Standardized High-Performance Liquid Chromatographic Method. J Chromatogr **646**: 417

292. Feige GB, Röser G, Lumbsch HT (1997) Chemotaxonomic Studies on European *Squamarina* Species (Ascomycotina, Lecanorales). Bibl Lichenol **67**: 25

293. Feige GB, Viethen B, Geyer M, Follmann G (1986) Untersuchungen zur Phytochemie der Flechtenfamilie Roccellaceae Chev. I. Über die Sekundärstoffe und Chemotypen von *Roccella hypomecha* (Ach.) Bory. J Hattori Bot Lab **60**: 143

294. Fernandez E, Quilhot W, Gonzalez I, Hidalgo ME, Molina X, Meneses I (1996) Lichen Metabolites as UV B Filters. Cosmetics and Toiletries **111**: 69

295. Fernholz E (1988) Von chiralen, nachwachsenden Rohstoffen zum Flechtenmakrolid (+)-Aspicilin. Inaugural-Dissertation, FB Chemie, Frankfurt/M.

296. Fiedler P, Gamboro V, Garbarino JA, Quilhot W (1986) Epiphorellic Acid 1 and 2, Two Diaryl Ethers from the Lichen *Cornicularia epiphorella*. Phytochemistry **25**: 461

297. Focella A, Teitel S, Brossi A (1977) A Simple and Practical Synthesis of Olivetol. J Org Chem **42**: 3456

298. Follmann G (1972) Zur Vergleichenden Phytochemie der Basidiolichenen. Willdenowia **6**: 427

299. Follmann G, Geyer M (1986) Vorarbeiten zu einer Monographie der Flechtenfamilie Roccellaceae Chev. VII. Inhaltsstoffe and Verwandschaftsbeziehungen der Gattungen *Combea* de Not. und *Schizopelte* T.M. Fries. Z Naturforsch **41c**: 1117

300. Follmann G, Huneck S, Schulz M, Sánchez-Pinto L (1993) Neue Untersuchungen zur Ausstattung der Roccellaceen mit sekundären Inhaltsstoffen. Herzogia **9**: 653

301. Forster MO, Saville WB (1922) Constitution of Picroroccellin, a Diketopiperazine Derivative from *Roccella fuciformis*. J Chem Soc **121**: 816

302. Fournet A, Ferreira M-E, Rojas de Arias A, Torres de Ortiz A, Inchausti A, Yaluff G, Quilhot W, Fernandez E, Hidalgo ME (1997) Activity of Compounds Isolated from Chilean Lichens Against Experimental Cutaneous Leishmaniasis. Comp Biochem physiol **C116**: 51

303. Fraser J (1996) Traditional Scottish Dyes and How to Make them. Canongate Book Ltd, Edinburgh

304. Fujikawa F, Hirayama T, Mori T, Nakamura N, Takagaki Y, Motoda Y, Mukainaka H, Ito K, Matsui R, Fijisawa M (1973) On the Free Amino Acids in Lichens of Japan. III. Yakugaku Zasshi **93**: 1558

305. Fujikawa F, Hirai K, Hirayama T, Toyota T, Nakamura T, Nishimaki T, Yoshikawa T, Yasuda S, Nishio S, Kojitani K, Nakai T, Ando T, Tsuji Y, Tomisaki K, Watanabe M, Fujisawa M, Nagai M, Koyama M, Matsuami N, Urasaki M, Takagawa M (1970) On the Free Amino Acids in Lichens of Japan. III. Yakugaku Zasshi **90**: 1267

306. Fujikawa F, Hirai K, Hirayama T, Toyota T, Urasaki M, Takagawa M, Fukuda M, Moritani K, Tomoike S, Harada H, Matoba C (1972) On the Free Amino Acids in Lichens of Japan. II. Yakugaku Zasshi **92**: 823

307. Fujiwara T, Ohashi T, Hamade R, Kinoshita Y, Yamamoto Y, Fukumori Y, Yoshimura I, Yamanaka T (1995) Cytochromes in the Cultured Mycobiont of a Lichen, *Cladonia vulcani.* Plant Cell Physiol **36**: 183

308. Fuzikawa F, Nakamura I (1942) Synthese der 3-Oxyorsellinsäureester und 5-Oxyorsellinsäureester (in Japanese). Yakugaku Zasshi **62**: 304

309. Gagnaire D, Marchessault RH, Vincendon M (1975) Nuclear Magnetic Resonance of Lichenin. Tetrahedron Letters 3953

310. Gagnaire D, Vincendon M (1977) Spectres de RMN de ^{13}C de la Lichenine, Terpolymere du Glucose; Comparaison avec les Deux Homo Glucanes Correspondants: Cellulose et Laminarin. Bull Soc Chim France 479

311. Galun M (ed) 1988 CRC Handbook of Lichenology, vol III, 3. CRC Press, Inc, Boca Raton

312. Galun M, Ronen R (1988) Interaction of Lichens and Pollutants. In: Galun M (ed) CRC Handbook of Lichenology, vol III. CRC Press, Inc, Boca Raton, p 55

313. Garbarino JA, Chamy MC, Gambaro V, Quilhot W, Naranjo O, Bolt E (1987) Studies on Chilean Lichens. X. The Phenolic Constituents of *Protousnea magellanica.* J Nat Prod **50**: 745

314. Garcia F, Espinoza A, Coltantes G, Rios V, Quilhot W (1982) Lichen Substances and the Plant Growth. III. The Effects of Roccellic Acid on the Growth of Germlings of *Ulva lactuca* L. J Hattori Bot Lab **53**: 443

315. Garcia-Junceda E, Cambon MC, Vicente C (1991) Enzymatic Hydrolysis of Physodic Acid in *Pseudevernia furfuracea.* Characterization of the Product of Depsidone Hydrolysis. Plant Science **77**: 197

316. Garcia-Junceda E, Gonzalez A, Vicente C (1987) Photosynthetical and Nutritional Implications in the Accumulation of Phenols in the Lichen *Pseudevernia furfuracea.* Biochem Syst Ecol **15**: 289

317. Gaskell SJ, Eglinton G (1973) Hydrocarbon Constituents of Three Species of Norwegian Lichens: *Cetraria nivalis, C. crispa, Siphula ceratites.* Phytochemistry **12**: 1174

318. Gaucher GM, Shepherd MG (1971) Orsellinic Acid. Biochem Prep **13**: 71

319. Gavin J, Nicollier G, Tabacchi R (1978) Composants Volatils de la "Mousse de Chêne" (*Evernia prunastri* (L.) Ach.). Helv Chim Acta **61**: 352

320. Gavin J, Tabacchi R (1975) Isolement et Identification de Composés Phénoliques et Monotérpeniques de la Mousse de Chêne" (*Evernia prunastri* (L.) Ach.). Helv Chim Acta **58**: 190

321. Geyer M (1985) Hochdruck-Flüssigkeits-Chromatographie (HPLC) von Flechten-Sekundärstoffen. Inauguraldiss, FB 9 Univ Essen

322. Geyer M, Feuerer T, Feige GB (1984) Chemie und Systematik in der Flechtengattung *Rhizocarpon*; Hochdruckflüssigkeitschromatographie (HPLC) der Flechten-sekundärstoffe der *Rhizocarpon superficiale*-Gruppe. Plant Syst Ecol **145**: 41

323. Ghogomu RT, Bodo B (1982) Structural Elucidation of 13-Acetoxylichesterinic Acid and 13-Acetoxyprotolichesterinic Acid, Two Aliphatic Lichen Metabolites from *Neuropogon trachycarpus*. Phytochemistry **21**: 2355

324. Giez I, Lange OL, Proksch P (1994) Growth Retarding Activity of Lichen Substances Against the Polyphagous Herbivorous Insect *Spodoptera littoralis*. Biochem Syst Ecol **22**: 113

325. Giles RGF, Sargent MV (1986) Naturally Occurring Dibenzofurans. X. A New Synthesis of Di-*O*-methylstrepsilin. Aust J Chem **39**: 2177

326. Girard M, Moir DB, Apsimon JW (1987) A Simple and Efficient Synthesis of 5′-($^2H_3$)-Olivetol. Can J Chem **65**: 189

327. Gollapudi SR, Telikepalli H, Jampani HB, Mirhom YW, Drake SD, Bhattiprolu KR, Vandervelde D, Mitscher LA (1994) Alectosarmentin, a New Antimicrobial Dibenzofuranoid Lactol from the Lichen, *Alectoria sarmentosa*. J Nat Prod **57**: 934

328. González AG, Barrera JB, Pérez EMR (1992) Synthesis of Hierridin, a Phenol from the Lichen *Ramalina hierrensis*. Phytochemistry **31**: 1436

329. González AG, Barrera JB, Pérez EMR, Padrón CEH (1991) Chemical Constituents of the Lichen *Cladina macaronesica*. Z Naturforsch **46c**: 12

330. González AG, Barrera JB, Péreze EMR, Padrón CEH (1992) Chemical Constituents of the Lichen *Ramalina hierrensis*. Planta Med **58**: 214

331. González A, Vicente C, Legaz ME (1984) A simple Assay Demonstrating the Effect of Rehydration on the Orsellinate Depside Hydrolase Activity of *Evernia prunastri*. J Plant Physiol **116**: 219

332. Gorin PA, Baron M, Silva MC, Teixeira AZA, Iacomini M (1993) Lichen Carbohydrates. Ciência e Cultura **45**: 27

333. Gorin PAJ, Iacomini M (1985) Structural Diversity of D-Galacto-D-mannan Components Isolated from Lichens Having Ascomycetous Mycobionts. Carbohydr Res **142**: 253

334. Gowan SP (1989) A Character Analysis of the Secondary Products of the Porpidiaceae (Lichenized Ascomycotina). Syst Bot **14**: 77

335. Griffin DA, Staunton J (1975) A Novel Biogenetic-type Synthesis of an Orsellinic Acid Derivative. J Chem Soc Chem Comm 675

336. Griffin FK (1993) Structure Determination and Synthesis of some β-Orcinol *para*-Depsides. Thesis, Austral Natl Univ, Canberra

337. Grigsby RD, Jamieson WD, McInnes AG, Maass WSG, Taylor A (1974) The Mass Spectra of Derivatives of Polyporic Acid. Can J Chem **52**: 4117

338. Gudjönsdottir GA, Ingolfsdottir K (1997) Quantitative Determination of Protolichesterinic and Fumarprotolichesterinic Acids in *Cetraria islandica* by High-performance Liquid Chromatography. J Chromatogr **757**: 303

339. Gunzinger J (1985) Etùde de la Composition Chimique de *Pseudevernia furfuracea* (L.) Zopf. Synthèse d` un Depside et d` une Depsidone. These Univ Neuchatel

340. Gunzinger J, Tabacchi R (1985) Isolement et Identification de l` Acide Furfurique, Nouvelle Depsidone du Lichen *Pseudevernia furfuracea* (L.) Ach. Helv Chim Acta **68**: 1936

341. Gunzinger J, Tabacchi R (1985) Synthese d' une Nouvelle Depsidone Derivée de l Acide Furfurique, le Dimethoxy-3,8-(dimethoxy-2,4-methoxycarbonyl-5-di-methyl-3,6-benzyl)-trimethyl-1,4,6-oxo-11,11H-dibenzo [b,e] carboxylate-7-demethyle. Helv Chim Acta **68**: 1940
342. Halvorsen R, Bendiksen E (1982) The Chemical Variation of *Pseudevernia furfuracea* in Norway. Nord J Bot **2**: 371
343. Hamada N (1988) Depside frome Isolated Mycobiont. II. Lichenologist **20**: 294
344. Hamada N (1989) The Effect of Various Culture Conditions on Depside Production by an Isolated Lichen Mycobiont. Bryologist **92**; 310
345. Hamada N, Tanahashi T, Goldsmith S, Nash III TH (1997) Introduction of Secondary Products in Isolated Mycobionts from North American Lichens. Symbiosis **23**: 219
346. Hamada N, Ueno T (1987) Depside from an Isolated Lichen Mycobiont. Agric Biol Chem **51**: 1705
347. Hamada N, Ueno T (1990) Lecanoric Acid from the Mycobiont of the Lichen *Stereocaulon curtatum*. Phytochemistry **29**: 678
348. Hamat ALB, Din LB, Samsudin MWB, Elix JA (1993) Two New Depsidones from the Lichen *Erioderma phaeorhizum* Vainio *sensu lato*. Aust J Chem **46**: 153
349. Handong S, Xiaoyu S, Zhong Wen L (1985) The Structure of Vermicularin. Acta Bot Yunnanica **7**: 109
350. Hanko B, Leuckert C, Ahti T (1985) Beiträge zur Chemotaxonomie der Gattung *Ochrolechia* (Lichenes) in Europa. Nova Hedwigia **42**: 165
351. Hanssen H-P, Schädler M (1985): Pflanzen in der traditionellen chinesischen Medizin. II. Flechten. Deutsche Apoth Z **125**: 1239
352. Hardman JT, Hale ME, Hardman PK, Beck ML (1983) Report on Activity of Lichen Lectins. Lichenologist **15**: 303
353. Hareed AM (1992) Phytochemical Investigations on *Usnea articulata* (L.) Hoffm., and *Usnea hesperina* Mot. Contribution to the Syntheses of Phthalides, Depsides and Depsidones. Thesis, Fac Sci Univ Neuchatel
354. Härmälä P, Hiltunen R, Oksman-Caldentey K-M, Kauppinen V (1992) Isolation and *In Vitro* Cultivation of Lichen Algae and their Antimicrobial Properties. Fitoterapia **58**: 217
355. Harris TM, Carney RL (1967) Synthesis of 3,5,7-Triketo Acids and Esters and Models of Biosynthesis of Phenolic Compounds. J Am Chem Soc **89**: 6734
356. Hase TA, Suokas E, McCoy K (1978) An Improved Synthesis of 2-Methyl-4,6-dihydroxybenzoic acid (Orsellinate) Esters and Homologues. Acta Chem Scand **B 32**: 701
357. Heide R, Provatoroff N, Traas PC, Valois PJ, Plasse N, Wobben HJ, Timmer R (1975) Qualitative Analysis of the Odoriferous Fraction of Oakmoss (*Evernia prunastri* (L.) Ach.). J Agr Food Chem **23**: 950
358. Heilman AS, Sharp AJ (1963) A Probable Antibiotic Effect of some Lichens on Bryophytes. Rev Bryol Lichenol **32**: 215
359. Hermann S, Leuckert C, Poelt J (1973) Zur Kenntnis der Flechtengruppe *Lecanora radiosa s.ampliss*. Willdenowia **7**: 9
360. Hertel H, Rambold G (1987) *Miriquidica* Genus Novum Lecanoracearum (Ascomycetes Lichenisati). Mitt Bot Staatssamml München **23**: 377
361. Herzig R (1990) Integriertes Biologisches Messystem der Luftverschmutzung mit Fletchten. Umwelttechnik **24**: 15

362. Herzig R, Liebendörfer L, Urech M (1987) Flechten als Bioindikatoren der Luftverschmutzung in der Schweiz: Methoden-Evaluation und Eichung mit wichtigen Luftschadstoffen. VDI Ber **609**: 619
363. Hesbacher S, Giez I, Embacher G, Fiedler K, Max W, Trawöger A, Türk R, Lange OL, Proksch P (1995) Sequestration of Lichen Compounds by Lichen-feeding Members of the Arctiidae (Lepidoptera). J Chem Ecol **21**: 2079
364. Hesse O (1900) Beitrag zur Kenntnis der Flechten und ihrer charakteristischen Bestandteile. 5. Mitt. J Prakt Chem **62**: 430
365. Hesse O (1904) Beitrag zur Kenntnis der Flechten und ihrer charakteristischen Bestandteile. 9. Mitt. J Prakt Chem **70**: 449
366. Hickey BJ, Lumsden AJ, Cole ALJ, Walker JRL (1990) Antibiotic Compounds from New Zealand Plants: Methyl Haematommate, an Anti-fungal Agent from *Stereocaulon ramulosum.* New Zealand Natural Sciences **17**: 49
367. Hidalgo ME, Fernández E, Quilhot W, Lissi E (1994) Antioxydant Activity of Depsides and Depsidones. Phytochemistry **37**: 1585
368. Higuchi M, Miura Y, Boohene J, Kinoshita Y, Yamamoto Y, Yoshimura I, Yamada Y (1993) Inhibition of Tyrosinase Activity by Cultured Lichen Tissues and Bionts. Planta Med **59**: 195
369. Himmelreich U, Huneck S (1994) Haemophaein, ein Dibenzofuran aus der Flechte *Phyllopsora haemophaea.* Z Naturforsch **49b**: 1292
370. Himmelreich U, Huneck S, Feige GB, Lumbsch HT (1994) Squamaron, ein Naphthochinon aus der Flechte *Squamarina cartilaginea.* Z Naturforsch **49b**: 1289
371. Hirabayashi K, Iwata S, Ito M, Shigeta S, Narui T, Mori T, Shibata S (1989) Inhibitory Effect of a Lichen Polysaccharide Sulfate, GE-3-S, on the Replication of Human Immunodeficiency Virus (HIV) *in vitro.* Chem Pharm Bull **37**: 2410
372. Hirayama T, Fujikawa F, Kasahara T, Otsuka M, Nishida N, Mizuno D (1980) Anti-tumor Activities of some Lichen Products and their Degradation Products. Yakugaku Zasshi **100**: 755
373. Hirose Y, Suehiro Y, Furukawa Y, Murakami T (1982) Chemische Studien über natürliche Anthrachinone. II. Synthese von Citreosin, Fallacinol und Fallacinal. Chem Pharm Bull **30**: 4186
374. Holland PT, Wilkins AL (1979) Mass Spectra of Some Naturally Occurring Stictane Triterpenoids and their Trimethylsilyl Derivatives. Org Mass Spectrom **14**: 160
375. Holzmann G, Leuckert C (1990) Application of Negative Fast Atom Bombardment and MS/MS to Screening of Lichen Compounds. Phytochemistry **29**: 2277
376. Höfle G (1977) ^{13}C-NMR-Spektroskopie chinoider Verbindungen. II. Substituierte 1,4-Naphthochinone und Anthrachinone. Tetrahedron **33**: 1963
377. Hudlicky T, Fleming A, Radesca L (1989) [2 + 3] and [3 + 4] Annulation of Enones. Enantiocontrolled Total Synthesis of (−)-Retigeranic Acid. J Am Chem Soc **111**: 6691
378. Hudlicky T, Radesca-Kwart L, Li L-Q, Bryant T (1988) Short, Enantioselective Synthesis of (−)-Retigeranic Acid via [2 + 3] Annulation. Tetrahedron Letters **29**: 3283
379. Hudlicky T, Short RP (1982) Terpenic Acids by Cyclopentane Annulation of Exocyclic Dienes. Synthesis of Triquinane Portion of Retigeranic Acid. J Org Chem **47**: 1522
380. Huneck S (1976) Inhaltsstoffe von *Pyxine coccifera.* Phytochemistry **15**: 799
381. Huneck S (1982) (+)-Isorangiformic Acid, a Lichen Substance from *Lecanora stenotropa.* Phytochemistry **21**: 2407

382. Huneck S (1984) Fortschritte der Chemie von Flechtenstoffen. Nova Hedwigia, Beiheft **79**: 794

383. Huneck S (1984) Pseudocyphellarins A and B, two Fully Substituted Depsides from the Lichen *Pseudocyphellaria endochrysea*. Phytochemistry **23**: 431

384. Huneck S (1984) *Tert*-butanolysis of Lichen Depsides. Phytochemistry **23**: 2697

385. Huneck S (1991) New Results in the Chemistry of Lichens. Symbiosis **11**: 225

386. Huneck S (1994) New Results in the Field of Lichen Chemistry. J Hattori Bot Lab **76**: 221

387. Huneck S (1995) The Chemistry and Chemotaxonomy of the Lichen Order Arthoniales, Especially the Family Roccellaceae. Crypt Bot **5**: 36

388. Huneck S (1999) The Significance of Lichens and their Metabolites. Naturwissenschaften **86**: 559

389. Huneck S, Akinniyi JA, Cameron AF, Connolly JD, Mulholland AG (1981) The Absolute Configurations of (+)-Usnic and (+)-Isousnic Acids. X-ray Analyses of the (−)-α-Phenyl-ethylamine Derivative of (+)-Usnic Acid and of (−)-Pseudoplacodiolic Acid, a New Dibenzofuran from the Lichen *Rhizoplaca chrysoleuca*. Tetrahedron Letters **22**: 351

390. Huneck S, Culberson CF, Culberson WL, Elix JA (1991) Haematommone, a Red Pigment from Apothecia of *Haematomma puniceum*. Phytochemistry **30**: 706

391. Huneck S, Elix JA (1993) The Chemistry of the Lichens *Anamylopsora pulcherrima* and *Tephromela armeniaca*. Herzogia **9**: 647

392. Huneck S, Elix JA, Naidu R, Follmann G (1993) 3-*O*-Demethylschizopeltic Acid, a New Dibenzofuran from the Lichen *Roccella hypomecha*. Aust J Chem **46**: 407

393. Huneck S, Feige GB, Lumbsch HT (1994) High Performance Liquid Chromatographic Analysis of Aliphatic Lichen Acids. Phytochem Anal **5**: 57

394. Huneck S, Follmann G (1969) Mitteilungen über Flechteninhaltsstoffe. LXIX. Zur Phytochemie und Chemotaxonomie der Arthoniaceae. Österr Bot Z **117**: 163

395. Huneck S, Himmelreich U, Nicholson G (1995) Arthogalin, a Cyclic Depsipeptide from the Lichen *Arthothelium galapagoense*. Z Naturforsch **50b**: 1101

396. Huneck S, Himmelreich U, Schmidt J, John V, Zeybek U (1994) Zur Chemie von Flechten aus der Türkei. Struktur von Nemetzon, dem Apothecienpigment von *Haematomma nemetzii*. Z Naturforsch **49b**: 1561

397. Huneck S, Jakupovic J, Follmann G (1991) 3-*O*-Methylpannaric Acid from the Lichen *Roccella capensis*. Z Naturforsch **46b**: 969

398. Huneck S, Jakupovic J, Follmann G (1992) The Final Structures of the Lichen Chromones Galapagin, Lobodirin, Mollin and Roccellin. Z Naturforsch **47b**: 449

399. Huneck S, Porzel A (1994) Synthese und spektroskopische Eigenschaften der Stereoisomeren Ester aus L- und D-*N*-Benzoylalanin und L- und D-*N*-Benzoylalaninol. Z Naturforsch **49b**: 569

400. Huneck S, Porzel A, Schmidt JC (1993) Structure and Synthesis of Arthonin, a Lichen Metabolite from *Arthonia endlicheri*. Tetrahedron Asymmetry **4**: 303

401. Huneck S, Porzel A, Schmidt J, Feige GB, Posner B (1993) Crustinic Acid, a Tridepside from *Umbilicaria crustulosa*. Phytochemistry **32**: 475

402. Huneck S, Porzel A, Schmidt J, Follmann G (1992) Hypothallin, ein weiterer Vertreter eines Aminosäure-aminoalkohol-esters aus der Krustenflechte *Schismatomma hypothallinum*. Z Naturforsch **47c**: 785

403. Huneck S, Preiss A, Schmidt J, Mendez AM (1983) 3β-Acetoxyhopan-1β,22-diol, a Triterpene from the Lichen *Pseudoparmelia texana*. Phytochemistry **22**: 2027

404. Huneck S, Schmidt J (1980) Lichen Substances. 126. Mass Spectroscopy of Natural Products. 10. Comparative Positive and Negative Ion Mass Spectroscopy of Usnic Acid and Related Compounds. Biomed Mass Spectr **7**: 301
405. Huneck S, Schmidt J (1995) Neue Flechtenanalysen aus verschiedenen Verwandtschaftskreisen. Flechten Follmann. Contrib to Lichenology in Honour of Gerhard Follmann. Geobot and Phytotaxonomical Study Group, Cologne 15
406. Huneck S, Schmidt J, Alstrup V (1995) Lichen Substances from Subfossil and Recent *Umbilicaria cylindrica*. Bibl Lichenol **57**: 231
407. Huneck S, Schmidt J, Porzel A (1994) Zur Chemie der Roccellsäure. Z Naturforsch **49b**: 561
408. Huneck S, Schmidt J, Tabacchi R (1989) Thermal Decomposition of Lichen Depsides Z Naturforsch **44b**: 1283
409. Huneck S, Schreiber K (1972) Wachtumsregulatorische Eigenschaften von Flechten-und Moos-Inhaltsstoffen. Phytochemistry **11**: 2429
410. Huneck S, Schreiber K, Steglich W (1973) Flechteninhaltsstoffe – XCVIII. Struktur des Aspicilins. Tetrahedron **29**: 3687
411. Huneck S, Steglich W (1983) Structure of (+)-Rangiformic Acid. Phytochemistry **22**: 2855
412. Huneck S, Tabacchi R (1987) Ψ-Esters of Depsidones with a Lactole Ring. Phytochemistry **26**: 1131
413. Huneck S, Takeda R (1992) Zur Chemie der Proto- und Allo-protolichesterinsäure. Z Naturforsch **47b**: 842
414. Huneck S, Tønsberg T, Bohlmann F (1986) (–)-Allo-pertusaric Acid and (–)-Dihydropertusaric Acid from the Lichen *Pertusaria albescens*. Phytochemistry **25**: 453
415. Huneck S, Yoshimura I (1996) Identification of Lichen Substances, Springer, Berlin, Heidelberg
416. Huovinen K (1985) Variation of Lichen Acids in *Cladina stellaris* and *Cladina rangiferina* in Finland and North Norway. Act Pharm Fenn **94**: 113
417. Huovinen K (1986) Chromatographic Studies on the Aromatic Lichen Substances in *Cladina* and *Cladonia*, Section Unciales. Thesis, Fac Sci Univ Helsinki
418. Huovinen K (1987) A Standard HPLC Method for the Analyses of Aromatic Lichen Compounds. Bibl Lichenol **25**: 457
419. Huovinen K, Ahti T (1986) The Composition and Contents of Aromatic Lichen Substances in the Genus *Cladina*. Ann Bot Fenn **23**: 93
420. Huovinen K, Ahti T (1986) The Composition and Contents of Aromatic Lichen Substances in *Cladonia*, Section Unciales. Ann Bot Fenn **23**: 173
421. Huovinen K, Ahti T (1988) The Composition and Contents of Aromatic Lichen Substances in *Cladonia*, Section Perviae. Ann Bot Fenn **25**: 371
422. Huovinen K, Ahti T, Stenroos S (1989) The Composition and Contents of Aromatic Lichen Substances in *Cladonia*, Section Cocciferae. Ann Bot Fenn **26**: 133
423. Huovinen K, Ahti T, Stenroos S (1990) The Composition and Contents of Aromatic Lichen Substances in *Cladonia*, Section Helopodium and Subsection Foliosae. Ann Bot Fenn **26**: 297
424. Huovinen K, Ahti T, Stenroos S (1990) The Composition and Contents of Aromatic Lichen Substances in *Cladonia*, Section Cladonia and Group Furcatae. Bibl Lichenol **38**: 209
425. Huovinen K, Härmälä P, Ottoila P, Hiltunen R, Von Schantz M (1987) Fast Atom Bombardment Mass Spectrometry in the Characterization of Lichen Depsides. Poster at the Univ Leiden

426. Huovinen K, Hiltunen R, Von Schantz M (1985) A High Performance Liquid Chromatographic Method for the Analysis of Lichen Compounds from the Genera *Cladina* and *Cladonia*. Acta Pharm Fenn **94**: 99
427. Huovinen K, Lampero M (1989) Usnic Acid as a Mitotic Inhibitor in the *Allium* Test. Planta Med **55**: 98
428. Hveding-Bergseth N, Bruun T, Kjøsen H (1983) Isolation of 30-Nor-21α-H-hopan-22-one (Iso-adiantone) from the Lichen *Platismatia glauca*. Phytochemistry **22**: 1826
429. Hylands PJ, Ingolfsdottir K (1985) The Isolation of Methyl β-Orcinol-carboxylate from *Stereocaulon alpinum* and Comments on the Isolation of 4,6-Dihydroxy-2-methoxy-3-methylacetophenone from *Stereocaulon* Species. Phytochemistry **24**: 127
430. Iacomini M, Gorin PAJ, Baron M, Tulloch AP, Mazurek M (1988) Novel D-Glucanes Obtained by Dimethyl Sulfoxide Extraction of the Lichens *Letharia vulpina*, *Actinogyra muehlenbergii*, and an *Usnea* sp. Carbohydrate Res **176**: 117
431. Iacomini M, Schneider CL, Gorin PAJ (1985) Comparative Studies on the Polysaccharides of *Cladonia alpestris* (Reindeer Moss), *Cladonia confusa* and *Cladonia amaurocraea*. Carbohydrate Res **142**: 237
432. Iacomini M, Zanin SMW, Fontana JD, Hogge L, Gorin PAJ (1987) Isolation and Characterization of β-D-Glucan, Heteropolysaccharide, and Trehalose Components of the Basidiomycetous Lichen *Cora pavonia*. Carbohydrate Res **168**: 55
433. Ichinose T, Miller M, Shibamoto T (1994) Inhibition of Malondialdehyde Formation from Liver Microsomes by a Lichen Constituent. Food Chem Toxicol **32**: 1167
434. Ihda T-A, Nakano T, Yoshimura I, Iwatsuki Z (1993) Phycobionts Isolated from Japanese Species of *Anzia* (Lichenes). Arch Protistenkunde **143**: 163
435. Inayma S, Hori H, Pang G-M, Nagasawa H, Ageta H (1989) Isolation of a Hopane-type Triterpenoid, Zeorin, from a Higher Plant, *Tripterygium regelii*. Chem Pharm Bull **37**: 2836
436. Ingolfsdottir K, Bloomfield SF, Hylands PJ (1985) *In vitro* Evaluation of the Antimicrobial Activity of Lichen Metabolites as Potential Preservatives. Antimicrob Agents and Chemotherapy **28**: 289
436a. Ingolfsdottir K, Chung GAC, Skúlason VG, Gissurarson SR, Vilhelmsdottir M (1988) Antimycobacterial Activity of Lichen Metabolites *in vitro*. Europ J Pharm Sci **6**: 141
436b. Ingolfsdottir K, Gissurarson SR, Nenninger A, Neszmelyi B, Wiedemann B, Wagner H (1997) Biologically Active Alkamide from the Lichen *Stereocaulon alpinum*. Phytomed **4**: 331
437. Ingolfsdottir K, Hjalmarsdottir MA, Sigurdson A, Gudjonsdottir GA, Brynjolfsdottir A, Steingrimsson O (1997) *In vitro* Susceptibility of *Helicobacter pylori* to Protolichesterinic Acid from the Lichen *Cetraria islandica*. Antimicrob Agents and Chemotherapy **41**: 215
438. Ingolfsdottir K, Hylands PJ, Solberg Y (1986) Structure of Vesuvianic Acid from *Stereocaulon vesuvianum*. Phytochemistry **25**: 550
439. Ingolfsdottir K, Wiedemann B, Birgisdottir M, Nenninger A, Jónsdóttir S, Wagner H (1997) Inhibitory Effects of Baeomycesic Acid from the Lichen *Thamnolia subuliformis* on 5-Lipoxygenase *in vitro*. Phytomed **4**: 125
440. Jacobsen JP, Reffstrup T, Boll PM (1977) ^{13}C Nuclear Magnetic Resonance of some 3-Acyltetronic Acids. Acta Chem Scand **B31**: 756
441. Jacques E (1990) A Stabilized Ethanolic PD-Reagent in Lichenology. Mém Soc Roy Bot Belg **12**: 148

442. Jakupovic J, Huneck S (1989) Zuordnung der ^{1}H-NMR-Spektren von Depsiden, Depsidonen, Depsonen and Dibenzofuranen aus Flechten durch NOE-Differenzspektroskopie. Z Naturforsch **44b**: 1117
443. Jaxa-Chamiec AA, Sammes PG, Kennewell PD (1980) A New Route to 5-Substituted Resorcinols and Related Systems. J Chem Soc Perkin I 170
444. Jerris PJ, Movkulich PM, Smith III AB (1979) A Facile Synthesis of Simple Tetronic Acids and Pulvinones. Tetrahedron Letters **45**: 4517
445. Jiang H (1991) Structure and Synthesis of Lichen Xanthones. Thesis, Austral Natl Univ, Canberra
446. John V (1998) Flechten als Bioindikatoren für Anthropogene Luftbelastungen. In: Barth HG, Köseoglu H (eds) Umweltqualität und Naturraum-Potentiale im Ballungsraum Izmir. Hannover, p 1
447. Joulain D (1999) Personal Communication to the Author
448. Kato T, Hozumi T (1972) Studies on Ketene and its Derivatives. XLIX. Reaction of Diketene with β-Ketoesters to Give Ethyl Orsellinate, Divarate, Olivetol Carboxylate and Sphaerophorol Carboxylate. Chem Pharm Bull **20**: 1574
449. Kauppi M, Verseghy-Patay K (1990) Determination of the Distribution of Lichen Substances in the Thallus by Fluorescence Microscopy. Ann Bot Fenn **27**: 189
450. Kawahara N, Nozawa K, Nakajima S, Kawai K-I (1988) Isolation and Structures of Novel Fungal Depsidones, Emeguisins A, B, and C, from *Emericella unguis*. J Chem Soc Perkin I 2611
451. Kinoshita K, Matsubara H, Koyama K, Takahashi K, Yoshimura I, Yamamoto Y (1993) A Higher Alcohol from *Solorina crocea*. Bibl Lichenol **53**: 129
452. Kinoshita K, Matsubara H, Koyama K, Takahashi K, Yoshimura I, Yamamoto Y, Higuchi M, Kinoshita Y, Kawai K-I (1994) New Phenolics from *Protusnea* Species. J Hattori Bot Lab **75**: 359
453. Kinoshita Y, Matsubara H, Koyama K, Takahashi K, Yoshimura I, Yamamoto Y, Miura Y, Kinoshita Y, Kawai K-I (1994) Topics in the Chemistry of Lichen Compounds. J Hattori Bot Lab **76**: 227
454. Kinoshita Y, Hayase S, Higuchi M, Yamamoto Y, Ahmadjian V, Yoshimura I, Yamada Y (1991) Improvement of Protoplast Isolation from Lichen Mycobionts. Agric Biol Chem **55**: 1891
455. Kinoshita Y, Hayase S, Yamamoto Y, Yoshimura I, Kurokawa T, Ahti T, Yamada Y (1993) Morphogenetic Capacity of the Mycobiont in *Usnea* (Lichenized Ascomycete). Proc Jap Acad **69**: 18
456. Kinoshita Y, Yamamoto Y, Yoshimura I, Kurokawa T, Huneck S (1997) Distribution of Optical Isomers of Usnic and Isousnic Acids Analyzed by High Performance Liquid Chromatography. J Hattori Bot Lab **83**: 173
457. Kinoshita Y, Yamamoto Y, Yoshimura I, Kurokawa T, Yamada Y (1993) Production of Usnic Acid in Cultured *Usnea hirta*. Bibl Lichenol **53**: 137
458. Kirschbaum U, Wirth V (1995) Flechten erkennen, Luftgüte bestimmen. Ulmer, Stuttgart
459. Kitahara N, Endo A, Furuya K, Takahashi S (1981) Thielavin A and B, New Inhibitors of Prostagladin Biosynthesis Produced by *Thielavia terricola*. J Antibiotics **34**: 1562
460. Kitahara N, Haruyama H, Hata T, Takahashi S (1983) The Structures of Thielavins A, B and C. Prostaglandin Synthetase Inhibitors from Fungi. J Antibiotics **36**: 599
461. Kjaer A, Kjaer DK (1985) Synthesis of Siphulin, a Naturally Occurring Homoflavone. Acta Chem Scand **B39**: 65

462. Kjølberg O, Kvernheim AL (1984) Studies on the Polysaccharides of Lichens. II. The Structure of Water-soluble Polysaccharides in *Umbilicaria pustulata* (L.) Hoffm. and *Umbilicaria spodochroa* (Ach.) Hoffm. Acta Chem Scand **B38**: 735

463. Kjølberg O, Kvernheim AL (1989) Studies on the Polysaccharides of Lichens. III. The Structure of Akali-soluble Polysaccharides in *Umbilicaria pustulata* (L.) Hoffm. and *Umbilicaria spodochroa* (Ach). Hoffm. Acta Chem Scand **43**: 280

464. Knoph J-G (1990) Untersuchungen an gesteinsbewohnenden xanthonhaltigen Sippen der Flechtengattung *Lecidella* (Lecanoraceae, Lecanorales) unter besonderer Berücksichtigung von außereuropäischen Proben exklusive Amerika. Bibl Lichenol **36**

465. Knoph J-G, Leuckert C (1994) Chemotaxonomic Studies in the Saxicolous Species of the Lichen Genus *Lecidella* (Lecanorales, Lecanoraceae) in America. Nova Hedwigia **59**: 455

466. Knoph J-G, Leuckert C, Hertel H (1995) Chemotypes and Distribution Patterns of Saxicolous Species of *Lecidella* (Lecanoraceae, Lecanorales). Crypt Bot **5**: 45

467. Knoph J-G, Schmidt R, Elix JA (1995) Untersuchungen einiger Arten der Gattung *Lecidella* mit Hochdruckflüssigkeitschromatographie unter besonderer Berücksichtigung von epiphytischen Proben. Bibl Lichenol **57**: 307

467a. Kobayashi Y, Nakano M, Okui H (1997) Chiral Synthesis of (+)-Aspicilin by Using a Furyl Group as the masked γ-Oxo-α,β-unsaturated Carboxylic acid. Tetrahedron Letters **38**: 8883

468. Kon Y, Iwashina T, Kashiwadani H, Wardlaw JH, Elix JA (1997) A New Dibenzofuran, Isostrepsilic Acid, Produced by Cultured Mycobiont of the Lichenized Ascomycete *Usnea orientalis*. J Jap Bot **72**: 67

469. Kon Y, Kashiwadani H, Wardlaw JH, Elix JA (1997) Effects of Culture Conditions on Dibenzofuran Production by Cultured Mycobionts of Lichens. Symbiosis **23**: 97

470. Korte F, Sieper H (1960) Zur chemischen Klassifizierung von Pflanzen, XX. Isolierung von Haschisch-Inhaltsstoffen aus *Cannabis sativa* non *indica*. Liebigs Ann Chem 71

471. Koyama M, Takahashi K, Chou T-C, Darsynkiewicz Z, Kapuscinski J, Kelly TR, Watanabe K (1989) Intercalating Agents with Covalent Bond Forming Capacity. A Novel Type of Potential Anticancer Agents. 2. Derivatives of Chrysophanol and Emodin. J Med Chem **32**: 1594

472. Kranner I, Grill D (1996) Determination of Glutathione and Glutathione Disulfide in Lichens: a Comparison of Frequently Used Methods. Phytochem Anal **7**: 24

473. Krause W (1976) Massenspektrometrische Untersuchungen an Flechten. Dissertat Friedrich-Alexander Univ Erlangen-Nürnberg

474. Krishnamurthy HG, Siva Prasad J (1975) New and Simple Syntheses of Olivetol. Tetrahedron Letters 2511

475. Krivoshchekova OE, Maximov OB, Stepanenko LS, Mishchenko NP (1982) Quinones of the Lichen *Cetraria cucullata*. Phytochemistry **21**: 193

476. Krivoshchekova OE, Mishchenko NP, Stepanenko LS, Maximov OB (1983) Aromatische Flechtenmetabolite der Familie Parmeliaceae. I. Depsidone (in Russian). Chim Prirodn Soedin 13

477. Krivoshchekova OE, Stepanenko LS, Mishchenko NP, Denisenko WA, Maximov OB (1983) Study of Aromatic Metabolites of Lichens from Parmeliaceae. 2. Pigments (in Russian). Chim Prirodn Soedin 283

478. Kulka K, Zazula T (1974) Methyl Orsellinate. US Patent 3,884,843

479. Kurokawa T, Yoshimura I (1993) Identification of C-1-Depsides Including their Phenol Carboxylic Acids and their Esters by Using Three dimensional High Performance Liquid Chromatography. Bull Kochi Gakuen Coll **24**: 827
480. Kutney JP, Leman JD, Salisbury PJ, Yee T, Sanchez IH (1984) Studies in the Usnic Acid Series. IX. The Biodegradation of (+)-Usnic Acid by *Mucor globosus*. Can J Chem **62**: 320
481. Kutney JP, Sanchez IH (1976) Studies in the Usnic Acid Series. I. The Condensation of (+)-Usnic Acid with Aliphatic and Aromatic Amines. Can J Chem **54**: 2795
482. Kutney JP, Sanchez IH, Yee T (1976) Studies in the Usnic Acid Series. II. The Condensation of (+)-Usnic Acid with Hydroxylamine. Can J Chem **54**: 3713
483. Kutney JP, Sanchez IH, Yee T (1977) Studies in the Usnic Acid Series. IV. The Base Catalyzed Usnic Acid – Isousnic Acid Rearrangement. Part II. An Improved Synthesis of (+)-Isousnic Acid. Can J Chem **55**: 1073
484. Kümmerling H, Leuckert C (1993) Chemische Flechtenanalysen VIII. *Lepraria lesdainii* (Hue) R.C. Harris. Nova Hedwigia **56**: 483
485. Kümmerling H, Leuckert C, Wirth V (1991) Chemische Flechtenanalysen VI. *Lepraria incana* (L.) Ach. Nova Hedwigia **53**: 507
486. Kümmerling H, Leuckert C, Wirth V (1993) Chemische Flechtenanalysen VII. *Lepraria lobificans* Nyl. Nova Hedwigia **56**: 211
487. Kümmerling H, Leuckert C, Wirth V (1994) Chemische Flechtenanalysen IX. *Lecanactis latebrarum* (Ach.) Arnold. Nova Hedwigia **58**: 437
488. Kümmerling H, Leuckert C, Wirth V (1994) Chemische Flechtenanalysen. *Lepraria jackii*. Nova Hedwigia **60**: 457
489. Kümmerling H, Leuckert C, Wirth V (1995) Chemische Flechtenanalysen X. *Lepraria rigidula* (B. de Lesd.) Tønsberg. Nova Hedwigia **60**: 233
489a. Land CJ, Lundström H (1989) Inhibition of Fungal Growth by Water Extracts from the Lichen *Nephroma arcticum*. Lichenologist **30**: 259
490. Lascéve G, Gaugain F (1990) Effects of Usnic Acid on Sunflower and Maize Plantlets. J Plant Physiol **136**: 723
491. Lauterwein M, Oethinger M, Belsner K, Peters T, Marre R (1995) *In Vitro* Activities of the Lichen Secondary Metabolites Vulpinic Acid, (+)-Usnic Acid, and (−)-Usnic Acid against Aerobic and Anaerobic Microorganisms. Antimicrob Agents and Chemotherapy **39**: 2541
492. Lawrey JD (1983) Vulpinic and Pinastric Acids as Lichen Antiherbivore Compounds: Contrary Evidence. Bryologist **86**: 365
493. Lawrey JD (1989) Lichen Secondary Compounds: Evidence for a Correspondence between Antiherbivore and Antimicrobial Function. Bryologist **92**: 326
494. Legaz M, Xavier Filho L, Vicente C (1992) Acciones Alelopaticas de los Liquenes. Bol Soc Broteriana **265**: 107
495. Leisner JMR, Bilger W, Czygan F-C, Lange OL (1993) Lipophilous Carotenoids of Cyanobacterial Lichens from different Habitats, including an Extreme Desert Site. Crypt Bot **4**: 74
496. Leuckert C (1985) Probleme der Flechten-Chemotaxonomie – Stoffkombinationen und ihre taxonomische Wertung. Ber Deutsch Bot Ges **98**: 401
497. Leuckert C, Ahmadjian V, Culberson CF, Johnson A (1990) Xanthones and Depsidones of the Lichen *Lecanora dispersa* in Nature and of its Mycobiont in Culture. Mycologia **82**: 370
498. Leuckert C, Knoph J-G (1992) European Taxa of Saxicolous *Lecidella* containing Chloroxanthones: Identification of Patterns Using Thin Layer Chromatography. Lichenologist **24**: 383

499. Leuckert C, Knoph J-G (1993) Secondary Compounds as Taxonomic Characters in the Genus *Lecidella* (Lecanoraceae, Lecanorales). Bibl Lichenol **53**: 161

500. Leuckert C, Knoph J-G, Hertel H (1992) Chemotaxonomische Studien in der Gattung *Lecidella* (Lecanorales, Lecanoraceae) II. Europäische Arten der *Lecidella asema*-Gruppe. Herzogia **9**: 1

501. Leuckert C, Knoph J-G, Ziegler H G, Hertel H (1990) Chemotaxonomische Studien in der Gattung *Lecidella* (Lecanorales, Lecanoraceae). I. *Lecidella carpathica* und *Lecidella viridans* – Untersuchungen an mittel- und südeuropäischen Proben. Herzogia **8**: 265

502. Leuckert C, Kümmerling H (1991) Chemotaxonomische Studien in der Gattung *Leproloma* Nyl. ex Crombie (Lichenes). Nova Hedwigia **52**: 17

503. Leuckert C, Kümmerling H, Wirth V (1995) Chemotaxonomy of *Lepraria* Ach. and *Leproloma* Nyl. ex Crombie, with Particular Reference to Central Europe. Bibl Lichenol **58**: 245

504. Leuckert C, Mathey A (1975) Beiträge zur Chemotaxonomie einiger Xanthon-haltiger Arten der Flechtengattung *Buellia*. Herzogia **3**: 461

505. Leuckert C, Mayrhofer H (1984) Beiträge zur Chemie der Flechtengattung *Rinodina* (Ach.) Gray II. Herzogia **6**: 373

506. Leuckert C, Mayrhofer H (1985) Chemische Flechtenanalysen IV. Herzogia **7**: 99

507. Leuckert C, Poelt J (1989) Studien über die *Lecanora rupicola*-Gruppe in Europa (Lecanoraceae). Nova Hedwigia **49**: 121

508. Leuckert C, Poelt J, Hähnel G (1976) Zur Chemotaxonomie der eurasischen Arten der Flechtengattung *Rhizoplaca*. Nova Hedwigia **28**: 71

509. Llera JM, Fraser-Reid B (1989) An Expeditious Route to the Northern Part of Retigeranic Acid a from (R)-(−)-Carvone. J Org Chem **54**: 5544

510. Lochmüller CH, Hill Jr WB, Porter RM, Hangac HH, Culberson CF, Ryall RR (1983) Separation of Lichen Metabolites, Pyridine Derivatives, and Pyrimidine Bases Using Microbore, Reversed-phase LC. J Chrom Sci **21**: 70

511. Lorimer SD, Barns G, Evans AC, Foster LM, May BCH, Perry NB, Tangney RS (1996) Cytotoxicity and Antimicrobial Activity of Plants from New Zealand's Subantarctic Islands **2**: 327

512. Löffler A, Pratt RD, Pucknat J, Gelbard G, Dreiding AS (1969) Préparation des α-Méthylène Butyrolactones par Reaction de Reformatsky; Synthèse de l' Acide Protolichestérinique. Chimia **23**: 413

513. Lumbsch HT (1994) Die *Lecanora subfusca*-Gruppe in Australasien. J Hattori Bot Lab **77**: 1

514. Lumbsch HT (1998) Taxonomic Use of Metabolic Data in Lichen-forming Fungi. In: Frisvad JC, Bridge PD, Arora DK (eds) Chemical Fungal Taxonomy, Marcel Dekker Inc, New York Basel Hongkong, p 345

515. Lumbsch HT, Elix JA (1993) Notes on the Circumscription of the Lichens *Lecanora leprosa* and *L. sulphurescens* (Lecanoraceae, Lichenized Ascomycotina). Tropical Biology **7**: 71

516. Lumbsch HT, Feige GB, Elix JA (1994) Chemical Variation in Two Species of the *Lecanora subfusca* Group (Lecanoraceae, Lichenized Ascomycotina). Plant Syst Evol **191**: 227

517. Lücking R, Lumbsch HT, Elix JA (1994) Chemistry, Anatomy and Morphology of Foliicolous Species of *Fellhanera* and *Badimia* (Lichenized Ascomycotina: Lecanorales). Bot Acta **107**: 393

518. Maass WSG, Hanson A (1986) Wrightiin, a New Chlorinated Depside from *Erioderma wrightii* Tuck. (Ascolichenes). Z Naturforsch **41b**: 1589

519. MacDonald AL, Rettig SJ, Trotter J (1974) Crystal and Molecular Structure of 2-Deacylusnic Acid. Can J Chem **52**: 723
520. Machado MJ, Guerrini M, Gorin PAJ, Torri G, Iacomini M (1997) A Galactosphingolipid from the Lichen, *Ramalina celastri*. Phytochemistry **45**: 651
521. MacMillan JG, Browne JC (1977) A Regiospecific Synthesis of Haematommic Acid. J Org Chem **42**: 2526
522. Mahato SB, Kundu AP (1994) ^{13}C-NMR Spectra of Pentacyclic Triterpenoids – a Compilation and some Salient Features. Phytochemistry **37**: 1517
523. Malterud KE, Anthonsen T (1987) Carbon-13-NMR Studies on Partially *O*-Methylated Phenols: Spin–Lattice Relaxation Time Measurements as a Tool for Signal Assignments. Acta Chem Scand **B41**: 6
524. Marcuccio SM, Elix JA (1983) A Structural Revision of Picroroccellin. Tetrahedron Letters **24**: 1445
525. Marcuccio SM, Elix JA (1985) Pyrazine Chemistry. V. Synthesis of Methylanhydropicroroccellin and Dimethylpicroroccellin. Aust J Chem **38**: 1785
526. Martin J, Watts PC, Johnson F (1974) Carboxylations of γ-Butyrolactones with Methyl Methoxymagnesium Carbonate. A New Synthesis of dl-Protolichesterinic Acid. J Org Chem **39**: 1676.
527. Martinez E, Mestres R (1972) Componentes de Liquenes. III. Espectros de Masas de Depsidonas. An Real Soc Esp Fis Quim, Quim **68**: 1321
528. Mathey A (1981) De In Situ Lichenum Investigatione. These, Univ P et M Curie, Paris
529. Mathey A (1981) LAMMA: New Perspectives for Lichenology? Fresenius Z Anal Chem **308**: 249
530. Mathey A, Hoder D (1978) Distribution of Lichen Substances by Means of Fluorescence Microscopy, Cathodoluminescence in Scanning Electron Microscope and X-ray Microanalysis in *Lecanora*-, *Buellia*-, *Laurera*- and *Trypethelium* Species. Nova Hedwigia **30**: 127
531. Mathey A, Van Roy W, Van Vaeck L, Eckhardt G, Steglich W (1994) *In situ* Analysis for a New Perylene Quinone in Lichens by Fourier-Transform Laser Microprobe Mass Spectrometry with External Source. Rapid Commun in Mass Spectrom **8**: 46
532. Mathey A, Van Vaeck L, Steglich W (1986) De l' Analyse *in situ* du Microlichen Crustacé Corticole Tropical *Lecidea russula* Ach. Papers of the 10th Internatl Symp on Microchem Techniques, Antwerpen
533. Mathey A, Van Vaeck L, Steglich W (1987) Investigation of Semi-thin Cryosections of Lichens by Laser Microprobe Mass Spectrometry. Anal Chim Acta **195**: 89
534. Matsubara H, Kinoshita K, Koyama K, Takahashi K, Yoshimura I, Yamamoto Y, Kawai K-I (1994) An Amino Acid from *Solorina crocea*. Phytochemistry **37**: 1209
535. Matsubara A, Kinoshita K, Koyama K, Yang Y, Takahashi K, Yoshimura I, Yamamoto Y, Miura Y, Kinoshita Y (1997) Anti-tyrosinase Activity of Lichen Metabolites and their Synthetic Analogues. J Hattori Bot Lab **83**: 179
536. Maximov OB, Gorschkova PP, Stepanenko LS, Mishchenko NP, Krivoshchekova OE (1990) Ein wasserlösliches lineares β-1, 6-D-Glukan und andere Komponenten der Flechte *Haematomma lapponicum* (in Russian). Chim Prirodn Soedin 400
537. Mayrhofer H, Leuckert C (1985) Beiträge zur Chemie der Flechtengattung *Rinodina* (Ach.) Gray III. Herzogia **7**: 117
538. McCune B (1987) Distribution of Chemotypes of *Rhizoplaca* in North America. Bryologist **90**: 6

539. McDonald IA, Simpson TJ, Sierakowski AF (1997) ^{13}C-NMR Spectral Studies of some Naturally Occurring Quinones and Related Compounds. Aust J Chem **30**: 1727
540. McLean SM, Mahler P, Nyburg SC, Sawyer JF, Webster CJ, Wong Ng W (1983) Structure of Rhodocladonic Acid. X-ray Crystal Structure Analysis of its Triacetate. Can J Chem **61**: 2055
541. Mendez AM, Lopez JL, Negron E (1986) Acido Fumarico, Aislamiento por Primera vez de Una Especie Liqueniqua. Rev Fac Farmac Univ de los Andes 1
542. Meyappan A, Neelakantan S, Ramesh P (1981) *t*-Butanolysis of Lecanoric Acid. Curr Sci **50**: 1028
543. Miceno AM, Gorin PAJ, Iacomini M (1991) Galactomannan and Isolichenan Components of the Carbohydrate-rich *Ramalina ecklonii* (Spreng.) Mey. et Flot. Agric Biol Chem **55**: 1391
544. Mietzsch E, Lumbsch HT, Elix JA (1993) Notice: a New Computer Program for the Identification of Lichen Substances. Mycotaxon **47**: 475
545. Mietzsch E, Lumbsch HT, Elix JA (1996) Wintabolites (Mactabolites for Windows). Users Manual, 3rd edn Univ Essen, FB 9/Botanik
546. Minami K (1944) Synthese des Glomellins (in Japanese). Yakugaku Zasshi **46**: 315
547. Mishchenko NP, Stepanenko LS, Krivoshchekova OE, Maximov OB (1980) The Anthraquinones from the Lichen *Asahinea chrysantha* (in Russian). Chim Prirodn Soedin 160
548. Miyagawa H, Hamada N, Sato M, Ueno T (1993) Hypostrepsilic Acid, a New Dibenzofuran from the Cultured Lichen Mycobiont of *Evernia esorediosa*. Phytochemistry **34**: 589
549. Miyagawa H, Hamada N, Sato M, Ueno T (1994) Pigments from the Cultured Lichen Mycobionts of *Graphis scripta* and *G. desquamescens*. Phytochemistry **36**: 1319
550. Miyagawa H, Yamashita M, Ueno T, Hamada N (1997) Hypostrepsilalic Acid from a Cultured Lichen Mycobiont of *Stereocaulon japonicum*. Phytochemistry **46**: 1289
551. Miyawaki H (1988) Studies on the *Lecanora subfusca* Group in Japan. J Hattori Bot Lab **64**: 271
552. Mulzer J, Salimi N, Hartl H (1993) First Asymmetric Synthesis of (+)- and (−)-Roccellaric Acid and Dihydroprotolichesterinic Acid. Tetrahedron Asymmetry **4**: 457
553. Mulzer J, Kattner L, Strecker AR, Schröder C, Buschmann J, Lehmann C, Luger P (1991) Highly Felkin-Anh Selective Hiyama Additions of Chiral Allylic Bromides to Aldehydes. Applications to the First Synthesis of Nephromopsinic Acid and its Enantiomer. J Amer Chem Soc **113**: 4218
554. Muralikrishna K, Manoharachary C, Prasad U (1989) Effect of Quinone Extracted from *Pyxine pefricola* Nyl. on Mitosis in *Allium cepa* L. Root Meristems. Curr Sci **58**: 1393
555. Murta MM, De Azevedo MBM, Greene AE (1993) Synthesis and Absolute Stereochemistry of (−)-Protolichesterinic Acid, Antitumour Lactone from *Cetraria islandica*. J Org Chem **58**: 7537
556. Narui T, Culberson CF, Culberson WL, Johnson A, Shibata S (1996) A Contribution to the Chemistry of the Lichen Family Umbilicariaceae (Ascomycotina). Bryologist **99**: 199

557. Narui T, Sawada K, Takatsuki S, Okuyama T, Culberson CF, Culberson WL, Shibata S (1998) NMR Assignments of Depsides and Tridepsides of the Lichen Family Umbilicariaceae. Phytochemistry **48**: 815
558. Narui T, Takatsuki S, Sawada K, Okuyama T, Culberson CF, Culberson WL, Shibata S (1996) Lasallic Acid, a Tridepside from the Lichen *Lasallia asiae-orientalis*. Phytochemistry **42**: 839
559. Neamati N, Hong H, Mazumder A, Wang S, Sunder S, Nicklaus MC, Milne GWA, Proksa B, Pommier Y (1997) Depsides and Depsidones as Inhibitors of HIV-1 Integrase: Discovery of Novel Inhibitors Through 3D Database Searching. J Med Chem **40**: 942
560. Nicoletti TM, Raston CL, Sargent MV (1988) A New Synthesis of Anthraquinones. J Chem Soc Chem Comm 1491
561. Nicollier G (1979) Isolement, Identification et Quelques Synthéses de Composés de la Mousse de Chêne (*Evernia prunastri* (L.) Ach.). Thesis Fac Sciences Neuchatel
562. Nicollier G, Tabacchi R, Gavin J, Breton JL, Gonzalez AG (1979) Triterpénes de la "Mousse de Chêne" (*Evernia prunastri* (L.) Ach.) Helv Chim Acta **62**: 807
563. Nimis PL, Pina D, Salvadori O (1992) Licheni e Conservazione di Monumenti. CLUEB, Bologna
564. Nishikawa Y, Tanaka M, Shibata S, Fukuoka F (1970) Polysaccharides of Lichens and Fungi. IV. Antitumour Active *O*-Acetylated Pustulan-type Glucans from the Lichens of *Umbilicaria* Species. Chem Pharm Bull **18**: 1431
565. Nishitoba Y, Nishimura H, Nishiyama T, Mizutami J (1987) Lichen Acids, Plant Growth Inhibitors from *Usnea longissima*. Phytochemistry **26**: 3181
566. Norrestam R, Von Glehn M, Wachtmeister CA (1974) Three-dimensional Structure of Usnic Acid. Acta Chem Scand **B28**: 1149
567. Okuyama E, Meyama K, Yamazaki M, Kinoshita Y, Yamamoto Y (1995) Usnic Acid and Diffractaic Acid as Analgesic and Antipyretic Components of *Usnea diffracta*. Planta Med **61**: 113
568. Orange A (1995) The British Species of *Lepraria* and *Leproloma*: Chemistry and Identification. British Lichen Soc Bull **76**: 1
569. Ott S, Zwoch I (1992) Ethylene Production by Lichens. Lichenologist **24**: 73
570. Paquette LA, Wright J, Drtina GJ, Roberts RA (1987) Enantio-specific Total Synthesis of Natural (−)-Retigeranic Acid A and two (−)-Retigeranic Acid B Candidates. J Org Chem **52**: 2960
571. Patra A, Mitra AK, Chatterjee TK, Barua AK (1981) Carbon-13 NMR Spectra of some Novel Migrated Hopanes and 21αH-Hopanes. Org Magn Res **17**: 148
572. Patra A, Mukhopadhyay AK, Mitra AK (1981) Carbon-13 Resonance Experiments of Some Friedelanes and Taraxasteranes. Org Magn Res **17**: 166
573. Pattenden G (1978) Natural 4-Ylidenebutenolides and 4-Ylenetetronic Acids. Progr Chem Org Nat Comp **35**: 133
574. Pengsuparp T, Cai L, Constant H, Fong HHS, Lin L-Z, Kinghorn AD, Pezzuto JM, Corell GA, Ingolfsdottir K, Wagner H, Hughes SH (1995) Mechanistic Evaluation of New Plant-derived Compounds that Inhibit HIV-1 Reverse Transcriptase. J Nat Prod **58**: 1024
575. Pereira EC, Nascimento SC, Lima RC, Silva NH, Oliveira AFM, Bandeira E, Boitard M, Beriel H, Vicente C, Legaz ME (1994) Analysis of *Usnea fasciata* Crude Extracts with Antineoplastic Activity. Tokai J Exp Med **19**: 47
576. Piervittori R, Alessio F, Maffei M (1994) Fatty Acid Variations in the Lichen, *Xanthoria parietina*. Phytochemistry **36**: 853

577. Poch GK, Gloer JB (1991) Aurantiacins A and B; Two New Depsidones from a Mangrove Isolate of the Fungus *Preussia aurantiaca*. J Nat Prod **54**: 213

578. Poelt J, Leuckert C, Roux C (1995) Die Arten der *Lecanora dispersa*-Gruppe (Lichenes, Lecanoraceae) auf kalkreichen Gesteinen im Bereich der Ostalpen – Eine Vorstudie. Bibl Lichenol **58**: 289

579. Polborn K, Steglich W, Connolly JD, Huneck S (1995) Structure of the Macrocyclic Bis-lactone Lepranthin from the Lichen *Arthonia impolita*; an X-ray Analysis. Z Naturforsch **50b**: 1111

580. Porwoll JP, Leete E (1985) Synthesis of [5,6-^{13}C, 1-^{14}C]-Olivetolic Acid, Methyl [1′-^{13}C] Olivetolate and [5,6-$^{13}C_2$, 1-^{14}C] Cannabigerolic Acid. J Labelled Comp Radiopharm **22**: 257

581. Posner B, Feige GB, Huneck S (1990) Phytochemische Untersuchungen an westeuropäischen *Lasallia*-Arten. Z Naturforsch **45c**: 161

582. Posner B, Feige GB, Huneck S (1991) Studies on the Chemistry of the Lichen Genus *Umbilicaria*. Z Naturforsch **47c**: 1

583. Posner B, Feige GB, Leuckert C (1991) Beiträge zur Chemie der Flechtengattung *Lasallia* Mérat. Z Naturforsch **46c**: 19

584. Proksa B, Adamcová J, Šturdíková M, Fuska J (1994) Metabolites of *Pseudevernia furfuracea* (L.) Zopf and their Inhibition Potential of Proteolytic Enzymes. Pharmazie **49**: 282

585. Proksa B, Šturdíková M, Prónayová N, Liptaj T (1996) (−)-Usnic Acid and its Derivatives. Their Inhibition of Fungal Growth and Enzyme Activity. Pharmazie **51**: 195

586. Pueyo G (1981) Constituants Glucidiques dans les Lichens. III. Osides. Rev Bibliographique (1). Ann Fals Exp Chim Toxicol **797**: 343

587. Pueyo G (1982) Carbohydrate Constituents in Lichens. V. Glycosides. Bibliographic Review. Ann Fals Exp Chim Toxicol **809**: 75

588. Pulgarin C (1989) Synthèse de Depsides et Depsidones. Approche Biomimetique. Thesis Fac Sci Univ Neuchatel

589. Pulgarin C, Gunzinger J, Tabacchi R (1985) Synthèse de l' Eriodermine. Helv Chim Acta **68**: 945

590. Pulgarin C, Gunzinger J, Tabacchi R (1985) Synthèse des Pseudocyphellarines A et B, Deux Depsides du Lichen *Pseudocyphellaria endochrysea*. Helv Chim Acta **68**: 1948

591. Pulgarin C, Tabacchi R (1988) Synthèse de l Acide Décarboxythamnolique. Helv Chim Acta **71**: 876

592. Puntambekar HM, Naik DG, Kapadi AH (1993) Mild, One-pot Conversion of Carboxylic Acids into Esters Using Phase Transfer Catalysis. Ind J Chem **32B**: 793

593. Purvis OW, Elix JA, Broomhead JA, Jones GC (1987) The Occurrence of Copper-Norstictic Acid in Lichens from Cupriferous Substrata. Lichenologist **19**: 193

594. Purvis OW, Elix JA, Gaul KL (1990) The Occurrence of Copper-Psoromic Acid in Lichens from Cupriferous Substrata. Lichenologist **22**: 345

595. Quinkert G, Döller U, Eichhorn M, Küber F, Nestler HP, Becker H, Bats JW, Zimmermann G, Dürner G (1990) Weitere Beiträge zur Synthese von (+)-Aspicilin. Helv Chim Acta **73**: 1999

596. Quinkert G, Fernholz E, Eckes P, Neumann D, Dürner G (1989) Synthese von (+)-Aspicilin mit Bausteinen aus nachwachsenden Rohstoffen. Helv Chim Acta **72**: 1753

597. Quinkert G, Heim N, Bats JW, Oschkinat H, Kessler H (1985) Die Struktur des Flechten-Makrolides (+)-Aspicilin. Angew Chem **97**: 985
598. Quinkert G, Heim N, Glenneberg J, Billhardt U-M, Bats JW, Dürner G (1987) Totalsynthese des enantiomeren-reinen Flechten-Makrolides (+)-Aspicilin. Angew Chem **99**: 363
599. Quinkert G, Heim N, Glenneberg J, Döller U, Eichhorn M, Billhardt U-M, Schwarz C, Zimmermann G, Bats JW, Dürner G (1988) Synthese des Flechtenmakrolides (+)-Aspicilin mit Photolactonisierung als Schlüsselreaktion. Helv Chim Acta **71**: 1719
600. Raghava Raju K, Rao PS (1985) Chemistry of Lichen Products. Part I. – Reaction of Hydrazine with Pulvinic Dilactone. Indian J Chem **24B**: 785
601. Raju KR, Appa Rao AVN, Rao PS (1985) Leprapinic Acid Derivatives with Antibacterial Activity. Fitoterapia **56**: 221
602. Raju KR, Rao PS (1986) Chemistry of Lichen Products. Part V – Synthesis and Antimicrobial Activity of some New 1,4-Benzoxazinones from Pulvinic Acid Dilactone. Indian J Chem **B 25**: 94
603. Raju KR, Rao PS (1986) Chemistry of Lichen Products. Part VI – Synthesis of Some New Benzimidazole Derivatives from Pulvinic Acid Lactone and their Fungicidal Activity. Indian J Chem **B 25**: 97
604. Rama M (1997) Étude Phytochimique de "*Bryoria fuscescens*" (Gyelnik) Brodo & D. Hawksw. Synthèse de Depsidones Comportant un Cycle Hydroxyphthalide. Thesis Fac Sci Univ Neuchatel
605. Rama NH, Turner E, Whalley WB (1981) The Chemistry of Fungi. Part 79. A New Synthesis of 3,5-Dihydroxy-4-methylphenylacetic Acid. J Chem Res (S) 149
606. Ramage R, Griffiths GJ, Sweeney JNA (1984) Dioxolanones as Synthetic Intermediates. Part 3. Biomimetic Synthesis of Pulvinic Acids. J Chem Soc Perkin I 1547
607. Ramaut JL, Brouers MB, Serusiaux E, Corvisier M (1978) Separation of Mixtures of Atranorin and Chloroatranorin by Thinlayer Chromatography. J Chromatogr **155**: 450
608. Ramaut JL, Corvisier M (1975) Effets Inhibiteurs des Extraits de *Cladonia impexa* Harm., *C. gracilis* (L.) Willd. et *Cornicularia muricata* (Ach.) Ach. sur la Germination des Graines de *Pinus sylvestris* L. Oecol Plant **10**: 295
609. Ramaut JL, Degraeve N, Moutschen J (1974) Modalités d' Action de Différentes Extraits d' *Evernia prunastri* et de *Cornicularia muricata* chez la *Nigella damascena*. Phyton **15**: 239
610. Ramaut JL, Thonar J (1972) Inhibition de la Germination de Differentes Graines d' Angiospermes par *Evernia prunastri* (L.) Ach. I. An Real Soc Espan Fis y Quim, Quim **68**: 575
611. Ramaut JL, Thonar J (1972) Inhibition de la Germination de Differentes Graines d' Angiospermes par *Evernia prunastri* (L.) Ach. An Real Soc Espan Fis y Quim, Quim **68**: 597
612. Ramesh P, Baig ESA (1992) A Convenient Synthesis of Depsides. Indian J Heterocycl Chem **2**: 1
613. Rao PS, Raju KR (1989) Chemistry of Lichen Products: Part VII. Some New Amide Derivatives from the Reaction of Acid Hydrazides with Pulvinic Dilactone. Current Sci (India) **58**: 444
614. Rapsch S, Ascaso C (1985) Effect of Evernic Acid on Structure of Spinach Chloroplasts. Ann Bot **56**: 467

615. Renner B, Gerstner E (1978) Anthrachinone aus der Mycobiontenkultur und dem Thallus von *Caloplaca ferruginea.* Naturwissenschaften **65**: 439

616. Renner B, Gerstner E (1982) Stoffwechselunterschiede zwischen lichenisierten und dem isolierten Mycobionten von *Baeomyces rufus* (Huds.) Rebent. Z Pflanzenphysiol **107**: 47

617. Renner B, Henssen A, Gerstner E (1982) Zur Phytochemie südamerikanischer *Nephroma*-Arten. Z Naturforsch **37c**: 739

618. Řezanka T, Dembitsky VM (1993) Occurrence of C_{40}–C_{130} Polyisoprenoid Alcohols in Lower Plants. Phytochemistry **34**: 1335

619. Řezanka T, Dembitsky V (1999) Brominated Fatty Acids from Lichen *Acarospora gobiensis.* Phytochemistry **50**: 97

620. Ribár B, Kapor A, Argay G, Engel P, Djarmati Z, Jankov RM (1993) Crystal Structure of Usnic Acid Sodium 2½ Hydrate. J Cryst Spectroscop Res **23**: 107

621. Roberts RA, Schüll V, Paquette LA (1983) Electrophile-initiated Ring-opening Reactions of 2-Methylene-6,6-dimethylbicyclo [3.1.0] hexanes. New Methodology for the Synthesis of Highly Functionalized 1,2,3-Trisubstituted Cyclopentenes. J Org Chem **48**: 2076

622. Rogers RW (1989) Chemical Variation and the Species Concept in Lichenized Ascomycetes. Bot J Linn Soc **101**: 229

623. Ruef BD (1990) Massenspektrometrische Untersuchungen an Flechtenxanthonen. Inauguraldiss, Phil.-naturwiss Fak Univ Bern

624. Rycroft DS, Cole WJ, Aslam N, Lamont YM, Gabriel R (1999) Killarniensolide, Methyl Orsellinates and 9,10-Dihydrophenanthrenes from the Liverwort *Plagiochila killarniensis* from Scotland and the Azores. Phytochemistry **50**: 1167

625. Rycroft DS, Connolly JD, Huneck S, Himmelreich U (1995) Revised Structure of Haemoventosin. Z Naturforsch **50b**: 1557

626. Safe S, Safe LM, Maass WSG (1975) Sterols of Three Lichen Species: *Lobaria pulmonaria, L. scrobiculata* and *Usnea longissima.* Phytochemistry **14**: 1821

627. Sakurai A, Goto Y (1987) Chemical Studies on the Lichen. I. The Structure of Isolecanoric Acid, a New *Ortho*-depside Isolated from *Parmelia tinctorum* Despr. Bull Chem Soc Japan **60**: 1917

628. Salimi N (1988) Stereokontrollierte Synthese von Verrucarinsäurelacton und Roccellarsäure. Inaugural Diss FB Chemie FU Berlin

629. Samsudin MW, Said M, Yusoff I, Zakaria Z, Elix JA (1995) The Isolation and Identification of a New Nortriterpenoid from the Lichen *Pseudocyphellaria sulphurea* (Schaerer). Science International (Lahore) **7**: 295

630. Sandifer RM, Harris TM (1979) Biogenetically Modelled Synthesis of Polyketide-type Xanthones. J Chem Soc Chem Comm 442

631. Santesson J (1970) Syntheses of Orsellinic Acid and Related Compounds. Acta Chem Scand **24**: 3373

632. Sargent MV (1987) Naturally Occurring Dibenzofurans. Part 9. A Convenient Synthesis of Phthalides: the Synthesis of Methyl Di-*O*-methylporphyrilate. J Chem Soc Perkin I 231

633. Sargent MV, Vogel P, Elix JA (1975) Structure of the Lichen Depsidone Gangaleoidin. J Chem Soc Perkin I 1986

634. Sassa T, Igarashi M (1990) Structure of (−)-Mycousnine, (+)-Isomycousnine and (+)-Oxymycousnine, New Usnic Acid Derivatives from Phytopathogenic *Mycosphaerella nawae.* Agr Biol Chem **54**: 2231

635. Schade A (1975) Über das Vorkommen von Calciumoxalat-Exkreten bei den Usneaceen (Lichenes) nebst Bemerkungen über Höhlungen der Achse, gelegentliche

Ölhyphen und Größenverhältnisse bei den *Usnea*-Arten Afrikas und mit einem Nachtrag über die Usneen Japans. Nova Hedwigia **26**: 45
636. Schimmer O, Lehner H (1973) Untersuchungen zur Wirkung von Usninsäure auf die Grünalge *Chlamydomonas reinhardii*. Arch Microbiol **93**: 145
637. Schmalz H-G, Volk T, Bernicke D, Huneck S (1997) On the Deprotonation of η^6-1,3-Dimethoxybenzene-$Cr(CO)_3$ Derivatives: Influence of the Reaction Conditions on the Regioselectivity. Tetrahedron **53**: 9219
638. Schindler H (1988) zur Geschichte der Anwendung von Flechten (Lichenes) in der Medizin. Carolinea **46**: 31
639. Schmidt J, Huneck S, Franke P (1981) Lichen Substances – 128. Mass Spectroscopy of Natural Products – 12. Comparative Positive and Negative Ion Mass Spectroscopy of Nitrogen-containing and Ring C Cleaved Usnic Acid Derivatives. Biomed Mass Spectr **8**: 293
640. Schöller H (1997) Flechten. Kleine Senckenberg-Reihe **27**: 1
641. Schulz H, Abroscheit G (1989) Characterization of Oakmoss Products Used in Perfumery by High-performance Liquid Chromatography. J Chromatogr **466**: 301
642. Scott FE, Simpson TJ, Trimble LA, Vederas JC (1986) Biosynthesis of the Meroterpenoid Austin, by *Aspergillus ustus*: Synthesis and Incorporation of $^{13}C,^{18}O$-Labelled Ethyl 3,5-Dimethyl Orsellinate. J Chem Soc Chem Comm 214
643. Seifert P, Bertram C (1995) Usnic Acid – Natural Preservation from Lichens. Seifen Öle Fette Wachse **121**: 480
644. Seriña E, Arroyo R, Manrique E, Sancho LG (1996) Lichen Substances and their Intraspecific Variability within Eleven *Umbilicaria* Species in Spain. Bryologist **99**: 335
645. Sharnoff SD, Sharnoff S (1997) Lichens. Natl Geographic **191**: 58
646. Shibamoto T, Wei C-I (1984) Mutagenicity of Lichen Constituents. Environmental Mutagenesis **6**: 757
647. Shibata S, Iitaka Y (1984) Renewed Studies on the Structure of Didymic Acid. Chem Pharm Bull **32**: 366
648. Shibuya M, Ebizuka Y, Noguchi H, Iitaka Y, Sankawa U (1983) Inhibition of Prostaglandin Biosynthesis by 4-*O*-Methylcryptochlorophaeic Acid; Synthesis of Monomeric Arylcarboxylic Acids for Inhibitory Activity Testing and X-ray Analysis of 4-*O*-Methylcryptochlorophaeic Acid. Chem Pharm Bull **31**: 407
649. Shin C-G, Nakano T, Sato Y, Kato H (1986) Syntheses of Picroroccellin Diastereomers and their Regiomers. Chem Letters 1453
650. Shiojima K, Arai Y, Masuda K, Takase Y, Ageta T, Ageta H (1992) Mass Spectra of Pentacyclic Triterpenoids. Chem Pharm Bull **40**: 1683
651. Sholichin M, Yamasaki K, Kasai R, Tanaka O (1980) ^{13}C Nuclear Magnetic Resonance of Lupane-type Triterpenes, Lupeol, Betulin and Betulinic Acid. Chem Pharm Bull **28**: 1006
652. Sinha SC, Keinan E (1994) Short Synthesis of (+)-Aspicilin *via* Asymmetric Hexahydroxylation of a Triene. J Org Chem **59**: 949
653. Sinha SC, Keinan E (1997) Total Synthesis of (+)-Aspicilin. The Naked Carbon Skeleton Strategy vs the Bioorganic Approach. J Org Chem **62**: 377
654. Smriga M, Saito H, Shibata S, Narui T, Okuyama T, Nishiyama N (1996) PC-2, Linear Homoglucan with α-Linkages, Peripherally Enhances the Hippocampal Long-term Potentiation. Pharmaceut Res **13**: 1322
655. Solberg Y (1975) Studies on the Chemistry of Lichens. Chemical Investigations of Five Norwegian *Alectoria* Species. Acta Chem Scand **B29**: 145

656. Solberg Y (1977) Studies on the Chemistry of Lichens, XV. New Synthetic Nitrogen Derivatives of Pulvinic Acid. Z Naturforsch **32c**: 292

657. Solberg Y (1979) Studies on the Chemistry of Lichens, XIX. New Amino Compounds from *Anaptychia fusca* and several *Stereocaulon* Species. Z Naturforsch **34c**: 493

658. Solberg Y (1986) Chemical Constituents of the Lichen Species *Cetraria islandica*. J Hattori Bot Lab **60**: 391

659. Solberg Y (1987) Chemical Constituents of the Lichens *Cetraria delisei, Lobaria pulmonaria, Stereocaulon tomentosum* and *Usnea hirta*. J Hattori Bot Lab **63**: 357

660. Solberg Y, Remedios G (1978) Studies on the Chemistry of Lichens, XVIII. Chemical Investigations of the Species *Letharia vulpina* (L.) Hue and New Derivatives of Vulpinic Acid. Z Naturforsch **33c**: 449

661. Solladie G, Fernandez I, Maestro C (1991) Asymmetric Synthesis of the C-3/C-9 Fragment of (−)-Aspicilin. Tetrahedron Letters **32**: 509

662. Solladie G, Fernandez I, Maestro C (1991) Asymmetric Synthesis of the Macrolide (−)-Aspicilin. Tetrahedron Asymmetry **2**: 801

663. Song-San S, Watanabe S, Saita T (1989) An Antifungal Triterpenoid from *Mollugo pentaphylla*. Phytochemistry **28**: 1767

664. Stach J, Möder M, Herzschuh R (1990) Tandem-Massenspektrometrie – Grundlagen und Applikationen. Z Chemie **30**: 157

665. Staiger B, Kalb K (1995) Haematomma-Studien. I. Die Flechtengattung *Haematomma*. Bibl Lichenol **59**

666. Steffen UA (1989) Asymmetrische Synthese von Paraconsäurederivaten. Inaugural Diss FB Chemie FU Berlin

667. Steffens JC, Robeson DJ (1987) Secalonic Acid A, a Vivotoxin in Pink Root-infected Onion. Phytochemistry **26**: 1599

668. Steglich W, Besl H, Zipfel K (1974) Pilzpigmente, XXIX. Festlegung der Struktur von Pulvinsäuren mit Hilfe der NMR-Spektroskopie. Z Naturforsch **29b**: 96

669. Steiner E (1974) Étude de la Synthèse et de la Biosynthèse de Quelques Pigments de Champignons Inferieurs. These Fac Sci Univ Geneve

670. Stenhouse J, Groves C (1877) Über Picroroccellin. Liebigs Ann Chem **185**: 14

671. Stepanenko LS, Krivoshchekova OE, Dmitrenok PS, Maximov OB (1997) Quinones of *Cetraria islandica*. Phytochemistry **46**: 565

672. Stepanenko LS, Krivoshchekova OA, Mishchenko NP (1985) Chemical Variations of *Asahinea chrysantha*. Phytochemistry **24**: 354

673. Stephenson NL, Rundel PW (1979) Quantitative Variation and the Ecological Role of Vulpinic Acid and Atranorin in the Thallus of *Letharia vulpina*. Biochem Syst Ecol **7**: 263

674. Stipanovic AJ, Giammatteo PJ, Robie SB (1986) Polysaccharides from Lichens. ^{13}C-NMR Studies on (1 → 6)-β-D-Glucan (Pustulan). Polymer Science Technol **33**: 41

675. Stübler D, Buchenauer H (1996) Antiviral Activity of the Glucan Lichenan (Poly-β-3,1 → 4D-anhydroglucose). 1. Biological Activity in Tobacco Plants. J Phytopathol **144**: 37

676. Stübler D, Buchenauer H (1996) Antiviral Activity of the Glucan Lichenan (Poly-β-3,1 → 4D-anhydroglucose) 2. Studies on the Mode of Action. J Phytopathol **144**: 45

677. Sucharita B, Malleshwar D, Manoharachary C, Mulaikrishna K (1983) Extraction of Lecanoric Acid from *Parmelia andina* Müll. Arg. and its Effect on Mitosis in *Allium cepa* Root Tips. Curr Sci (India) **52**: 21

678. Sugawara H, Kasuya A, Iitaka Y, Shibata S (1991) Further Studies on the Structure of Retigeranic Acid. Chem Pharm Bull **39**: 3051
679. Sugawara T, Strobel S, Strobel G, Larsen RD, Berglund DL, Gray G, Takahashi N, Coval SJ, Stout TJ, Clardy J (1991) The Structure and Biological Activity of Cercosporamide from *Cercosporidium henningsii.* J Org Chem **56**: 909
680. Sundholm EG (1978) Total Syntheses of Lichen Xanthones. Tetrahedron **34**: 577
681. Sundholm EG (1979) Syntheses and ^{13}C-NMR Spectra of some 5-Chloro-substituted Lichen Xanthones. Acta Chem Scand **B33**: 475
682. Sundholm EG (1974) Synthesis of 2,2′, 4,4′, 6′-Pentahydroxy-6-methyl benzophenone. Acta Chem Scand **B28**: 1102
683. Tabacchi R, Allemand P, Tsoupras G (1991) Direct Analysis of Lichens by Tandem Mass Spectromertry. Symbiosis **11**: 193
684. Tabacchi R, Hareed AM (1993) The Chemistry of *Usnea articulata* (L.) Hoffm. and *Usnea hesperina* Motyka. Bibl Lichenol **53**: 259
684a. Tabacchi R, Tsoupras G (1986) Sur la Presence de Triterpenes dans Quelques Lichens. Proc Conf Terpene Chem, Grasse 293
685. Tabacchi R, Tsoupras G (1987) The Chemical Composition of *Letharia vulpina.* Bibl Lichenol **25**: 475
686. Tabacchi R, Tsoupras G, Allemand P (1995) Identification of Triterpenes from Lichens by Tandem Mass Spectrometry (MS-MS). Bibl Lichenol **57**: 429
687. Tabacchi R, Tsoupras G, Huneck S (1987) Steroids and Nortriterpenoids from Lichens. J Hattori Bot Lab **63**: 351
688. Takahashi K, Kon T, Yokota I, Shibata S (1981) Chemotaxonomic Studies on the Polysaccharides of Lichens. Polysaccharides of Stereocaulaceous Lichens. Carbohydrate Res **89**: 166
689. Takai M, Uehara Y, Beisler JA (1979) Usnic Acid Derivatives as Potential Antineoplastic Agents. J Med Chem **22**: 1380
690. Takani M, Takahashi K (1985) Usnic acid. XVIII. The Photolysis of Usnic Acid and its Derivatives (1). Chem Pharm Bull **33**: 2772
691. Takenaka S, Ojima N, Seto S (1972) The Isolation of 2,4-Dihydroxy-3,6-dimethylbenzoic Acid (3-Methylorsellinic Acid) from a Culture of *Aspergillus terreus.* J Chem Soc Chem Comm 391
692. Takeo K, Suzuki T (1986) Synthesis of the Tri- and Tetra-saccharides Related to the Fine Structures of Lichenan and Cereal β-D-Glucanes. Carbohydrate Res **147**: 265
693. Tanahashi T, Kuroishi M, Kuwahara A, Nagukura N, Hamada N (1997) Four Phenolics from the Cultured Lichen Mycobiont of *Graphis scripta* var. *pulverulenta.* Chem Pharm Bull **45**: 1183
694. Texieira AZA, Iacomini M, Gorin PAJ (1992) An Unusual Glucomannan from *Tornabenia intricata.* Phytochemistry **31**: 3467
695. Teixeira AZA, Iacomini M, McCune B, Gorin PAJ (1994) Heteropolysaccharides of the Lichen *Evernia prunastri.* Carbohydrate Res **264**: 63
696. Thune PO, Solberg YJ (1980) Photosensitivity and Allergy to Aromatic Lichen acids, Composite Oleoresins and other Plant Substances. Contact Dermatitis **6**: 81
697. Timdale E (1989) The Production of Rhodocladonic Acid in *Cladonia bacilliformis* and *C. norvegica* Triggered by the Presence of Lichenicolous Mite. Graphis scripta **2**: 125
698. Titze E (1987) Sekikaic Acid Methyl Ester Found in *Protousnea* (Mot.). Krog. Bibl Lichenol **25**: 469

699. Tsuda Y, Nakajiama S, Udagawa S-I, Uzawa J (1980) The Isolation, from *Myrothecium* Species, and Long-range Selective Proton Decoupling ^{13}C-NMR of Rhizonic Acid. J Nat Prod **43**: 467

700. Tvaroska I, Ogawa K, Deslandes Y, Marchessault RH (1983) Crystalline Conformation and Structure of Lichenan and Barley β-Glucan. Can J Chem **61**: 1608

701. Tyman JHP, Durrani AA (1973) A Novel Synthesis of Homologous Orsellinic Acids and 5-Alkylresorcinols. Tetrahedron Letters **73**: 4839

702. Unverricht A, Pfützner P, Stecher B, Schütte HR (1972) Synthesen radioaktiver Verbindungen; Darstellung von Orsellinsäure-(2-^{14}C). Z Chem **12**: 289

703. Vainshtein EA (1985) Lichen Acids and Permeability of Alga *Trebouxia erici* Cells (in Russian). Fiziol Rast **32**: 1153

704. Vainshtein EA, Ravinskaja AP, Shapiro IA (1990) Guide to Lichen Chemotaxonomy. USSR Acad Sci, Leningrad

705. Van Roy W, Mathey A, Van Vaeck L (1996) *In-situ* Analysis of Lichen Pigments by Fourier Transform Laser Microprobe Mass Spectrometry with External Ion Source. Rapid Comm Mass Spectrom **10**: 562

706. Vänskä H (1984) The Identity of the Lichens *Leconora frustulosa* and *L. argopholis.* Ann Bot Fenn **21**: 391

707. Venkataramana D (1992) High-performance Liquid Chromatographic Determination of Usnic Acid in Plasma. J Chromatogr **575**: 167

708. Vicente C, Gonzalez A (1986) Purification and Properties of D-Usnic Acid Dehydrogenase of *Evernia prunastri.* Biochem **1**: 47

709. Vicente C, Gonzalez A, Legaz ME (1986) Factors Affecting Two Different Usnic Acid Dehydrogenases in the Lichen *Evernia prunastri.* Biochem Syst Ecol **14**: 283

710. Vicente C, Mateos JL, Pedrosa MM, Legaz ME (1991) High-performance Liquid Chromatographic Determination of Sugars and Polyols in Extracts of Lichens and Sugarcane Juice. J Chromatogr **553**: 271

711. Vinet C, Quilhot W, Gambaro V, Garbarino JA (1990) Studies on Chilean Lichens, XIII. Polysubstituted Depsides from *Lecania brialmontii.* J Nat Prod **53**: 500

712. Vinet C, Quilhot W, Garbarino JA (1990) Studies on Chilean Lichens, XIV. 2′-*O*-Methylhiascic Acid, a New Tridepside in *Catillaria corymbosa.* J Nat Prod **53**: 1025

713. Vogt H, Heinen HJ, Meier S, Wechsung R (1981) LAMMA 500 Principle and Technical Description of the Instrument. Fresenius Z Anal Chem **308**: 195

714. Waanders PP, Thijs L, Zwanenburg B (1987) Stereocontrolled Total Synthesis of the Macrocyclic Lactone (−)-Aspicilin. Tetrahedron Letters **28**: 2409

715. Wadstein T, Moberg R (1985) Calcium Oxalate Hydrates on the Surface of Lichens. Lichenologist **17**: 239

716. Watanabe M, Iwai K, Shibata S, Takahashi K, Narui T, Tashiro T (1986) Purification and Characterization of Mouse α_1-Acid Glycoprotein and its Possible Role in the Antitumor Activity of some Lichen Polysaccharides. Chem Pharm Bull **34**: 2532

717. Weinstock J, Blank JE, Oh H-J, Sutton BM (1979) A Regiospecific Synthesis of Substituted Vulpinic Acids. J Org Chem **44**: 673

718. Wender PA, Singh SK (1990) Synthetic Studies on Arene-olefine Cycloaddition. 11. Total Synthesis of (−)-Retigeranic Acid. Tetrahedron Letters **31**: 2517

719. Wenkert E, Baddeley GV, Burfitt IR, Moreno LN (1978) Carbon-13 Nuclear Magnetic Resonance Spectroscopy of Naturally-occurring Substances. LVII. Triterpenes Related to Lupane and Hopane. Org Magn Res **11**: 337

720. Whalley WB (1949) The Application of the Modified Gattermann Reaction to Methyl and Ethyl Orsellinate. J Chem Soc 3278
721. White FJ, James PW (1985) A New Guide to Microchemical Techniques for the Identification of Lichen Substances. British Lichen Soc Bull **57**: 1
722. White JD, Johnson AT (1990) Synthesis of the Lichen Metabolite (+)-Bourgeanic Acid and Conformational Analysis of its Dilactone. J Org Chem **55**: 5938
723. White JD, Johnson AT (1994) Synthesis of the Aliphatic Depside (+)-Bourgeanic acid. J Org Chem **59**: 3347
724. Whiton JC, Lawrey JD (1984) Inhibition of *Cladonia cristatella* and *Sordaria fimicola* Ascospore Germination by Lichen Acids. Bryologist **85**: 222
725. Whiton JC, Lawrey JD (1984) Inhibition of Crustose Lichen Spore Germination by Lichen Acids. Bryologist **87**: 42
726. Wilkins AL (1980) Nephrin: Structure and Occurrence in *Nephroma* Species. Phytochemistry **19**: 696
727. Wilkins AL, Elix JA (1990) New Fernene Triterpenes from the Lichen *Pseudocyphellaria aurata.* Aust J Chem **43**: 623
728. Wilkins AL, Elix JA, Gaul KL, Moberg R (1989) New Hopane Triterpenoids from Lichens in the Family Physciaceae. Aust J Chem **42**: 1415
729. Wilkins AL, Elix JA, Gonzalez AG, Perez C (1989) A One- and Two-dimensional ^{1}H and ^{13}C-NMR Study of some Fernene Triterpenoids. Aust J Chem **42**: 1185
730. Wilkins AL, Elix JA, Whitton AA (1990) A One- and Two-Dimensional ^{1}H and ^{13}C-NMR Study of some Lichen Triterpenoids of the Pyxinol Group. Aust J Chem **43**: 411
731. Wilkins AL, Ronaldson KJ, Jager PM, Bird PW (1987) A ^{13}C-NMR Study of some Oxygenated Hopane Triterpenes. Aust J Chem **40**: 1713
732. Williams DE, Bombuwala K, Lobkovsky E, Dilip de Silva E, Karunaratne V, Allen TM, Clardy J, Andersen RJ (1998) Ambewelamides A and B, Antineoplastic Epidithiapiperazinediones Isolated from the Lichen *Usnea* sp. Tetrahedron Letters **39**: 9579
733. Wilson MJ, Jones D (1984) The Occurrence and Significance of Manganese Oxalate in *Pertusaria corallina* (Lichenes). Pedobiologia **26**: 373
734. Wiseman JR, Pendery JJ, Otto CA, Chiong KG (1980) Regioselective Synthesis of Islandicin and Digitopurpone. J Org Chem **45**: 516
735. Wollenweber E, Arriaga-Giner FJ (1991) Grammitic Acid, a Long-chain Substituted Benzoic Acid Derivative from the Fern *Grammitis argyrata.* Phytochemistry **30**: 2307
736. Wolseley PA (1995) A Global Perspective on the Status of Lichens and their Conservation. Mitt Eidgen Forschungsanstalt Wald Schnee Landschaft **70**: 11
737. Wong S-M, Oshima Y, Pezzuto JM, Fong HHS, Farnsworth NR (1986) Plant Anticancer Agents. XXXIX: Triterpenes from *Iris missouriensis* (Iridaceae). J Pharmaceut Sci **75**: 317
738. Wood S, Huffman J, Weber N, Andersen D, North J, Murray B, Sidwell R, Hughes B (1990) Antiviral Activity of Naturally Occurring Anthraquinones and Anthraquinone Derivatives. Planta Medica **56**: 651
739. Wright D (1995) A Simplified TLC Method. Bull Calif Lichen Soc **2**: 7
740. Wright J, Drtina GJ, Roberts RA, Paquette LA (1988) A Convergent Synthesis of Triquinane Sesterterpenes. Enantioselective Synthesis of (−)-Retigeranic Acid A. J Am Chem Soc **110**: 5806

741. Yamamoto Y (1991) Production of Lichen Substances. In: Komamine K, Misawa M, Dicosmo F (eds) Plant Cell Culture in Japan. Progress in Production of Useful Plant Metabolites by Japanese Enterprises Using Plant Cell Culture Technology. CMC Co, Ltd., Tokyo, p 58
742. Yamamoto Y, Kinoshita Y, Kurokawa T, Yoshimura I, Ahmadjian V, Yamada Y (1995) Cell growth and Pigment Production is Suspension Cultures of a Mycobiont Isolated from the Lichen *Cladonia cristatella.* Can J Chem **73** (Suppl 1): 590
743. Yamamoto Y, Kinoshita Y, Matsubara H, Kinoshita K, Takahashi K, Kurokawa T, Yoshimura I (1998) Screening of Biological Activities and Isolation of Biological-active Compounds from Lichens. Recent Res Developm in Phytochemistry **2**: 23
744. Yamamoto Y, Matsubara H, Kinoshita Y, Kinoshita K, Koyama K, Takahashi K, Ahmadjian V, Kurokawa T, Yoshimura I (1996) Naphthazarin Derivatives from Cultures of the Lichen *Cladonia cristatella.* Phytochemistry **43**: 1239
745. Yamamoto Y, Miura Y, Higuchi M, Kinoshita Y, Yoshimura I (1993) Using Lichen Tissue Cultures in Modern Biology. Bryologist **96**: 384
746. Yamamoto Y, Miura Y, Kinoshita Y, Higuchi M, Yamada Y, Murakami A, Ohigashi H, Koshimizu K (1995) Screening of Tissue Cultures and Thalli of some Lichens and some of their Active Constituents for Inhibition of Tumor Promoter-induced Epstein-Barr Virus Activation. Chem Pharm Bull **43**: 1388
747. Yamamoto Y, Mizuguchi R, Takayama S, Yamada Y (1987) Effects of Culture Conditions on the Growth of Usneaceae Lichen Tissue Cultures. Plant Cell Physiol **28**: 1421
748. Yamamoto Y, Mizuguchi R, Yamada Y (1985) Tissue Cultures of *Usnea rubescens* and *Ramalina yasudae* and Production of Usnic Acid in their Cultures. Agric Biol Chem **49**: 3347
749. Yamamoto Y, Yoshimura I, Yamada Y (1987) Cultures of Usneaceae Species and Growth Factors in their Cultured Tissues, Bibl Lichenol **25**: 163
750. Yang X, Shimizu Y, Steiner JR, Clardy J (1993) Nostoclide I and II, Extracellular Metabolites from a Symbiotic Cyanobacterium, *Nostoc* sp., from the Lichen *Peltigera canina.* Tetrahedron Letters **34**: 761
751. Yoshida T, Hashimoto T, Takaoka S, Kan Y, Asakawa Y, Pezzuto JM, Pengsuparp T, Cordell GA (1996) Phenolic Constituents of the Liverwort: Novel Cyclic Bis-benzyl Dimers from *Blasia pusilla.* Tetrahedron **52**: 14487
752. Yoshimura I, Kinoshita Y, Yamamoto Y, Huneck S, Yamada Y (1994) Analysis of Secondary Metabolites from Lichen High Performance Liquid Chromatography with a Photodiode Array Detector. Phytochem Analysis **5**: 197
753. Yoshimura I, Kurokawa T (1989) Tissue Cultures of Some *Umbilicaria* Species, Lichenized Fungi. Bull Kochi Gakuen Coll **20**: 527
754. Yoshimura I, Kurokawa T (1991) Biogenesis of Lecanoric Acid in Lichen Thalli and its Related Substances. Bull Kochi Gakuen Coll **22**: 663
755. Yoshimura I, Kurokawa T, Hamade R, Kinoshita YH (1992) Secondary Metabolic Substances of *Usnea diffracta* Vain. Bull Kochi Gakuen Coll **23**: 687
756. Yoshimura I, Kurokawa T, Kanda H (1990) Tissue Culture of some Antarctic Lichens Preserved in the Refrigerator. Proc NIPR Symp Polar Biol **3**: 224
757. Yoshimura I, Kurokawa T, Kinoshita Y, Yamamoto Y, Miyakawi H (1994) Lichen Substances in Cultured Lichens. J Hattori Bot Lab **76**: 249
758. Yoshimura I, Kurokawa T, Nakano T, Yamamoto Y (1989) Production of Secondary Metabolic Substances by Cultured Tissues of *Usnea flexilis.* Bull Kochi Gakuen Coll **20**: 535

759. Yoshimura I, Kurokawa T, Yamamoto Y, Kinoshita Y (1993) Development of Lichen Thalli *in vitro*. Bryologist **96**: 412

760. Yoshimura I, Kurokawa T, Yamamoto Y, Kinoshita Y (1994) *In vitro* Development of the Lichen Thallus of some Species of *Peltigera*. Crypt Bot **4**: 314

761. Yoshimura I, Yamamoto Y (1991) Development of *Peltigera praetextata* Lichen Thalli in Culture. Symbiosis **11**: 109

762. Zeybek U, Lumbsch HT, Feige GB, Elix JA, John V (1993) Chemosyndromic Variation in *Hypogymnia* Species, mainly from Turkey (Lichenized Ascomycotina). Crypt Bot **3**: 260

763. Zopf W (1904) Zur Kenntnis der Flechtenstoffe. 12. Mitt. Liebigs Ann Chem **336**: 46

764. Zopf W (1904) Die Flechtenstoffe in chemischer, botanischer, pharmakologischer und technischer Beziehung. G Fischer Jena, p 332

765. Zygadlo JA, Pignata ML, Gonzalez CM, Levin A (1993) Alkanes in Lichens. Phytochemistry **32**: 1453

(*Received February 3*, 2000)

Author Index

Subject Index

Springer-Verlag and the Environment

MIX
Papier aus verantwortungsvollen Quellen
Paper from responsible sources
FSC® C105338

If you have any concerns about our products,
you can contact us on
ProductSafety@springernature.com

In case Publisher is established outside the EU,
the EU authorized representative is:
Springer Nature Customer Service Center GmbH
Europaplatz 3, 69115 Heidelberg, Germany

Printed by Libri Plureos GmbH
in Hamburg, Germany